WHY WE EAT, HOW WE EAT

Critical Food Studies

Series Editor
Michael K. Goodman, Kings College London, UK

The study of food has seldom been more pressing or prescient. From the intensifying globalization of food, a world-wide food crisis and the continuing inequalities of its production and consumption, to food's exploding media presence, and its growing re-connections to places and people through 'alternative food movements', this series promotes critical explorations of contemporary food cultures and politics. Building on previous but disparate scholarship, its overall aims are to develop innovative and theoretical lenses and empirical material in order to contribute to – but also begin to more fully delineate – the confines and confluences of an agenda of critical food research and writing.

Of particular concern are original theoretical and empirical treatments of the materializations of food politics, meanings and representations, the shifting political economies and ecologies of food production and consumption and the growing transgressions between alternative and corporatist food networks.

Other titles in the series include:

Embodied Food Politics
Michael S. Carolan
9781409422099

Liquid Materialities
A History of Milk, Science and the Law
Peter Atkins
9780754679219

Why We Eat, How We Eat

Contemporary Encounters between Foods and Bodies

Edited by

EMMA-JAYNE ABBOTS
University of Wales Trinity St David and
SOAS, University of London, UK

ANNA LAVIS
University of Birmingham and University of Oxford, UK

ASHGATE

Published by
Ashgate Publishing Limited
Wey Court East
Union Road
Farnham
Surrey, GU9 7PT
England

Ashgate Publishing Company
110 Cherry Street
Suite 3-1
Burlington, VT 05401-3818
USA

www.ashgate.com

British Library Cataloguing in Publication Data
Why we eat, how we eat : contemporary encounters between foods and bodies. -- (Critical food studies)
1. Food habits. 2. Nutrition--Psychological aspects.
3. Food preferences--Psychological aspects. 4. Food--Symbolic aspects.
I. Series II. Abbots, Emma-Jayne. III. Lavis, Anna.
613.2'019-dc23

The Library of Congress has cataloged the printed edition as follows:
Abbots, Emma-Jayne.
Why we eat, how we eat : contemporary encounters between foods and bodies / by Emma-Jayne Abbots and Anna Lavis.
p. cm. -- (Critical food studies)
Includes bibliographical references and index.
ISBN 978-1-4094-4725-2 (hardback) -- ISBN 978-1-4094-4726-9 (ebook) -- ISBN 978-1-4094-7379-4 (epub) 1. Food habits. 2. Food preferences. 3. Food--Symbolic aspects. I. Lavis, Anna. II. Title.
GT2850.A34 2013
394.1'2--dc23

2012039400

ISBN 9781409447252 (hbk)
ISBN 9781409447269 (ebk – PDF)
ISBN 9781409473794 (ebk – ePUB)

Printed and bound in Great Britain
by MPG PRINTGROUP

Contents

PART III: CONTRADICTIONS AND COEXISTENCES: WHAT WE SHOULD AND SHOULD NOT EAT

PART IV: ENTANGLEMENTS AND MOBILIZATIONS: THE MULTIPLE SITES OF EATING ENCOUNTERS

List of Figures, Tables and Boxes

Figures

Tables

Boxes

Notes on Contributors

Emma-Jayne Abbots is a Lecturer in Social/Cultural Anthropology and Heritage at the University of Wales Trinity St David and a Research Associate at The Food Studies Centre, SOAS, University of London. A political and economic anthropologist, her research centres on the politics and practices of food and eating in the context of migration and demographic change, with a particular focus on the interplay between cultural heritage, discourses of 'sustainability' and rural livelihoods. She has further interests in care, kinship, consumption and the brokerage of food and body knowledges.

Lucy Aphramor is a Dietician who is the Director of Well Founded Ltd. and an honorary research fellow at Coventry University, England. Her research interests centre on social justice and health.

Kim Baker holds an Art MA (Norwich) and a PhD in Visual Anthropology (University of London). Her writing, visual research interests and lecturing activities focus on rural lives and the human-animal relations engendered by contemporary agriculture in Britain.

Jennifer Brady is a PhD student at Queen's University, Canada. Her research interests include feminist perspectives in critical fat studies, critical dietetics and food studies.

Sally Brooks is a Teaching Fellow in the Department of Social Policy and Social Work at the University of York. She also teaches on online Masters programmes in public policy and management and international development. Sally is the author of *Rice Biofortification: Lessons from Global Science and Development.*

Duika Burges Watson is based at Durham University in the School of Medicine and Health. A Lecturer in the Evaluation of Policy Interventions within FUSE, The Centre for Translational Research in Public Health, Burges Watson is a social scientist with interests in qualitative methodologies, post-structuralism, food policy and translational research.

Simon Cohn is a Medical Anthropologist and Senior University Lecturer in the Department of Public Health and Primary Care, Cambridge University. Working within a multidisciplinary context on a range of applied topics, his current research is focused on finding ways to conceive of medical issues that offer a new and critical perspective.

Benjamin Coles is a Lecturer in Political and Economic Geography at the University of Leicester. His research interests centre on place, space and scale, with a particular interest in the relationship(s) between place-making, commodity culture and the production and consumption of food.

Alizon Draper is a Reader in Public Health Nutrition at the University of Westminster. She has a background in social anthropology and her work has focused on the social and policy aspects of nutrition. Draper has long been interested in choice and its rhetorics as used in public health and policy.

Jacqui Gingras is an Associate Professor in the School of Nutrition at Ryerson University, Canada. Jacqui is editor of the *Journal of Critical Dietetics* and applies feminist methodology to transform dietetic research, practice, teaching and policy.

Michael Goodman is a Senior Lecturer in Geography at King's College, London. His current research interests are in the cultural politics of 'alternative' foods, environment and development, celebrity politics and the shifting geographies, moralities and politics of consumer cultures.

Jon Holtzman is an Associate Professor in Cultural Anthropology at Western Michigan University. His work centres on Samburu pastoralists in northern Kenya and Nuer (Sudanese) refugees in Minnesota. He is the author of *Uncertain Tastes: Memory, Ambivalence, and the Politics of Eating in Samburu, Northern Kenya.* Jon is currently developing a new research project on food and memory in Japan.

Samantha Hurn is a Lecturer in Anthropology in the Department of Sociology and Philosophy at the University of Exeter where she also convenes an award winning MA in Anthrozoology. Samantha has conducted fieldwork in Wales, Spain, South Africa and Swaziland. Her book *Humans and Other Animals* is published by Pluto Press.

Rachael Kendrick recently completed her PhD, on obesity and the material ontology of metabolism, at the University of Melbourne. Her research interests include systems biology, feminist materialisms, medicine and the practice of doping in high performance sport.

Heidi Kvalvaag is a PhD Candidate at the Department of Sociology and Political Science at NTNU, Norwegian University of Science and Technology, Trondheim and an Assistant Professor in the Department of Nursing & Health at Diakonhjemmet University College. Heidi's research explores how understandings of action, especially eating, are related to understandings of the body and food.

Anna Lavis is a Medical Anthropologist and Research Fellow at the University of Birmingham. Within the multidisciplinary context of the Department of Primary Care, she conducts applied research into first episode psychosis. Also with a focus on individuals' lived experiences and subjectivities of mental illness, Anna's doctorate explored pro-anorexia and she continues to work on eating, food and eating disorders. As a Research Associate in the Institute of Social and Cultural Anthropology at the University of Oxford, she is currently collaborating on projects investigating media representations of obesity.

Kaori O'Connor is a Research Fellow in the Department of Anthropology at UCL, where she specializes in material culture and food. She won the Sophie Coe Prize for Food History in 2009. Kaori is the author of *English Breakfast: The Cultural Biography of a National Meal*, and of *Lycra: How a Fiber Shaped America*.

Jim Ormond is currently completing his PhD in Geography at King's College London. This examines the introduction of 'Product Carbon Footprinting', specifically the micro-politics of its implementation within the UK food/value chain. Jim grew up on a dairy farm in Oxfordshire.

Anne Murcott is a Professorial Research Associate at the Food Studies Centre, SOAS, University of London and has career-long interests in the cultural conceptions of food and eating. Anne is the editor of *The Nation's Diet: The Social Science of Food Choice*, and *The Sociology of Food and Eating*, as well as co-editor of *The Sociology of Food: Diet, Eating and Culture*.

Elspeth Probyn is Professor of Gender & Cultural Studies at the University of Sydney, and Adjunct SA Research Professor and co-convenor of the Research in Regions Cluster at the University of South Australia. She is the author of several monographs as well as over a hundred articles and chapters. Her current research (funded by an ARC Discovery Project) focuses on place and taste within the transglobal food system.

Elizabeth Saleh is a PhD candidate in the Anthropology Department of Goldsmiths, University of Lond, and has conducted fieldwork in Lebanon since 2006. The working title of her PhD thesis is 'Trade-marking Tradition: An Ethnography of the Lebanese Wine Industry', and her interests are in elites, economic enterprising, agriculture, food studies and the senses, memory and history.

Wendy Wills is a Reader in Adolescent and Child Health at the University of Hertfordshire's Centre for Research in Primary and Community Care (CRIPACC). She is a sociologist and public health nutritionist and her research focuses on everyday food, eating and weight management practices and public health/nutrition policy discourses

Emily Yates-Doerr is a Postdoctoral Researcher at the University of Amsterdam on Annemarie Mol's *Eating Bodies* project, where she is studying global health and agricultural debates surrounding meat consumption while revising her book manuscript, *The Weight of the Body: Fatness, Nourishment, and Health in Translation.*

Maria Yotova is a Visiting Researcher at the National Museum of Ethnology (Japan) and she also lectures in the Department of Anthropology, Shiga University. Her research interests lie at the intersection of food and business anthropology, with a focus on Eastern-Europe and Japan. She is currently working on her book manuscript, *The Narratives of Yogurt in Socialist and Postsocialist Bulgaria.*

Acknowledgements

This volume began as a selection of papers presented at the conference *Why We Eat, How We Eat* that took place at Goldsmiths, University of London, in February 2011. When we put out the call for papers we were pleasantly overwhelmed by the variety, depth and originality of the responses. The large number of abstracts and level of interest suggested that the time had indeed arrived for a considered discussion of 'eating'. Selecting the papers was a difficult task and we chose those which would foster an interdisciplinary dialogue on eating in order to map out its contours and elucidate both its current depths and lacks. To this end we organized the panels not by topic but by conceptual theme, and selected contributions engaged with these broader frameworks. It was that structure which laid the foundations for this edited collection. Sadly, not all the papers presented at the conference could make it into this collection, but our thanks go to all those who contributed on the day; whether as presenters, chairs or participants, they all directly or indirectly informed the contents of this book. Thanks, also, to Emma Felber, Tim Martindale, Hannah Roberson, Elizabeth Saleh, Jessie Sklair and Max Waterman who not only provided invaluable support during the conference, but whose insightful feedback on the papers also helped us during the process of selecting chapters for this volume. Our gratitude also goes to the Anthropology Department and Postgraduate Office at Goldsmiths and to the SOAS Food Studies Centre, all part of the University of London, for their financial support for the conference. We should especially like to thank Les Back, Hugh Macnicol, Frances Pine and Harry G. West.

In the course of convening the conference and editing this collection, both editors have received encouragement, support and advice from a number of colleagues. Our appreciation therefore goes to: Catherine Alexander, Max Birchwood, Simon Cohn, Sophie Day, Karin Eli, Linda Everard, Rachel Foskett-Tharby, Sarah Hinton, Lizzie Hull, Jakob Klein, Helen Lester, Tatum Matharu, Amy McLennan, Anne Murcott, Johan Pottier, Emilia Sanabria, Stanley Ulijaszek, Megan Warin and Sami Zubaida. We should like to thank our series editor, Michael Goodman, and also Katy Crossan, Val Rose, Lianne Sherlock and Caroline Spender at Ashgate for their enthusiasm for the book. We would also like to express our appreciation to all the authors and 'interluders' for their hard work, openness to critiques and suggestions, and their ability to bear – and deliver work of a high standard to – tight deadlines.

Finally, Emma-Jayne Abbots would personally like to thank David for his patience, love and encouragement, and for letting this book invade our lives. Thanks, also, to Margaret and John Birt for their ceaseless love and support.

Anna Lavis would like to thank David Martin and Pamela and David Lavis for their invaluable support in all its many forms, both during the production of this book and beyond its pages.

Introduction
Contours of Eating: Mapping the Terrain of Body/Food Encounters

Emma-Jayne Abbots and Anna Lavis

Ensuing from an international multi-disciplinary one-day conference that took place in London in February 2011, this volume explores the multidimensional act of eating, and turns its attention to the many spatial, cultural and biological moments at which commensality encounters materiality. Recognizing that food has multiple 'regimes of value' (Appadurai 1986) and is simultaneously symbolic, economic, political, material and nutritional, the conference purposefully sought contributions from a wide range of disciplines, drawing into dialogue nutritionists, activists and social scientists. This edited collection takes its lead from that interdisciplinary approach and looks to reflect on, and extend, the scholarly debates that emerged there. Our focus on eating, as a site in which both the symbolic resonances and the material properties of food are brought to the fore, engages with the conceptual crossing points in those discussions. Asking not what, but rather *why* and *how*, we eat offers new avenues for theoretical exploration and interdisciplinary engagement. As all the contributors to this volume elucidate in varying and nuanced ways, why and how are terms more complex and, perhaps, fraught than they might at first appear. They allow us to ask what eating is and, moreover, what it does. Eating, as a conceptual and a physical act, brings both foods and bodies into view; food does not remain on a supermarket shelf, in the kitchen or on a plate, but is placed in the mouth, chewed, tasted, swallowed and digested. Its solidity is thus broken down and rendered into fragments that both pass through, and become, the eater's body. This is a process that concomitantly establishes and ruptures social relations between bodies, whether those of the food's producers, retailers, micro-biological components or even of the original animal sources. Unpacking the encounters between foods and (human and non-human) bodies then, offers a way to take account of the many networks and relations embedded in and performed by eating; these may be obscured by more established paradigms that consider food inside *or* outside the body, rather than inside *and* outside the body. The theoretical 'golden thread' that runs through this volume, therefore, is a simultaneous reflection on foods both within and beyond bodies, and the signifying of how an exploration of this dual perspective can illuminate the oft-hidden relations of eating. These relations, as well as possibly 'partial' (Strathern 2004) and transient, are, as all the contributors in this volume demonstrate, often unplanned or unknown. Hence our use of the term 'encounters' signifies a way in which to avoid assumptions of intentionality and to engage with its frequent lack, offering instead a tracing of dissipated agency and contingency. Some consumers may purposefully seek out specific relations of eating, as is the case with 'ethical' foods, whereas other meetings between foods, bodies and personhoods may be more haphazard and inchoate. Thus, whilst concurring with Jack Goody's statement that eating signifies 'a way of placing oneself in relation to others' (1982: 37), the chapters in this volume illuminate that in the act of eating, agency cannot, and should not, be located entirely with the consumer; it is dispersed across multiple actors and within both individual and social bodies. Moreover if food, as Elspeth Probyn argues, 'reminds us of others' (2000: 12) then eating *makes* those others; it is an act that (re)orders, constructs, destroys and mobilizes across cultural and geographical distances.

Thus, this volume employs eating as a tool – a novel way of looking – whilst also drawing attention to the term 'eating' itself, and the multiple ways in which it can be constituted. What centrally emerges from this layered analysis is a complex and nuanced portrait of relationality. We suggest that much scholarship of food to date has privileged the symbolic meanings, commensality and political economy of food. The foundations laid by Douglas (2002), Douglas and Isherwood (1979) and Mintz (1985) continue to provide innumerable valuable insights into relationships between food, personhood and belonging through a range of analytical lenses, as well as expounding the historical and contemporary, and local and global, processes through which our food preferences and options are shaped.[1] Debates over the structural mechanisms and political and economic relations through which foods are made available to consumers (cf. Hawkes et al. 2009; Reardon and Berdegué 2002; Ritzer 2010) and the ways they, and their spaces, are in turn appropriated, localized and domesticated (Caldwell 2004; Miller 1998; Turner 2003; Watson 1997) have drawn attention to the individual agency and creativity of purchasing, preparing, and eating foods. However, these discussions have rarely overtly engaged with bodies, embodiment and biologies, and the breadth of work across the humanities and social sciences exploring theses facets of human existence; we rarely gain a sense of eating as bodily and affective – of how that piece of fried chicken tastes, smells, and feels as it enters the body and leaves its residue on sticky fingers. The visceral, haptic and sensorial experiences of eating have been discursively shifted into the background and thus our 'understanding of the political importance of food may be hindered by an inability to specify the link between the materiality of food choice (bodily experiences of food) and ideologies regarding food and eating' (Hayes-Conroy and Hayes-Conroy 2008: 462).

1 Prolific examples with which the contributors to this volume engage include Bestor (2000), Carsten (1997), Pilcher (2002) and Wilk (1999, 2002, 2006).

Consequently, with a few notable theorists (cf. Bennett 2010; Lupton 1996; Mol 2008; Probyn 2000), scholarship has tended to privilege either the symbolic *or* the material dimensions of food, and to focus on either the social *or* the individual body. What might be termed a dislocation between foods and mouths has perhaps become entrenched as reflections on the movements of food through the spaces and socialities outside bodies have blinded us to its traversals of those within them and, importantly, to the relationships between these. As such, the ways in which chewing, ingesting and digesting are profoundly social as well as individual, have become lost from view. We suggest that paying attention to eating offers insights into how foods and bodies both haphazardly encounter, and actively engage with, one another in ways that are *simultaneously* material, social and political. Moreover, unfurling from these body/food encounters is, to borrow from Jane Bennett, a 'complicated web of dissonant connections between bodies' (Bennett 2010: 4) and also between geographical and biological spaces and a multiplicity of actors.

While located in diverse geographical, cultural and disciplinary contexts, all the chapters in this volume therefore explore eating as a participatory practice not just in the sense of what is eaten, where it is eaten and with whom it is eaten (or not), but also in terms of individual eating bodies and their relations with other social actors, both human and non-human. In different but complementary ways, all demonstrate that eating draws together people, places and objects that may never tangibly meet and show how relationships between these are made and unmade with every mouthful. They thereby trace the multiple scales in which eating bodies are entangled and show how eating bodies become coterminous with other bodies over geographic, affective and social distances. Across the volume, these connections engendered by eating emerge as rhizomatic, resembling a 'map that is always detachable, connectable, reversible, modifiable, and has multiple entryways and exits and its own lines of flight' (Deleuze and Guattari 2004: 23). Moreover, given that 'any point in a rhizome can be connected to anything other, and must be' (ibid.: 7), eating itself is also a process in constant movement, always changing its contours and shape as it enfolds, and is folded into, ever-changing actors. As such, eating draws our attention to the 'complex assemblages that come to compose bodies and worlds simultaneously' (Seigworth and Gregg 2010: 6, see also Collier and Ong 2005; Latour 2005). Mapping contradictions and coexistences across these scales, folds and 'assemblages' illuminates how these connections are played out within the mouths and stomachs of eaters themselves, thereby bringing the visceral back into an examination of food and commensality; taking account of the visceral – to the 'sensations, moods, and ways of being that emerge from our sensory engagement with the material and discursive environments in which we live' (Longhurst, Johnston and Ho 2009: 334) – broadens the concept of, and challenges current assumptions about, socialized eating. We suggest that individual acts of eating are always inherently social and relational, and that the socialities they enact or disrupt are as viscerally corporeal as they are political.

Paying attention to the simultaneity of commensality *and* viscerality, of the symbolic *and* material, allows the contributors to this volume to reflect critically on what eating is – what it performs and silences, what it produces and destroys, and what it makes present and absent. In tracing these ever-shifting contours of eating, each chapter in this volume expounds eating's inherent paradoxes by highlighting both the creative and destructive nature of the act; they explore the fundamental tensions and interplays between choice, risk and necessity, and life and death, that are embedded in this quotidian and apparently mundane activity. Eating, as Deane Curtin has argued, highlights that 'some aspects of our experience are valuable just because they are physical, transitory, and completely ordinary' (Curtin 1992: 4).

To this volume there are 13 case studies framed by four theoretical interludes. Mirroring the 'rhizomatic' (see Deleuze and Guattari 2004) connections that are engendered by the moment of placing food in the mouth, the form of this book is also based on rhizomatic resonances that see its arguments unfolding not only across chapters and sections, but also across the many disciplines from which its authors write. We have intentionally resisted any easy narratives about eating, not grouping the chapters into clearly defined 'topics' that might be assumed to 'fit' naturally together. Rather, knowing how ways of looking may become too-easily accepted and solidified, and not seeking to perpetuate established blinkers around eating, we have sought juxtaposition rather than neatness within each of the four sections of the text. We have aimed to put together chapters that touch edges rather than align or dialogue in any neat way, seeking out partiality (see Haraway 1991), tropes and resonances through which to forge new pathways of reflection. Within each section thus, chapters with differing approaches and disciplinary backgrounds are drawn into a dialogue which allows each to illuminate the others. The conceptually-led section headings have been designed not only to reflect key concepts of eating, but also to open up spaces in which chapter authors consider their own topics in relation to common, but productively ambivalent, conceptual hinges. This has allowed them to go beyond and even challenge our own thinking about what eating may be and the ways in which it can be theoretically and methodologically addressed. Each section's interlude also teases out and cements the theoretical synergies and intersections between its chapters. These interludes offer a secondary layer to this exploration of eating, allowing complexities to emerge in clear but nuanced ways throughout. Not only engaging with their own sections, but also echoing the discussions in others, the interludes thereby weave together diverse perspectives into a cohesive multi-disciplinary discussion of the symbolic and material encounters between foods and bodies.

The explorations in this volume intersect with a variety of topical and timely theoretical and applied debates. These wider themes take many guises and engender different, even contradictory, critical reflections on the part of authors, as eating emerges as an act through which outsides and insides, bodies and landscapes, foods and imaginings, are drawn into moments of encounter. This multidimensionality resonates across the volume as the contributors explore shifting intersections of meaning and materiality. That eating appears to be an 'assemblage' (Collier and Ong 2005; Deleuze and Guattari 2004) and can be reflected on in relation to wider theoretical explorations of networks (Callon 1999; Latour 2005), flows (Deleuze 1998; Grosz 1994; Martin 1994) and multiplicity (Law and Mol 2002; Mol 2002) is most tangibly articulated through accounts of circulations of food's material culture: O'Connor's seaweed snacks, Coles's coffee cup and 'travel brochures' and the packaging of Baker's pork sausages are just a few of the examples found in this volume. Yet, networks of many actors are also apparent in other, less tangible ways. Ormond, for example, illustrates how the materiality of milk, such as whether or not it tastes of garlic, is constituted through a bewildering array of actors and agencies positioned temporally and spatially between producer and consumer. In turn, by reflecting on metabolism, Kendrick opens our vision to the many actors of digestion as well as ingestion, suggesting that to understand how and why we eat, it is necessary to take account of connections between the insides and outsides of bodies. Hurn too draws our attention to multiplicity and entanglements, illustrating how in fox hunting and meat eating, sheep, fox and human bodies are triangulated into a network through which each tangibly constitutes the other. Yet, multiplicities can also be collapsed into a shadowy singularity through eating; Baker's unpacking of a sausage illustrates how in the journey to the plate certain people and places become visible, whilst others are obscured and hidden.

Complementing these tropes is an exploration of how, in the act of placing food in the mouth, landscapes, people, objects and imaginings not only juxtapose with and fold into one another, but are also reconstituted and reordered. As foods circulate and are shaped into edible and affective commodities, they also processually shape the worlds both from which they originate and through which they move. Through these ways that eating itself acts, it always intertwines affective engagements with ethical choices and ramifications. It is in Coles's chapter that this sense of the processes of eating – or rather, perhaps – eating as process is most explored. By charting how the environment in which it grows is brought to the fore in the marketing of coffee, he argues that the bodies tasting and ingesting the drink reshape, and abstract, its landscape of production. In a very different context, Lavis too traces how eating (and not eating) (re)order, and even constitute, social worlds, by reflecting on ways in which the illness of anorexia may be processually remade through relationships of absence with food. In so doing, she, like O'Connor, illustrates that the absence as well as the presence of eating can be generative; its lack may also make connections. In exploring the not eating of seaweed, through the lens of blue rather than green issues, O'Connor further demonstrates that eating can be a destructive, as well as generative, act. Shaping the world is not always 'to the good', and a number of the contributors highlight the ways in which eating can be both detrimental and favourable, not only to material worlds but also to social, economic and cultural worlds; this is suggested by Abbots, Brooks et al., Lavis, Saleh and Yotova, to name but a few. While O'Connor asks how we can eat sustainably in relation to marine resources, Ormond explicitly reflects on what eating sustainably can mean; this is a question that is further raised, perhaps more obliquely, and widened, throughout the volume through the charting of the ways in which eating can sustain culture, social networks, and economic and political relations.

Sustainable and ethical food practices then do not just relate to the well-trodden paths of food security, scarcity, environmental conservation and 'fairer' relations of production (cf. Altieri 1995, 1998; Barrientos and Dolan 2006; Pottier 1999; Raynolds et al. 2007), but also to the sustainability or destruction of traditions, knowledges, family, community, nations, and ultimately, bodies. As such, eating emerges across the chapters as an everyday enactment of particular, even contradictory, and utterly 'messy' (Probyn 2010), ethical concerns, whether that is deliberate and active on the part of the eater or not. Eating is perhaps, therefore, always constituted by a plethora of macro and micro, relational and affectively intimate, dilemmas. In particular, together the authors in this volume raise questions not only regarding what we should and should not eat, but also of *whether* or not we should eat. Paying attention to how eating is thus, in many ways, problematic, illuminates why, as Abbots, Baker, Brooks et al., Hurn, Lavis and Ormond all stress in diverse ways, eating may require management, governance and regulation. As the authors show, this can take many forms and animal welfare provisions; European Union legislation; healthy eating guidelines; Corporate Social Responsibility frameworks or individual ideological practices are just a few examples addressed in the following chapters.

These discussions chart the ways in which national and supra-national agencies and institutions, markets, and knowledge brokers manipulate and inform people's eating practices – in terms of their palates and understandings of 'good' and 'bad' foods, and their selection of foods from the 'choices' available. Yet the manner in which eating may express individual agency, creativity and action is also a central thread of many of the chapters, which look to expound nuanced and complex dynamics between individual eating bodies and the social bodies in which they are embedded. At their heart is the question most famously posed by Mary Douglas (2002), that of what is edible and what is inedible. As Murcott points out in her interlude, concerns over what food is, let alone what food does, remain central and should not be assumed; before it is eaten it must become edible. It is perhaps O'Connor who most clearly elucidates this process by exploring how seaweed is and is not eaten in two contrasting geographical contexts of Wales and Japan. Exemplified by O'Connor's attention to the cultural reframings of *laver*, throughout the volume is an attention to how food is 'wrapped' both conceptually and materially, and on a number of different scales. Thus, imagined and marketed relations with producers – and other consumers – can be a source of reassurance for consumers, as Baker's discussion of welfare standards and Yotova's account of Bulgarian yogurt consumption indicate. Yet, they can also be disquieting, especially as imaginings become a lived, embodied reality, as Abbots's account of privileged migrants' experiences in Ecuadorean markets demonstrates. In this context, markets, and the physical proximity of producers, are constituted as problematic, and require the construction and maintenance of bodily and social boundaries. Abbots consequently draws attention to the viscerality of markets, and the ways in which they are experienced, sometimes in deeply unpleasant ways, through the body.

Resonating across the volume, at times implicitly and at others explicitly, thus, are multiple notions of the risk that eating poses to individual and social bodies, landscapes and cultures, and the betterment of risk positions (cf. Beck 1992; Giddens 1991). The viscerality of these diverse paradigms of risk highlights how eating is continually framed and reframed by many, often competing, discursive paradigms within which bodies become entangled. Bodily boundaries themselves may also become fluid (see Longhurst 2001), contested and re-sited when, as the contributors show, these knowledge constructions themselves are weighed, digested and embodied.

In her chapter, Yates-Doerr conjures up the smell of freshly-baked bread to explore how, as this object moves from Guatemala to New York, its meaning – and indeed materiality – remain intact, whilst in Guatemala competing value systems frame and alter the meanings of this materiality. We are thereby reminded of Appadurai's (1986, 1996) observation that ideas, ideologies and knowledges travel with, against and are caught up with, objects and social actors as they all move through different spatial and temporal dimensions. In examining the discursive, as well as material, flows of foods and the ways in which they produce places and people, this volume shows how, just as eating acts to shape social worlds and geographical landscapes, it too is perpetually acted on and reshaped (cf. Coveney 2006; Foucault 1977, 2002; Lupton 1996; Mennell 1987). Whether formalized in nutrition policies and practices; bound up in concerns over animal welfare, environmental protection, fairtrade and industrialization or contributing to the construction of 'imagined communities' (Anderson 1983), the chapters all elucidate how knowledges about eating and eating bodies serve to inform, 'order' (see Bowker and Star 2000), regulate or legitimize eating practices. As such, many also illuminate the multiple ways in which both human and animal bodies become entangled within, sometimes competing, ideologies.

Questions are thus raised over the legitimacy, authenticity and manipulation of competing paradigms, and the issue of whose voices are heard the loudest over the cacophony of claims and counter-claims comes to the fore (cf. Goodman 1999). Aphramor et al., Brooks et al., Cohn and Yates-Doerr tackle these questions directly in the context of diverging constructions of 'healthy eating' and the, often state-sponsored and medicalized, knowledges that are promoted in the UK and Guatemala. All, in their own way, draw attention to knowledge brokers – in the form of nutritionists, doctors, retailers, and the state – and offer a nuanced discussion of the power dynamics and political relations that are produced and maintained, as well as those that are challenged, by 'alternative' approaches and frameworks. Engaging with eating as a participatory practice that may be externally framed, Aphramor et al., show us how 'eaters' are persuaded to eat some foods while also being dissuaded from others and this is a process further elucidated by Yotova in her account of the international branding of 'national' yogurt in Bulgaria. Across these accounts, discursive paradigms may overlap as eaters draw together competing knowledges in one mouthful, or they may be incommensurate and invoke an assessment – by producers, policy makers and consumers – of priorities and the negotiation of potential 'trade-offs': this kind of discursive layering emerges from the discussions of Brooks et al., Yates-Doerr and Ormond in particular, with the latter pointing to broader concerns and subjective interpretations within such processes. As such, it is not just in the realm of nutrition and 'healthy' eating that competing knowledge paradigms emerge. Saleh's discussion of intergenerational transmission of food knowledges, in both its narrative and embodied forms, elucidates how mobilizations of knowledge can sustain a sense of being and belonging. In tracing the ways in which the embodied experiences of taste and smell communicate values, she highlights ruptures as well as continuities.

Thus, eating, as Holtzman reminds us in his interlude, is inherently fraught and ambivalent; food is a source of comfort and nourishment, and also of guilt and tension. All of these moreover, as much as the swallowing of food itself, are profoundly embodied (Csordas 1990, 1994) and affective (Clough and Halley 2007; Massumi 2002). Eating, as Cohn's interlude elucidates, cannot be measured as embodied subjectivities do not fit neatly into quantifiable ordered categories or even moments. This is most demonstrated by both Yates-Doerr's and Aphramor et al.'s critiques of the 'body as machine' metaphor. Yet, both Kendrick and Lavis, albeit very differently, also illustrate that the affective, hungry, viscerally-experiencing body cannot be dislocated from explorations of the many agential and unagential ways in which foods and bodies encounter one another. In so doing, Kendrick's chapter demonstrates that relationships between eating and obesity may themselves be problematically tangential. Taking account of, but going beyond, everyday intimacies of taste, pleasure and revulsion, this sense of the haptic in the volume's explorations of body/food encounters is also, we suggest, of theoretical importance. It is particularly in relation to debates around 'globesity' (Delpeutch et al. 2009, see also Gilman 2010) that this is most clear. By elucidating tensions between the sensory and the quantifiable in discussions of 'fat' (see Kulick and Meneley 2005), contributors to this volume challenge narrow and linear rhetoric of individual responsibility for obesity and bring the materiality of the body back into view alongside an attention to the 'topographical' complexities of subjectivity (see Haraway 1991: 193) in eating.

As many of the authors illustrate, the politics of eating and encounters between foods and bodies are at times explicit, visible and enacted on a global stage. Yet they are also intimate, micro, and bodily. They may be actively sought and produced or unexpected and unplanned. They may also be a source of comfort and sustenance – 'doing good in the world' – to communities and to individuals, or disquieting, even harmful and damaging to both social and individual bodies; assessments that are, in themselves, inherently subjective. In its emphasis on the multiplicity of objects, relations, biologies and affects folded into, and enacted by, eating, this volume addresses the 'mess' (Law 2004) that we risk not seeing if we take account only of foods outside *or* inside bodies. It suggests that, despite the forms food/body encounters take and in whatever mess they are entangled, eating is concomitantly always both ideological and material (see Conroy-Hayes and Conroy-Hayes 2008). As all the contributors demonstrate, thus, asking why and how we eat offers important insights into how every mouthful is political, visceral and, centrally, always relational.

References

Altieri, M. (ed.) 1995. *Agroecology: The Science of Sustainable Agriculture*. Boulder, Colorado: Westview Press.

Altieri, M. 1998. Ecological impacts of industrial agriculture and the possibilities for truly sustainable farming. *Monthly Review* 50(3), 60–71.

Anderson, B. 1983. *Imagined Communities: Reflections on the Origins and Spread of Nationalism*. London: Verso.

Appadurai, A. 1986. Introduction: Commodities and the politics of value, in *The Social Life of Things: Commodities in Cultural Perspective*, edited by A. Appadurai. Cambridge: Cambridge University Press, 3–63.

Appadurai, A. 1996. *Modernity at Large: Cultural Dimensions of Globalization*. Minneapolis: University of Minnesota Press.

Barrientos, S. and Dolan, C. (eds) 2006. *Ethical Sourcing in the Global Food System*. London: Earthscan.

Beck, U. 1992 [1986]. *Risk Society: Towards a New Modernity*. London: Sage Publications.

Bennett, J. 2010. *Vibrant Matter: A Political Ecology of Things*. Durham, NC: Duke University Press.

Bestor, T. 2000. How sushi went global. *Foreign Policy*, Dec. 2000, 54–63.

Bowker, G.C. and Star, S.L. (eds) 1999. *Sorting Things Out: Classification and its Consequences*. Cambridge, Mass.: MIT.

Caldwell, M. 2004. Domesticating the French fry: McDonalds and consumerism in Moscow. *Journal of Consumer Culture* 4(1), 5–26.

Callon, M. 1999. Some elements of a sociology of translation: Domestication of the scallops and the fisherman of St Brienc Bay, in *The Science Studies Reader*, edited by M. Biagioli. New York and London: Routledge, 67–84.

Carsten, J. 1997. *The Heat of the Hearth: The Process of Kinship in a Malay Fishing Community*. Oxford: Clarendon Press.

Clough, P.T. and Halley, J. (eds) 2007. *The Affective Turn: Theorizing the Social*. Durham and London: Duke University Press.

Collier, S.J. and Ong, A. 2005. Global assemblages, anthropological problems, in *Global Assemblages: Technology, Politics and Ethics as Anthropological Problems*, edited by S.J. Collier and A. Ong. Oxford: Blackwell Publishing.

Coveney, J. 2006. *Food, Morals and Meaning: The Pleasure and Anxiety of Eating*. London: Routledge.

Csordas, T.J. 1990. Embodiment as a paradigm for anthropology. *Ethos* 18(1), 5–47.

Csordas, T.J. (ed.) 1994. *Embodiment and Experience: The Existential Ground of Culture and Self.* Cambridge: Cambridge University Press.

Curtin, D.W. 1992. Food/Body/Person, in *Cooking, Eating, Thinking: Transformative Philosophies of Food*, edited by D.W. Curtin and L.M. Heldke. Bloomington: Indiana University Press.

Deleuze, G. 1998. *Essays Critical and Clinical.* Translated by D.W. Smith and M.A. Greco. London: Verso.

Deleuze, G. and Guattari, F. 2004. *A Thousand Plateaus: Capitalism and Schizophrenia.* London and New York: Continuum.

Delpeuch, F., Maire, B., Monnier, E. and Holdsworth, M. 2009. *Globesity: A Planet Out of Control?* London: Earthscan.

Douglas, M. 2002 [1966]. *Purity and Danger: An Analysis of Concepts of Pollution and Taboo.* London: Routledge and K. Paul.

Douglas, M. and Isherwood, B. 1979. *The World of Goods: Towards an Anthropology of Consumption.* New York: Basic Books.

Foucault, M. 1977. *Discipline and Punish: The Birth of the Prison.* London: Allen Lane.

Foucault, M. 2002. *The Order of Things: An Archaeology of the Human Sciences.* London: Routledge.

Giddens, A. 1991. *Modernity and Self-identity: Self and Society in the Late Modern Age.* Cambridge: Polity Press.

Gilman, S. 2010. *Obesity: The Biography.* Oxford: Oxford University Press.

Goodman, D. 1999. Agro-food studies in the 'Age of Ecology': Nature, corporeality, bio-politics. *Sociologia Ruralis* 39(1), 17–38.

Goody, J. 1982. *Cooking, Cuisine and Class: A Study in Comparative Sociology.* Cambridge and New York: Cambridge University Press.

Grosz, E. 1994. *Volatile Bodies: Towards a Corporeal Feminism.* Bloomington and Indianapolis: Indiana University Press.

Haraway, D.J. 1991. *Simians, Cyborgs, and Women: The Reinvention of Nature.* London: Free Association Books.

Hawkes, C., Blouin, C., Henson, S., Drager, N. and Dubé, L. (eds) 2009. *Trade, Food, Diet and Health: Perspectives and Policy Options.* Oxford: Wiley-Blackwell.

Hayes-Conroy, A. and Hayes-Conroy, J. 2008. Taking back taste: Feminism, food and visceral politics. *Gender, Place & Culture: A Journal of Feminist Geography* 15(5), 461–73.

Kulick, D. and Meneley, A. 2005. *Fat: The Anthropology of an Obsession.* London: Penguin.

Latour, B. 2005. *Reassembling the Social: An Introduction to Actor-Network-Theory.* Oxford: Oxford University Press.

Law, J. 2004. *After Method: Mess in Social Science Research.* Oxford and New York: Routledge.

Law, J. and Mol, A. (eds) 2002. *Complexities: Social Studies of Knowledge Practices*. Durham and London: Duke University Press.

Longhurst, R. 2001. *Bodies: Exploring Fluid Boundaries*. London: Routledge.

Longhurst, R., Johnston, L. and Ho, E. 2009. A visceral approach: Cooking 'at home' with migrant women in Hamilton, New Zealand. *Transactions of the Institute of British Geographers* 34(3), 333–45.

Lupton, D. 1996. *Food, the Body and the Self*. London, Thousand Oaks and New Delhi: SAGE.

Martin, E. 1994. *Flexible Bodies: Tracking Immunity in American Culture from the Days of Polio to the Age of AIDS*. Boston: Beacon Press.

Massumi, B. 2002. *Parables for the Virtual: Movement, Affect, Sensation*. Durham and London: Duke University Press.

Mennell, S. 1987. On the civilising of appetite. *Theory, Culture & Society* 4, 373–403.

Miller, D. 1998. Coca-Cola: A black sweet drink from Trinidad, in *Material Cultures: Why Some Things Matter*, edited by D. Miller. Chicago: Chicago University Press, 169–87.

Mintz, S. 1985. *Sweetness and Power: The Place of Sugar in Modern History*. New York: Viking.

Mol, A. 2002. *The Body Multiple: Ontology in Medical Practice*. Durham and London: Duke University Press.

Mol, A. 2008. I eat an apple: On theorizing subjectivities. *Subjectivity* 22, 28–37.

Pilcher, J. 2002. Industrial tortillas and folkloric Pepsi: The nutritional consequences of hybrid cuisines in Mexico, in *Food Nations: Selling Taste in Consumer Societies*, edited by W. Belasco and P. Scranton. New York and London: Routledge, 222–39.

Pottier, J. 1999. *Anthropology of Food: The Social Dynamics of Food Security*. Cambridge: Polity Press.

Probyn, E. 2000. *Carnal Appetites: FoodSexIdentities*. London and New York: Routledge.

Probyn, E. 2010. Feeding the world: Towards a messy ethics of eating, in *A Critical Introduction to Consumption*, edited by E. Potter and T. Lewis. London: Palgrave.

Raynolds, L., Murray, D. and Wilkinson, J. 2007. *Fair Trade: The Challenges of Transforming Globalization*. London: Routledge.

Reardon, T. and Berdegué, J.A. 2002. The rapid rise of supermarkets in Latin America: Challenges and opportunities for development. *Development Policy Review* 20(4), 371–88.

Ritzer, G. 2010. *McDonaldization: The Reader 4th Edition*. LA: Pine Forge Press.

Seigworth G.J. and Gregg, M. 2010. An inventory of shimmers, in *The Affect Theory Reader*, edited by M. Gregg and G.J. Seigworth. Durham and London: Duke University Press.

Strathern, M. 2004. *Partial Connections, Updated Edition*. Walnut Creek, Lanham, Oxford: Alta Mira Press: Rowman and Littlefield.

Turner B. 2003. McDonaldization: Linearity and liquidity in consumer cultures, in *American Behavioral Scientist* 47(2), 137–53.

Watson, J. (ed.) 1997. *Golden Arches East: McDonalds in East Asia.* Stanford, CA: Stanford University Press.

Wilk, R. 1999. 'Real Belizean food': Building local identity in the transnational Caribbean. *American Anthropologist* 101(2), 244–55.

Wilk, R. 2002. Food and nationalism: the origins of 'Belizean' food, in *Food Nations: Selling Taste to Consumer Societies*, edited by W. Belasco and P. Scranton. New York and London: Routledge, 67–89.

Wilk, R. 2006. *Home Cooking in the Global Village: Caribbean Food from Buccaneers to Ecotourists*. Oxford and New York: Berg.

PART I
Absences and Presences: How We (Do Not) Eat What (We Think) We Eat

This opening section explores what is made present or absent, visible or obscure, by eating and not eating across three different contexts. In unpacking the complex interplays of eating and (in)visibility, the chapters share a united interest in the concomitant visceral materiality and yet slippery intangibility of food. They trace how both knowing and unknowing absences are enacted by eating, as this seemingly mundane activity makes material certain relationships and overlooks, unsettles or even destroys others. The section highlights physical and conceptual gaps that lie not only between producers and consumers, personhoods and foods, but also within the same foods as they move between diverging contexts. As such, the chapters investigate how both eating and not eating can be transformative and generative processes, which are performed both with and against the agency of eaters and eating bodies. Beginning with an examination, through seaweed, of how and why certain foods become absent from our plates, the section then explores how resisting eating may produce relationships of absence with food through which the presence of anorexia is processually remade. Finally, by reflecting on sausages and pigmen, the section moves beyond material absences in, and of, eating to explore how the act of eating can engender ruptures as well as connections, as relations that constitute the substance consumed become obscured. In the coming together of these three chapters, thus, this section's central concern is to make visible a plethora of relationships and broken links that are continually (re-)enacted by food's trajectory not only from producers to plates, but also from mouths to embodied selves. Accompanied by an interlude by Simon Cohn, as well as visual imagery, these chapters thereby open discussion of the many actors embedded in eating. In so doing, they offer important insights into the ways in which complexities of eating cannot, and should not, be quantified in any simple ways.

Chapter 1
Invisible Foodscapes: Into the Blue

Kaori O'Connor

Introduction

Plunging into the Blue, this paper explores the invisible foodscapes that are central to a critical understanding of why we eat, how we eat. Food is a uniquely problematic form of material culture – ephemeral, elusive, always in the process of transformation. By 'foodscapes' I mean what Sahlins (1976: 170) called the 'cultural reason in our food habits', which defines the consumption practices of a society or group and allows us to better understand them through what they eat and drink; and just as importantly, what they *do not* eat and drink – what they 'see' and do not 'see' as edible. This chapter suggests that invisible foodscapes are as fundamental to human life as the visible ones that anthropologists usually study. Invisible foodscapes have tended to be overlooked and it is only through a systematic study of cultural 'blindness' regarding certain foods that we can begin to engage with pressing issues of health, sustainability and food security. Foodscapes can be invisible in several ways:

- Culturally invisible: People do not 'see' them, either as foods or as food producing zones.
- Physically invisible: Low profile industry or production, barely visible on land or sea.
- Judicially invisible: Production and practices largely unrecognized in and unprotected by law.
- Administratively invisible: Not administered or regulated in a formal way.
- Commercially invisible: Not promoted as a foodstuff.
- Nutritionally invisible: Nutritional benefits not recognized or publicized.
- Culinarily invisible: Does not play a central or recognized and valued role in cuisine.

My focus here is on the invisible Blue, the marine and coastal places that people do not 'see' or care for, the sea foods that people choose not to eat – yet which they may consume without realizing it – and the ocean resources that people ignore when they focus on 'Green' or terrestrial products and issues. This invisibility has serious consequences. The negative effects of the standard western diet on human health and on the terrestrial environment are among the most pressing concerns of the twenty-first century (Cordain et al. 2005; FAO 2002, 2003).

To limit the damage, organizations such as the WHO and the FAO are calling for modifications to dietary habits that involve eating less meat and it has been suggested that a reduction of even 10% in the human consumption of animal protein in the developed world would be beneficial. What is needed is a food for humans that can be produced without despoiling the terrestrial environment and whose consumption would allow a decrease in animal protein intake without nutritional loss. That food is edible seaweed; it already exists in abundance but in Britain and Europe little is done with it because it is culturally invisible.

Eating the Sea

The fundamental relationship between humans and their environment has always been one of alimentary exploitation and consumption. This is often lost from view in the developed world where people have become alienated from the production and harvesting of food and have come to see nature as something for contemplation and conservation rather than use. In the UK, 'fish' is what most people think of as *the* marine food, especially such popular species as Atlantic cod, plaice, wild halibut, wild salmon and haddock. These are severely over-fished and are often harvested using destructive fishing practices such as trawling, which result in a high level of discards in which netted fish of the 'wrong' species are simply dumped back into the sea, dead (Cefas-Defra 2011; Fishfight 2011; Smithers 2011). The depletion of the world's fish stocks epitomizes the 'Blue' predicament: beyond the limits of territorial waters (at present 12 nautical miles from the low-water mark of the shore) the high seas are the last frontier, where regulation is highly problematic even within voluntary marine economic zones. If harvesting cannot be controlled at sea, the alternative is to start at the other end of the chain. Realizing that consumption and demand are dictated by culture, numerous initiatives have been launched, often with the aid of celebrity chefs and supermarket chains, which aim to preserve fish stocks by changing the way people think about and eat fish. Examples of these in the UK include Marks & Spencer's 'Forever Fish' and Sainsbury's 'Switch the Fish' campaigns, and the 'Fish Fight' initiative led by the British celebrity chef and food writer Hugh Fearnley-Whittingstall. These involve encouraging the consumption of less well-known and non-endangered fish such as megrim and pouting, and promoting labels and logos that give assurances that the fish have been 'responsibly' and 'sustainably' sourced. However, such initiatives have become victim to what the American sociologist Robert K. Merton (1968) called 'the law of unintended consequences', whereby an intervention in a complex system can have wholly unexpected and undesirable outcomes. In this case, the promotion of obscure fish can stimulate the demand for fish in general, increasing pressure on depleted popular species, to which retailers and wholesalers must respond if they are to stay in business. Moreover, however responsibly fish are harvested within territorial waters, once they swim outside these, they can no longer be protected.

Fish are not the only marine foods, and for several years I have been conducting Blue ethnography in coastal zones within UK territorial waters, focusing on a less-mobile sea food – seaweed. While the depletion of fish stocks and overfishing are now global problems, on a local level seaweed – if properly promoted and responsibly exploited – can greatly benefit individual and social bodies by contributing to healthy eating, and thereby improving nutrition, and by sustaining micro- and larger economies. Seaweeds certainly do this in other parts of the world, notably in Asia and the Pacific, so why not in the UK where they grow naturally in abundance, can be cultivated through mariculture, and were used and consumed extensively in the past?

There are thousands of species of seaweed – marine algae – of which the most popular edible kind world-wide is porphyra, which was described by Rachel Carson with delicious accuracy as resembling 'nothing so much as little pieces of brown transparent plastic cut out of someone's raincoat' (1964: 443). It flourishes on rocky coasts and, despite its unprepossessing appearance, the 'human relationship with porphyra (Rhodophyta) is perhaps closer than with any other alga' (Mumford and Miura 1988: 88). This seaweed has traditionally been used for food in all the places around the world where it is found, valued for its taste and its nutritive qualities. It is high in protein (1.7% higher than beef by weight) and rich in essential vitamins A, B, B2 and C and minerals such as iodine. It also has jellifying or colloidal properties, which emerge after boiling, and it is both extremely digestible and full of dietary fibre. At the same time, there is concern about public health in Wales with the Welsh Government actively promoting healthy eating (Welsh Assembly Government 2006) in which increased consumption of seaweed could play a part. The Welsh economy is also in need of more local industries, which an expanded production of seaweed could meet. Having seen porphyra growing on British coasts and being familiar with its use in Japan and the Pacific, where it is highly esteemed, I set out to see how this seaweed was eaten in the UK, focusing on coastal South Wales, the one place in Britain still known for seaweed-eating.

Seaweed in Wales

'*Terroir*' – a term referring to the juxtaposition of a particular place and produce – applies perfectly to porphyra, locally called *laver*, and the coast of South Wales where it is gathered on the rocks. The preparation of Welsh *laver* traditionally begins with thorough cleaning, followed by boiling for six hours or more, until it becomes a dark and gelatinous sludge, highly digestible and full of the taste of the sea. It is customarily eaten with another local food, cockles – small saltwater clams – which have been boiled and shelled, and are usually purchased in that form. The cooked *laver* is warmed with the cockles in bacon fat, and served garnished with bacon. Sometimes the *laver* is mixed with oatmeal before frying, when it is called *laverbread.* The combination of *laver* or *laverbread*, cockles and bacon is popularly known as a 'Welsh breakfast', although it can be eaten at any time of day and the term has now been expanded to include eggs and fried bread (see O'Connor 2013).

As I have written elsewhere (O'Connor 2009, 2012), despite its renown as a distinctive local food, in South Wales *laver* is difficult to find. Or rather, it is visible to locals but invisible to outsiders. *Laver* is rarely offered in restaurants or seen for sale in supermarkets and consumption is usually in the home, with the *laver* purchased ready-cooked from a relatively small number of outlets well-known to Welsh consumers, prime among them the long-established municipal markets of Cardiff, Swansea, Llanelli and Carmarthen. As Mary Douglas (in Douglas and Isherwood 1979) observed, goods – of which food is a primary form – make visible the categories of culture, and an ethnographic study of Welsh seaweed consumption and non-consumption reveals differences of identity, affinity and generation that do not manifest in other ways. Seen from the outside as a 'national food' of Wales as a whole, inside Wales *laver* is considered a food specifically emblematic of South Wales, and within that two strands were discernible. Originally, *laver* was a defining food of the small South Wales coastal communities of fishermen and small farmers, which were transformed by the urban industrialization of the area in the nineteenth century, which turned Llanelli into 'Tinopolis' and Swansea into 'Copperopolis'. Labourers streaming into South Wales from elsewhere to work in mines and factories took up the consumption of *laver*, already established locally, because they appreciated it as both a cheap, nutritious food and also as a kind of prophylactic against the illnesses connected with their employment. People whose coastal family roots predate the industrial influx or whose antecedents were early incomers tend to associate *laver* with place – with South Wales – and speak of it as 'our food', while descendants of later incomers tend to think of *laver* in terms of period, connecting it to the nineteenth and early twentieth centuries, and the working men's life of the late industrial age. A gender element is still apparent; *laver* with or without cockles or bacon is considered more a 'man's meal' than a woman's, although women eat it too, and it is one of the few things that traditional Welsh men of the 1950s and 1960s would cook for themselves, especially on Sunday mornings. In those days, many older men and women remember, the only bacon used was the salty, smoky and, as one person put it wistfully, 'wonderfully fatty' kind; a side of Welsh bacon was commonly hung in the kitchen and slices were cut off to fry with the *laver*. Cooked *laver* has a distinctive consistency – gooey and viscid. It sticks to the plate and frying pan like a snail's trail, but to aficionados this is part of its appeal – 'That's goodness, that is' they would say. As to why people eat *laver*, there was a vague idea that it is 'good for you', and has 'lots of iodine', the latter being particularly mentioned by older people who remembered the disfiguring condition goitre, connected with iodine deficiency. Mainly, however, the reasons given were personal and nostalgic – 'We always had it when I was growing up' was a frequent response that cut across the boundaries of gender and age. Similarly, those who now lived away, often spoke of it as the food they liked to eat on their visits back – 'When I eat it, I know I'm home'.

More recently, *laver* has been caught up in generational change and the emerging identities of political devolution. It is eaten more by older people than by the young and it figures in the ongoing dialogue in the Welsh press between traditionalists and modernizers that surfaces every year around 1st March on St David's Day, which honours the patron saint of Wales. Those who want a new, progressive image for Wales seek to disassociate the nation from the past and its old emblematic foods in order to show that Wales is thoroughly modern and 'not just about leeks and *laver*'. Among many, *laver* is also seen as a 'poor person's food', associated with the hard industrial times. When I asked advocates of change – all under the age of 40 – what they thought of as a 'modern' and desirable food, the answer was invariably '*sushi*'. Would they eat *laver*, I asked. The answer was a dependable and emphatic 'no'. In some cases, the response was almost visceral, particularly among those who had never tasted it; the idea alone was repellent, and 'old-fashioned', 'slimy', 'looks weird' were the most common reasons given, accompanied by suitable facial expressions. Equally dependable was their shock and disbelief when I told them that the *sushi* they ate frequently at great expense was wrapped in Japanese *nori,* which is dried porphyra or *laver*. Seen through a cultural lens, one food was perceived as sublime and the other as slime (see Teas 2002), but the seaweed itself was exactly the same. Or to put it another way, in one context the seaweed was visible, in the other invisible.

My ethnography was expanded to explore further culinary and cultural transformations of seaweed in Wales and around Britain. The people who sold seaweed on the Welsh markets told me that in addition to conventional cookery, fresh *laver* was purchased for two reasons – health and beauty. By 'health', they did not mean general prophylactic use but, rather, the usually macrobiotic dietary therapies used in complementary or alternative medicine for cancer care. These regimes often specify regular consumption of seaweed and people came from considerable distances to stock up on freshly-prepared Welsh *laver*, which they would freeze and use as necessary. In other parts of Britain, I found that seaweed was spoken of more in medicinal than culinary terms, and anecdotal evidence suggests that its use as a 'country' or 'folk' remedy persists, although a precise identification of species used is hard to come by. One of the most commonly-reported uses is as a poultice for sprains or bruises, a treatment that involves rehydrating dried seaweed, applying it to the affected area, fixing it in place with a bandage and leaving overnight. Back in Wales, the second, and most unexpected, use to which fresh *laver* was put was as a beauty treatment, with people applying freshly-prepared *laver* paste directly to the face as a facial. 'Much cheaper than buying a commercial facial' I was told on several occasions, 'and completely natural. No additives, you know it's pure'. Also recounted was the application of *laver* to the hair as a conditioning treatment, combed or massaged in after shampooing and left to sit for about half an hour under a shower cap before being rinsed out. This corresponds to the fact that, in terms of the cultural mainstream in Britain and Europe today, the most visible use of seaweed is in the 'beauty and wellness' sector.

Thalassotherapy and Phycocolloids

Seaweed and the sea have been commodified and repackaged, with 'thalassotherapy' now a broad term for many types of beauty and relaxation treatments that claim to be based on the therapeutic properties of sea water and seaweed. There are seaweed facials and body masks, scrubs, wraps, oils, lotions, bath salts, exfoliants, shampoos and soaps. These are the cosmetic legacy of the marine therapies and 'sea cures' involving salt water baths and applications of seaweed and sea mud, which were popular in Britain and Europe in the nineteenth and early twentieth centuries. As Hassan (2003: 31) put it, 'the sea was conceived as some gigantic receptacle of mineral water, which impregnated coastal air. Thereby it was transformed into the elixir of life'. Seaweed and sea mud were believed to share in these beneficial qualities, and it is in this context that the logic of applying Welsh *laver* to the face and hair becomes apparent. But if people believe in the health benefits of seaweeds used externally, why won't they use them internally – by eating them as enthusiastically as they put them on their face or in their hair? Actually, in another example of culinary irony and invisibility, many people *do* eat seaweeds daily, in substantial quantities, but without realizing it.

'Phycocolloids' are seaweed gums – extracts of seaweed that have natural jellifying and emulsifying properties, as seen in the gelatinous consistency of Welsh *laver* after cooking or in the carrageen jellies that are part of traditional Irish cuisine. Extracted and added to solid foods or liquids, phycocolloids contribute viscosity and stability. Cheesecake, ice cream, cheese spreads, artificial whipped toppings, ketchup, bottled mayonnaise and salad dressings, prepared sauces and ready-made soups, microwave and frozen meals are just a few of the modern dietary mainstays that are literally held together by phycocolloids. The quivery glossiness of a strawberry trifle, the smoothness of Caesar salad dressing, the unctuous richness of prepared lasagne, the succulence of crabsticks and 'shaped' breaded cutlets, the melting comfort of meat pies with gravy and the silky liquidity of chocolate milk are all achieved through the use of phycocolloids. They are also used to enhance 'organoleptics', the name given to the study of sensory perception, which in the case of food means factors like body, texture, flavour, smell and taste, all vitally important in the competitive commercial market. Seaweed was the source of the most significant culinary discovery of recent times – *umami*. This so-called 'fifth sense' or 'fifth taste', which is now recognized along with sweet, sour, bitter and salt, was discovered by accident by Kikuenae Ikeda in 1908 when a cooking pan of *konbu* seaweed scorched, leaving a residue with the flavour now recognized as *umami*. Patented with the brand name 'Ajinomoto' and generically known as monosodium glutamate, this gave the taste of protein to non- or low-protein foods, greatly enhancing the appeal of processed foods generally as the market for them expanded. Today, monosodium glutamate is a mainstay of the global processed food market. The theory is that the more pleasant the sensory experience, the more food or drink will be consumed; the palate must be seduced. In London, in the company of food professionals who specialize in the organolepics of food and drink, I have observed many exhaustive brand and supermarket tastings of different versions of the same kind of product to select the most appealing one. The irony is that while fresh seaweed is healthy and nutritious, the phycocolloids extracted from them are largely stripped of their nutrients before they enter the food chain, where they become the foundation of the highly-processed, high-fat, high-carbohydrate, high-sugar and low-fibre foods that bedevil the standard western diet and contribute to chronic health problems. Because phycocolloids are 'natural', they don't get flagged on labels as artificial additives. Moreover, although extracted from healthy seaweed, the *umami* flavour can encourage unhealthy eating practices. In another case of cultural invisibility, people who would never knowingly eat seaweed do so every day, but with negative rather than positive consequences.

In Britain and Europe generally, the remaining elective method of seaweed consumption is nutritional supplements, usually as pills or capsules and less often in dried form. Mainly sold in health food shops or by mail order, what the claims of producers and packagers lack in detail they make up for in enthusiasm. 'Metabolism' is a term employed frequently and vaguely, and a recurrent theme is the need to restore 'balance' by adding the minerals and vitamins that may be 'lacking from the daily diet' or 'missing from the food supply' to use two phrases that regularly appear in magazine advertisements and promotional literature for seaweed supplements. The irony of using one form of seaweed to make up for the inadequacies facilitated by another is lost on the producers and largely unknown to the consumers and, leaving aside the question of efficacy, compared to the wide use of phycocolloids the supplement market is too tiny to reverse any damage. Despite this high level of use, seaweed remains culturally invisible in much of Britain and Europe.

Seaweed in Japan

For comparative purposes, I now turn to Japan, where I carried out fieldwork, and also to Japanese diaspora communities in Hawaii and North America, where my ethnography is ongoing. Japan is the country most people think of when seaweed – or at least *nori* – is mentioned. The consumption of porphyra seaweed in Japan dates from prehistoric times (Nisizawa et al. 1987), the earliest written reference being in the *Daihoritsuryo*, Japan's first known written legal code issued in 701CE. This designated 30 types of marine produce on which an annual tribute tax had to be paid to the nobility by their vassals. Of these, porphyra was described as the choicest and best, thus entering the Japanese food system as a commodity with high cultural value and visibility. A wild product gathered under difficult conditions, porphyra was a food for the elite, valued for its fragrance and savoury taste. By the medieval Kamakura period (1185–1333 CE), changes in Japanese food culture, which included the adoption of vegetarianism and a trend towards culinary simplicity as a result of the introduction of Buddhism (Ashkenazi and Jacob 2000; Ishige 2001; Matsuyama 2003), saw porphyra increase in popularity. But as a wild food in relatively short supply, its consumption was still restricted to the nobility.

The managed cultivation of porphyra began in the Edo period (1603–1868 CE). In many societies, the development of a food that is considered culturally significant is associated with an iconic figure (Flandrin and Montanari 1999) whose association with it increases its cultural value and visibility. In this case, that figure was the Shogun Ieyasu (1543–1616 CE), founder of the Tokugawa Shogunate, the feudal military dictatorship which ruled Japan for some 250 years. The common belief is that the Shogun commanded local fishermen to bring fresh fish to him daily at his new capital of Edo, later Tokyo. To ensure a constant supply, fishermen built fish-holding pens with bamboo fences in the waters of what is now called Tokyo Bay. The discovery that porphyra seaweed grew on the bamboo fences led to commercial cultivation. The procedure was to sink bunches of twigs in tidal waters, ideally located around the mouths of rivers, and to collect the seaweed that grew on them. Seaweed was soon being produced in sufficient quantity to be consumed by a wider sector of the population for the first time, although it was still expensive. Originally, seaweed had simply been boiled and made into a paste, but after papermaking was introduced to Japan, the seaweed producers of Akasuka, now a part of greater Tokyo, used the basic techniques of papermaking to develop a method of boiling, shredding and then drying seaweed on frames. Through this they made the dried paper-like rectangles of seaweed called *nori*, which became the most popular form of edible seaweed and were renowned as a regional food speciality of Edo.

Although seaweed had become well-established in the Japanese diet and in the elaborate Japanese cycle of gift-giving (see below) by the Meiji period (1868–1912), demand always outstripped supply and it remained a luxury. Seaweed cultivation was an uncertain business since the life cycle of seaweed was not fully understood and the successful propagation of new crops seemed to be as much a matter of luck as of skill. It was not until after World War II that a British phycologist and algologist, Kathleen Drew-Baker (1901–57), discovered the complex diploid life cycle of porphyra by studying seaweed or *laver* found on the west coast of Anglesey, North Wales. The Welsh porphyra (*laver*) was a species related to the Japanese porphyra (*nori*). Drew-Baker passed her findings to Japanese colleagues who succeeded in seeding *laver* artificially in 1953. At that point, the Japanese *nori* industry was in serious decline after a succession of poor harvests exacerbated by the after-effects of the war. As a result of Drew-Baker's discovery, artificial seeding techniques and the technology of systematic *laver* farming became possible for the first time, leading to the mass production of *nori*, the establishment of a stable market for it and increasing demand. Although Drew-Baker never visited Japan, Japanese seaweed producers commissioned a statue and a memorial shrine in tribute to her research, which revolutionized the industry and paved the way for modern *nori* production. Japanese seaweed producers have also celebrated a Drew Festival in her honour every year on 14th April, the anniversary of her birth, since 1963. The ceremony involves placing Drew-Baker's academic cap and gown on her monument overlooking Tokyo Bay while a British flag is raised, and then laying a tribute of *nori* from the current crop before the shrine, a unique example of ritualized culinary visibility.

After World War II and the revival of the industry, *nori* cultivators organized into cooperative unions and today Japanese seaweed beds can only be harvested by members of these cooperative associations. Members are responsible for managing and protecting 'their' seaweed resources while not encroaching upon those of others, although boundaries can be negotiated by agreement in times of poor harvest, an arrangement which has been described as a common property regime (Delaney 2003). *Nori* production is now highly mechanized; the propagation of seaweed spores is carried out onshore before the spores or 'germlings' are transferred to nets and lines that are attached to frames placed in the sea. The growing season lasts from November through to March. When it has matured, the seaweed is picked, washed with seawater, pounded, cut finely by machine, placed in *nori* frames, dried in a dehydrator to 16% moisture and then packed to await sorting. A small amount of wild *nori* is still gathered and sun-dried for the luxury market. *Nori* professionals distinguish between many different sub-species and grades of porphyra, differences that are invisible to outsiders. Processed *nori* is taken from the producers to cooperative headquarters where it is inspected and graded according to criteria that include fragrance, colour, gloss, thickness, growing location, subspecies, texture and the way it melts in the mouth. In the case of large production cooperatives, sorting can result in over a thousand different categories. The sorted *nori* is inspected by representatives of trading houses from all over Japan, who register their bids on 'hand boards', which are then submitted like sealed bids. The *laver* is then 'finished' according to the requirements of the successful bidder, and shipped or conveyed to warehouses. Depending on what product it will be made into, the *nori* will be subjected to further processing when it comes into the hands of the buyers.

The trading houses supply a range of domestic and export clients including department stores, markets such as Tsukiji market in Tokyo, retailers and restaurants. *Nori* is eaten every day in some form by Japanese of all ages, and is one of the standard components of a Japanese breakfast. For the export market, *nori* is processed according to differing national tastes. Teriyaki steak-flavoured *nori* is very popular in America, while in Korea the favourite flavours for *nori* are sesame and *kim chee* (the spicy pickled cabbage considered one of Korea's national foods). Currently, the range of edible *nori* products for the home export markets includes the following: The main form is *nori* sheets and strips – roasted, plain or seasoned – in various grades and sizes, organic and non-organic, used for rice balls (*musube/onigiri*), *nori-maki zushi* or rolled *sushi*, *nigiri-sushi* and *temaki* wraps, or eaten alone, as a snack. Condiments consist of powdered, chopped or shredded *nori*, both natural and flavoured, to sprinkle over salads, plain rice and other foods as a visible ingredient. Sheet and powdered *nori* is also used as an ingredient in instant foods, restaurant cooking and popular snacks, such as *nori*-flavoured rice crackers (*nori make arare*) and savoury biscuits. Seaweed facials are not popular in Japan, where seaweed is still seen as something for internal consumption not external application. The mucilaginous quality of many seaweed preparations and the rubbery consistency of some seaweeds used in soups and stews are entirely agreeable to the Japanese palate. The jellifying properties of seaweed have long been recognized in Japan, playing a central role in the preparation of Japanese sweets, especially *kanten*, a jellied sweet made from seaweed-derived agar. Organoleptically, *kanten* is one of those foods that divide the western palate from the Japanese: *kanten* is so firm that it is chewy, a consistency that most westerners accustomed to wobbly jellies do not find appealing. Although *nori* is the favourite, the Japanese use some twenty species of seaweed in their cookery, the other principal ones being *kombu* (*laminaria*), and *wakame* (*Undaria pinnatifida*), both forms of kelp. Taken together, seaweeds are estimated to make up about 10% of the Japanese diet (Guiry 2007). In Japanese traditional medicine (*kampo*) the food-as-medicine system is well developed, and seaweed has always been regarded as an essential food for good health and nutrition, part of an extensive marine pharmacopeia (Chapman 1970; Okazaki 1971).

In addition to direct consumption, *nori* plays a key role in the Japanese practice of gift-giving, through which different kinds of social networks are maintained (Creighton 1991; Mauss 1966; Rupp 2003). In addition to the special-occasion gifts tied to particular events such as marriage and graduation, there are two cyclical gift seasons, *oseibo* or winter (in December) and *chugen* or summer (in July). The former may be seen as an end-of-year thank you, and a way of beginning the new year with all accounts settled. The latter coincides roughly with the *o-bon* festival, when the ancestors are honoured. *Oseibo* and *chugen* gifts are exchanged between people in family, friendship, patronage or business networks to express appreciation for help, business or loyalty given in the past, as well as the desire to keep the relationship active. These gifts should be perishable, requiring periodic replacement that will keep the gifting cycle operative. But they should also be long-lasting perishables, such as preserved, dried or bottled consumables that can be displayed, stored for later use, or passed on as gifts to others, for which *nori* is ideal. Although the type of gifts considered appropriate is standard, differences in brands, regional origins, location purchased and ornamental packaging allow for the expression of the fine distinctions of status that are materialized through gift exchange. Ensuring that the gift accurately reflects the status, obligations, relationship and aspirations of giver and receiver is so important that department stores have gift departments and gift advisors and there are special gift websites. Department stores, both up-market establishments like Takashimaya and Mitsukoshi and mid-market chains like Seibu, compete vigorously for *oseibo* customers, and all the major Japanese *nori* producers have extensive ranges of gift-packaged *nori* in all its forms – sheets, strips, powdered, shredded and paste – for presentation. This reinforces *nori's* cultural value and visibility in Japan and in Japanese communities abroad.

While it has always been highly valued in Japan, *nori* only became a global commodity and item of international popular consumption as a direct result of the globalization of *sushi*. Before World War II, when *nori* was still a luxury good, its export to the west was minimal and mainly to the large communities of ethnic Japanese established in Hawaii and on the west coast of the United States, and this persisted after the mass production of *nori* became possible. Japan's economic success in the 1970s precipitated a vogue for all things Japanese, and Edo-style or *nigiri sushi* (Ishige 2001), which combined aesthetics and novelty with the speed of fast food, became highly popular among westerners as a marker of class, educational standing, cultural savvy and cool (Bestor 2001, 2005; Cwiertka 2006). *Sushi* also resonated with nascent Western concerns about wellbeing, fitness and healthy eating. *Sushi* bars proliferated, becoming a setting for new patterns of consumption, presentation and sociality, and boosting the Japanese export market. In 1966, the National Federation of *Nori*, Shellfish and Fishing Industry Co-operative Associations initiated the annual observation of a '*Nori* Day' on 6th February, the date the Daiho Code that first mentioned seaweed was issued in 701CE, to mark the continuing importance of *nori* in Japanese life. It is impossible – at least at present – to imagine such a thing happening in Wales, Britain or Europe.

Making the Invisible Visible

Returning to the invisibilities set out at the beginning of this chapter, the contrast between porphyra and *nori* in Japan and Wales is striking. In Japan, seaweed is culturally, physically, judicially, administratively, commercially, nutritionally and culinarily highly visible. In South Wales it is invisible in all regards outside the small, socially- and historically-specific laver-eating community, and even there it does not have a high profile. This difference has significant implications. On the macro-scale, world-wide degradation of the land through intensive terrestrial farming, along with over-fishing of the seas, both driven directly and indirectly by the high demand for animal protein associated with the problematic standard western diet, could be alleviated if more protein in the human diet came from seaweed. On the micro-level, human and economic health in Wales could be substantially improved through seaweed. But that will only happen if invisible foodscapes are made visible; if the consumption of seaweed in a visible and healthy form is encouraged; and if coastal and marine environments are protected and exploited responsibly and sustainably.

Making invisible foodscapes visible returns us to the question asked earlier about Welsh and European consumers – if people believe in the efficacy of seaweeds used externally, why will they not use them internally – by eating them? The receptivity to external applications is found all over the world stretching back into antiquity, a modern medical use being transdermal patches that deliver a variety of drugs through the skin on a time-release basis. Generally, all cultures see the skin as a body boundary, but one through which positive elements may pass in a less intrusive and invasive way than taking these in by mouth. As for ingestion, it all goes back to culture. Taste is in the mind, not the palate. What we eat, how we eat, is social construction from the inside out, and resistance can be fierce once established. Take, for example, snails: many English and American people would reject them out of hand without tasting them, yet they are widely regarded as a delicacy in Europe. Televisual and literary genres of 'extreme eating' play on these cultural boundaries. The embodied experience of smell, taste, consistency, and appearance are culturally determined, and an organoleptic map can reveal cultural divides not otherwise apparent. From a distance, the consumption of rice in the Far East may appear seamless; on the ground, there are sharp distinctions between the consumers of sticky and of non-sticky rice, which reach back into prehistory (Fuller and Rowlands 2011). And, of course, this evokes the divide in Wales between those who will eat laver and those who will not. I am convinced that culinary and medicinal seaweed use was once well-established in coastal Britain and Europe, until it was eclipsed by the introduction of Graeco-Roman medicine and socio-cultural attitudes that did not value marine products or therapies and did not consider seaweed edible, at least not by civilized peoples. As Virgil (1922: 82) put it, '*nihil vilior alga*' – 'nothing is more vile than seaweed'. As a result, seaweed use and consumption were marginalized, surviving in peripheral areas that escaped the dominant ideology. This is the kind of sweeping anthropological generalization as yet unsupported by conclusive archaeological evidence that is likely to outrage historians of medicine. Nonetheless, the ethnography is suggestive. Along the Atlantic coast of Britain today, from Cornwall to the Scottish islands, going up through Scandinavia to Greenland, and down through Normandy and Brittany to the Galician coast of Spain and over to Ireland, I have found consistent patterns of seaweed use. In addition to the use of seaweed poultices for bruises and sprains previously mentioned, pickled or fermented seaweed is nearly universal, along with the use of dried or powdered seaweed to thicken and enrich stews and soups. On this basis, and confident that my theory will be confirmed by isotope analysis and archaeobotany in due course, I posit the past existence of a widespread marine pharmacopeia and a wide range of culinary uses for seaweed in Europe. The area in question corresponds to Cunliffe's (2001, 2009) construct of 'Atlantic Europe', a geographical, cultural and economic entity that stretched from Iceland to Gibraltar and across to Ireland, and whose heyday was 9000BC–1000AD. Atlantic Europe predated the 'Mediterranean Europe' that now dominates the common perception of the European past. The heart of Atlantic Europe lay in the coastal zones that have long been considered marginal to the main land mass, but which now appear to have been central to the early settlement and later development of Europe and other parts of the world (Dillehay 2008; Fiedel 2000) because of their rich natural resources, including seaweed, and the relative ease of mobility provided by sea and estuaries compared to travelling on land. Looking out to sea, the Atlantic World view was Blue. People eat or do not eat seaweed because of culture, not nature. To make seaweed visible, it is necessary to reintroduce it into mainstream culture, and there are signs that this is happening.

In Wales, while *laver* producers value the iconic status and traditional associations of the 'weed', as they call it, they are also keen to move it into the present and to introduce *laver* to a wider public. Welsh chefs are devising recipes for *laver* quiche, *laver* and seafood chowder, *laver* sauce for lamb and a host of vegetarian and vegan dishes for which it is ideal. There are also innovative products like seaweed crisps intended for the healthy snack market (see Figure 1.1). One Welsh producer, building on the fact that Welsh seaweed is wild-harvested in waters known for their purity, has even exported Welsh seaweed to Japan for the luxury connoisseur trade. The packaging combines Japanese characters with a picture of the famed Gower Peninsula in South Wales, long associated with *laver* (see Figure 1.2).

Looking ahead, the waters off Wales are particularly suitable for the autotrophic or natural mariculture of seaweed, which depends only on solar radiation and nutrients in seawater. Amenable to mechanization, it can be carried out on a small scale requiring minimal investment, making it accessible to small communities, and it can also improve the marine environment and enhance biodiversity generally. There are producers in Ireland who are packaging seaweed for European consumption, including products such as 'sea spaghetti', and specialists in Galicia on the northwest coast of Spain, where there is a long tradition of seaweed use, are adapting it for use in modern dishes like seaweed pesto for pasta. Here we have seaweed becoming visible in a new culinary context, and these are interesting contrasts with Japan, where seaweed use is changing too.

Foodscapes are always shifting and sushi bars and sushi are continuing to evolve (Walraven 2002), to the point where the Japanese government has expressed concern about the loss of authenticity as sushi goes global. Are 'California roll' and the new wraps with filling encased in *nori* but with no rice really sushi? In the process of globalization, *nori* in the west has lost its original Japanese cultural connotations and configurations, epitomizing the way in which commodities change their meaning (Mansfield 2003; Mintz 2007) when they migrate, but meanings can also travel in the opposite direction. In Japan, as throughout Asia, customarily there was little distinction between food and medicine, and a healthy, well-balanced diet was considered a prophylactic that made the taking of additional supplements unnecessary. However, since the 1970s – when sushi began its global conquest – *nori* in Japan has increasingly gone from being perceived by the Japanese solely as a traditional and culture-specific food to being seen as a contemporary health food in the western sense.

Figure 1.1 New Welsh seaweed snack
Source: Courtesy of Selwyn's Penclawdd Seafoods.

Figure 1.2 Promotional literature for selling Welsh seaweed to Japan
Source: Courtesy of Selwyn's Penclawdd Seafoods.

Increasingly *nori* is being prepared and sold in Japan in capsule and tablet form, a development that parallels the increase in the consumption of western-style food in Japan, including foods with phycocolloids. And in areas of long-established Japanese cosmopolitanism like Hawaii, which has had a substantial Japanese population since the 1880s, there is a whole genre of East/West sushi wrapped in *nori* – hot dog sushi, garlic chicken sushi, salmon and cream cheese sushi, and a particular island favourite – Spam sushi or *musube.* Hawaii is renowned for its fresh tuna sushi, which coexists with a Native Hawaiian specialty called *poke* fish – chopped raw tuna mixed with chopped raw seaweed – which many people refuse to eat, even though they happily consume the sushi version. All are further transformations in the narrative of seaweed, and new shifts of visibility and invisibility.

What I have wanted to show here is that to further develop a critical anthropology of food, it is necessary to go beyond the visible to the invisible, and to deal not only with things that are, but things that are not – or at least things that are not 'seen' – and also with discontinuities and transformations that are not part of normal commodity chains. In this case, invisibility leads to the otherwise hidden ironies of modern consumption: the heavy but unknowing eating of seaweed extracts in the West, the 'unhealthy' uses to which 'healthy' foods are put, the juxtapositions of popular beliefs about the internal and external consumption of seaweed, the way in which the same seaweed can be simultaneously visible and invisible, the way cuisine transforms over time – all of which take place in the context of ongoing social, economic and culinary change. The technique of cultural biography pioneered by Kopytoff (1986) – following the trajectory of a single commodity across time, space and cultures – is now well-known in the study of food; by adding the dimension of the invisible to that of the visible, we can learn much more about how we eat what we eat.

In addition to showing how culture shapes consumption in the shifting arena of sand and sea, this study is an implicit critique of the values of 'Green'. The name and colour refer to the multi-faceted movement that has steadily gained ground among the public and within the academy since the 1970's, catalyzed by the 1962 publication of Rachel Carson's *The Silent Spring*, a founding text of modern environmentalism. Today, in the west, popular and academic debates about food policy and security, pollution, biodiversity, biotechnology, sustainability, self-sufficiency, conservation and water rights are all coloured Green – but is this a good thing?

While the profound importance of Carson's work on the dangers of the human destruction of the planet's ecosystem in the name of 'scientific progress' is undeniable, *The Silent Spring* has become another victim of 'the law of unintended consequences' (Merton 1968). So persuasive were Carson, her supporters and successors about the destructive effect on the land of pesticides and unregulated 'scientific' practices that Green interest – public, popular and academic – became firmly focused on the terrestrial, thereby rendering the coasts and seas, and their inhabitants and products, largely invisible. This invisibility means that the Green movement has tended to gain ground at the expense of the marine environment, sewage disposal and land run-off into the sea being but two examples of many. The particular irony here is that Carson was a marine biologist who wrote three highly-regarded books on marine and coastal ecosystems, later published together as *The Sea* (Carson 1964), before the work for which she is remembered today, and no one knew better than she the centrality of the sea and its wellbeing to the earth's holistic health. The ocean covers 71% of the planet, and with the world's lands and populations under unprecedented pressure of many kinds, the sea and its resources have never been more important. Instead of just thinking and eating Green, it is now essential to also think and eat Blue.

Acknowledgements

Thanks are due to Selwyn's Penclawdd Seafoods (www.selwynsseafoods.co.uk) and the Glamorgan County History Trust for their assistance in the preparation of this work.

References

Ashkenazi, M. and Jacob, J. 2000. *The Essence of Japanese Cuisine*. Philadelphia: University of Pennsylvania Press.

Bestor, T.C. 2001. Supply-side sushi: Commodity, market and the global city. *American Anthropologist New Series* 103(1), 76–95.

Bestor, T.C. 2005. How sushi went global, in *The Cultural Politics of Food and Eating*, edited by J.L. Watson and M.L. Caldwell. Oxford: Blackwell.

Carson, R. 1962. *The Silent Spring*. New York: Houghton Mifflin and Company.

Carson, R. 1964. *The Sea*. London: MacGibbon & Kee.

Cefas/Defra 2011. Under-utilised species. [Online]. Available at: http://www.cefas.defra.gov.uk/our-science/fisheries-information/under-utilised-species.aspx [accessed: May 2012].

Chapman, V.J. 1970. *Seaweeds and Their Uses*. London: Methuen and Company.

Cordain, L. with Eaton, S.B., Sebastian, A., Mann, N., Lindeberg, S., Watkins, B.A., O'Keefe, J.H. and Brand-Miller, J. 2005. Origins and evolution of the western diet: Health implications for the 21st century. *American Journal of Clinical Nutrition* 81(2), 341–54.

Creighton, M.R. 1991. Maintaining cultural boundaries in retailing. *Modern Asian Studies* 25(4), 675–709.

Cunliffe, B. 2001. *Facing the Ocean: The Atlantic and its Peoples 8000BC–1500 AD*. Oxford: Oxford University Press.

Cunliffe, B. 2009. *Europe Between the Oceans 9000BC–1000AD*. New Haven: Yale University Press.

Cwiertka, K. 2006. *Modern Japanese Cuisine: Food, Power and National Identity.* London: Reaktion Books.

Delaney, A. 2003. Setting nets in troubled waters: Environment, economics, and autonomy among nori cultivating households in a Japanese fishing cooperative. [Online]. Available at: http://dlc.dlib.indiana.edu/dlc/handle/10535/3575 [accessed: 8 May 2012].

Dillehay, T.D. 2008. Monte Verde: Seaweed, food, medicine and the peopling of South America. *Science* 320(5877), 784–6.

Douglas, M. with Isherwood, B. 1979. *The World of Goods: Towards an Anthropology of Consumption.* London: Penguin Books.

FAO 2002 (McHugh, D.J.). *Prospects for Seaweed Use in Developing Countries.* FAO Circ. No 968. Rome: FAO.

FAO 2003. (McHugh, D.J.). *A Guide to the Seaweed Industry*. FAO Fisheries Technical Paper 441. [Online] Available at: http://www.fao.org/docrep/006/y4765e/y4765e00.htm [accessed: 5 May 2012].

Fiedel, S.J. 2000. The peopling of the New World: Present evidence, new theories and future directions. *Journal of Archaeological Research* 8(1) 39–104.

Fishfight. 2011. [Online]. Available at: http://www.fishfight.net/the-campaign [accessed: 7 May 2012].

Flandrin, J.-L. and Montanari, M. 1999. *Food: A Culinary History from Antiquity to the Present.* New York: Columbia University Press.

Fuller, D. and Rowlands, M. 2011. Ingestion and food technologies: Maintaining differences over the long-term in West, South and East Asia, in *Interweaving Worlds – systematic interactions in Eurasia, 7th to 1st Millennia BC. Essays From a Conference in Memory of Professor Andrew Sherratt*, edited by J. Bennet, S. Sherratt and T.C. Wilkinson. Oxford: Oxbow Books, 37–60.

Guiry, M. 2007. *Porphyra Life History. Algae-L.* University College Galway. [Online]. Available at: http//www:seaweed.ucg.ie [accessed: 12 May 2012].

Hassan, J. 2003. *The Seaside, Health and the Environment in England and Wales since 1800.* Aldershot: Ashgate.

Ishige, N. 2001. *The History and Culture of Japanese Food.* London: Kegan Paul.

Kopytoff, I. 1986. The cultural biography of things: Commoditization as process, in *The Social Life of Things: Commodities in Cultural Process*, edited by A. Appadurai. Cambridge and New York: Cambridge University Press, 64–91.

Mansfield, B. 2003. 'Imitation crab' and the material culture of commodity production. *Cultural Geographies* 10, 176–95.

Matsuyama, A. 2003. *The Traditional Dietary Culture of SE Asia.* London: Kegan Paul.

Mauss, M. 1966. *The Gift*, translated by Ian Cunnison. London: Cohen and West.

Merton, R.K. 1968. *Social Theory and Social Structure*. New York: Free Press.

Mintz, S.W. 2007. Asia's contribution to world food: A beginning inquiry, in *Food and Foodways in Asia: Resource, Tradition and Cooking*, edited by S.C.H. Cheung and T. Chee-Beng. Abingdon and New York: Routledge, 201–10.

Mumford, T.F. Jr. and Miura, A. 1988. *Porphyra* as food: Cultivation and economics, in *Algae and Human Affairs*, edited by C.A. Lembi and J.R. Waaland. Cambridge and New York: Cambridge University Press, 87–117.

Nisizawa, K. with Noda, H., Kikuchi, R. and Watanabe, T. 1987. The main seaweed foods in Japan. *Hydrobiologia* 151–2(1), 5–29.

O'Connor, K. 2009. The secret history of the 'Weed of Hiraeth': Laverbread, identity and museums in Wales. *Journal of Museum Ethnography* 22, 82–101.

O'Connor, K. 2012. Imagining and consuming the coast: Anthropology, archaeology, 'heritage' and 'conservation' on the Gower in South Wales, in *Imagining Landscapes*, edited by M. Janowski and T. Ingold. Farnham: Ashgate, 121–42.

O'Connor, K. 2013. *The English Breakfast: the Biography of a National Meal, with Recipes*. London: Bloomsbury.

Okazaki, A. 1971. *Seaweeds and Their Uses in Japan*. Japan: Tokai University Press.

Rupp, K. 2003. *Gift Giving in Japan*. Palo Alto: Stanford University Press.

Sahlins, M. 1976. *Culture and Practical Reason*. Chicago: University of Chicago Press.

Smithers, R. 2011. Sales of sustainable fish soar in UK supermarkets. *The Guardian* [Online: 17th January 2011]. Available at: http://www.guardian.co.uk/environment/2011/jan/17/sustainable-seafood-supermarkets-fish-fight [accessed: 10 May 2012].

Teas, J. 2002. *The Cultural Construction of Seaweed: From Slime to the Macrobiotic Sublime*. Paper to the Society for Applied Anthropology Conference: Atlanta GA. 6th–10th March 2002.

Virgil (Publius Vergilius Maro). 1922. *The Eclogues, Bucolics and Pastorals of Virgil*. Translated by Thomas Fletcher Royds. Oxford: Basil Blackwell.

Walraven, B. 2002. Wild mushroom sushi? In *Asian Food: The Global and the Local*, edited by K. Cwiertka and B. Walraven. London: Curzon, 95–115.

Welsh Assembly Government. 2006. *Food and Fitness Plan*. [Online]. Available at: http://www.wales.nhs.uk/sitesplus/888/page/43758 [accessed: 11 May 2012].

Chapter 2
The Substance of Absence: Exploring Eating and Anorexia

Anna Lavis

Introduction

Gillian[1] had been anorexic for over 20 years when she was involuntarily admitted under the Mental Health Act (1983, 2007) to the inpatient eating disorders unit of a large NHS psychiatric hospital during the damp English summer of 2007. When Gillian arrived I was some months into a year of full-time anthropological fieldwork on the unit and, as we came to know each other, our conversations and research interviews offered insights into the experience of living through many years of anorexia and numerous hospital admissions. Echoing those previous hospitalizations, having limited her food intake to a point deemed potentially fatal by her outpatient team, Gillian suddenly found herself that July with no choice but to eat. Over the ensuing months, and with frequent articulations of both pain and anger, Gillian sat in the unit's dining room every day to eat three meals and three snacks, amounting to over 3,000 re-feeding calories. Arguably the object at the centre of inpatient treatment for anorexia (see Treasure and Schmidt 2005: 95) around which all other therapy is built, food both punctuated and produced the time of Gillian's admission. Her daily activities and freedoms depended on her having eaten her meals. During all of these Gillian held a fork in each hand; with one she reluctantly lifted food to her lips whilst the other, grasped in clenched fist, disappeared up her sleeve to scratch into her arm. When we were chatting during the last escorted cigarette break before ward lockdown one evening I asked Gillian about the deep scores that had appeared during dinner. She replied that digging the metal prongs into her arm made eating less painful, frightening and guilt-inducing; the physical pain, as the second fork entered her body through broken skin rather than open mouth, distracted her from the first fork, which transferred food from plate to lips. Other patients in the unit held ice cubes whilst eating, and some self-harmed afterwards to assuage their feelings of guilt. This distress at eating was also often discussed in interviews. Yet, these narratives illustrated that ice cubes, knives and forks are only partially effective at soothing distress. Gillian explained this partiality by contrasting the transience of such practices with the day-to-day incessancy of eating. Any relief gained, she argued, is ruptured by the looming

1 All names are pseudonyms.

necessity of eating again and Gillian echoed many other informants in describing herself as 'caught' between foods, always dreading the next mouthful as much as regretting the last. Yet, she also explained that this sense of entrapment not only arose from the eating disorders unit's enforcement of meals. It was also engendered by a tension that permeated her life more widely, which was encapsulated by her statement, 'I don't want to eat, but I don't want to die'. The tension between these desires lies, arguably, not solely in their binary opposition, but in their overlap; to (continue to) be anorexic, one must both eat and yet not eat.

Drawing on data from participant observation and semi-structured interviews undertaken during anthropological fieldwork in both the English inpatient eating disorders unit (EDU) (2007–2008) and on pro-anorexia websites (2005–2013), it is on subjectivities and lived experiences of eating amongst anorexic informants that this chapter focuses. I explore eating not only as absent to anorexia, the lack of which produces a disappearing body. Rather, I reflect on the substance of that absence, tracing informants' engagements with the materiality and meanings of food inside and outside their bodies. An attention to eating reveals complexity and nuance to informants' many and various food-centred practices; these engender active relationships of absence with food through which anorexia is maintained. (Not) eating thereby emerges as integral to how 'the everyday is produced' (Tucker 2010: 526) both within, and with, their illness by informants. Ensuing from interview narratives, central to this chapter is a focus on relationality as self-starvation cannot be delinked from informants' 'intersubjective fusion' (Jackson 2002: 340) with their anorexia. I suggest that anorexia both *is* not eating and is also *maintained by* not eating; to informants it is a presence both reliant on, but greater than, starvation practices. Recognizing how the illness is sometimes regarded as a part of self and, at others times, as both external and internal 'other', this chapter explores personhood as 'an unfolding *process*, with identity conditions which evolve over time' (Curtin and Heldke 1992: xiv, emphasis in the original) and in which '"the doer" is variably constituted in and through the deed' (Butler 1990: 142). Practices of eating and not eating mediate personhood and anorexia as these 'unfold' together and sometimes also enfold one another.

An attention to informants' (not) eating practices therefore offers an under-utilized way of taking account of subjectivities of anorexia, one that pays attention to the embodied and processual present moment lived with, through and inside the illness. It highlights the necessity of thinking beyond anorexia as primarily a future-orientated quest for thinness because, unlike food, thinness was not widely discussed by informants during fieldwork. When it was mentioned thinness was framed not as a goal of starvation, but rather as a visual indexical marker – to oneself and others – of the continuing presence of the illness.[2] My informants' accounts therefore intersect with a wider broadening of analysis beyond thinness in recent explorations of anorexia (see Allen 2008; Becker 2009; Gooldin 2008;

2 For more exploration of the relationship between thinness and anorexia and for wider discussions of the maintenance of, and desire for, the illness see Lavis 2011.

Warin 2010). Yet, reflecting on eating rather than thinness does not signify an attempt to dislocate anorexia from cultural context. Rather it considers that there may be other ways in which culture and anorexia touch edges, as cultural and affective spaces and contexts render anorexia valuable in and of itself, for what it *does*. This attention to what anorexic informants are doing and desiring, feeling and producing, when they (do not) eat therefore constitutes a 'micro-analysis' (see Deleuze 2007), which brings 'into view the immanent fields that people, in all their ambiguity, invent and live by' (Biehl and Locke 2010: 317).

However, whilst taking issue with the claim that 'the anorexic has no story to tell' (O'Connor and Van Esterik 2008: 9) and 'what happens makes no sense' (ibid.: 9), this chapter also acknowledges that 'to see the illness as anything less than cruel is to do an injustice to those who struggle with an eating disorder' (Allison 2009: 20). To recognize that anorexia is valued and even actively maintained by some informants is not to ignore the suffering and ambivalence in their narratives. Nor is it to suggest that anorexia is not a severe illness. Moreover, an ethnographic ethics of taking account of what matters to informants signifies 'an anthropological approach that includes from the start the possibility of diverse motivations for human action and diverse grounds for, and forms of, personal agency' (Desjarlais 1997: 202). Yet it also does not assume informants' agency to be unaffected by illness. Rather, it allows us to see that in the everyday complexities of why and how informants (do not) eat, 'different aspects of knowing, feeling, listening, and acting [circle] back on other aspects' (ibid.: 244).

The first half of the chapter explores *how* informants do not eat, tracing ways in which many position themselves against eating. To their (not) eating practices is an interplay between materiality and virtuality as actively-maintained relationships of absence with food are present, and even become more important, when informants 'actually' eat, such as in treatment. This chapter half, therefore, does not explore ways in which the act of eating is negotiated and contained when it takes place, such as through extreme calorie counting/'restricting' or purging (see APA 1994). Rather, it asks what embodied subjectivities of personhood and anorexia underpin informants' desires to position eating as 'other'. The second half of the chapter continues this reflection by exploring *why* informants do not eat. This is not to ask why people develop anorexia. Rather, by focusing on informants' relationships with their illness, it traces how these are maintained and mediated by (not) eating. The juxtapositions and mirrors between the two chapter halves illustrate that in exploring eating and anorexia there are no unambiguously straight lines, either between cause and effect or in terms of temporality. Encounters between eating and anorexia are multiple and varied and at these points of contact both are acted on and act.

(Not) Eating: Starvation Practices and Relationships of Absence

During fieldwork in the EDU, patients granted weekend leave were advised to plan this in writing, being careful to schedule in meals and snacks to exactly mirror those on the unit. This exercise was intended to make feeding oneself easier by removing some of the agency, and thus guilt, of eating. When Raja, an informant whom I had come to know well, was first allowed off the unit for weekends, she asked me to help her with this planning. So, on a number of Fridays Raja and I sat on a bench in the hospital grounds with a blank piece of paper and a pen. Raja's leave planning oscillated between tears, humour, evasion and frank honesty and I am grateful that she allowed me to record these. Raja would often tell me, laughing, that there was 'no point' writing meals into her days because she 'would not eat them'. She said with humour, 'I'll "forget" to go to the supermarket and then find the fridge is empty again – oops there's nothing to eat!' When charting hour-by-hour her future weekend activities, Raja also 'forgot' to leave any space for meals. By literally writing eating out of her life she demarcated it as outside her own time, which formed a contrast with its centrality to the temporal rhythms of the inpatient unit. When I would gently point this out, Raja would say with irony and a sad smile: 'So you noticed that again, huh? Damn!' During these planning sessions Raja described how her 'terror' at being near food 'did not allow' her to keep any food at home that would not be cooked, eaten or thrown away within a couple of days, and supermarkets were 'horrifying'. Such articulations of fear suggest that to reflect on experiences, subjectivities and practices of eating among anorexic informants, we need to take account of what Elspeth Probyn has termed the 'brute physicality' (Probyn 2000: 216) of food – its presence as well as absence.

Alongside fear, informants in both field sites described food with horror and disgust. In her interview Kate offered a description of mealtimes in treatment when eating was enforced; she said:

> Eating just feels wrong. It just doesn't feel right. It doesn't feel natural. Everything … like … my mind is saying 'no'. Sometimes I actually physically feel like something's holding my arms down and my legs there and I just feel held still because I'm so not wanting to do it and my body's pulling away, pulling away and I'm having to override my nat … what I perceive as my natural response and really sit on it and really stay really rigid and still so that I can get through it. It's thinking about what's going on, you know, it's like you're breaking down food. There's food in my mouth – food and saliva – and I think about what's in the food and it's all being broken down. I just think it's horrible to have something in your … eurgh, I can't really explain it but I think it's horrible to have something in your mouth and you're kind of moving it around and you swallow it and it's going into your digestive system … eurgh … I feel so greedy … I just feel like, when I've got something in my mouth, no matter how small the piece is, it just expands and it fills my mouth completely.

Kate's description of eating imbues food with an unstable and even perhaps, untrustworthy, materiality and its movements through her body are threatening. By attuning to this viscerally experiential horror, we can see how for informants such as Kate and Raja 'eating is always, more or less, a form of pollution' (Giordano 2005: 127). The threat and contamination of eating have also been highlighted by recent anthropological discussions of anorexia (see Warin 2003, 2010). Yet, in arguing that we need to take account of food's materiality to understand self-starvation, I also suggest that it is imperative to carefully trace the multi-directionality and temporality of cause and effect here; this at once demonstrates why food's physical presence is central to informants' productions of its absence whilst also showing how that centrality is complex.

We might suggest that the illness of anorexia itself causes informants to experience eating as contaminating and food as threatening; in line with clinical discourses, we could regard the 'anorexia [as] talking' (Tan 2003), as having altered informants' values (Tan 2005; Tan, Stewart and Hope 2009) and thereby their experience of food. A loss of agency to anorexia as a constituent part of informants' fear of food, which would support this perspective, is illustrated by statements such as this one from Abigail: 'You're too scared to eat it. You can't eat it'. Likewise, in her interview Eva said:

> I'll sit there and I know I'm hungry and I'll feel my stomach rumbling but then I'll sit there and think, 'ok, so what do you want to eat then?' and you think, 'an apple? No I can't eat that. No, no. Oh my God no'. And then you think of anything: 'Yoghurt? No'. You go through lists and lists in your head and the more you, like, realize there's nothing you feel like you can eat.

Moreover, that anorexia alters, and at least partly controls, conceptualizations and experiences of food was illustrated by another informant, Miriam. In her interview she said:

> I think it would be totally a lie if I said it [anorexia] hadn't changed me, it has. On a day-to-day basis, not necessarily now but before I came in [to the EDU], I would have screaming, ranting, slanging matches with my parents. I actually got quite hoarse and I'd be getting to the point where I'd almost be being sick because I was getting so angry and it would be over two tubes of pasta. Seriously, two of those little penne tube would set me off.

It is therefore clear that eating experiences are refracted through the agency exerted by anorexia, and also that such 'food beliefs and representations exist materially in the body' (Hayes-Conroy and Hayes-Conroy 2008: 461). Yet, this is also not the end of the story. To assume linearity of cause and effect in which anorexia's agency is unquestionably causally positioned between informants and food would frame their food refusal as *always* or *entirely* unagential and reactive. This would not encapsulate subjectivities of eating expressed in informants' narratives, in which loss of agency is only one element amongst many. We therefore need to hold in one analytical space an awareness of how anorexia may perform food and eating in certain ways, whilst also engaging with how food and eating practices may, likewise, perform anorexia; these 'hang together' (Mol 2002) in informants' modalities of being anorexic, and, even, of 'doing' anorexia.

Above, Kate described eating with abject horror. Yet, in her interview she also recounted how, when surviving on 100 calories a day just before admission to the EDU, she would collect menus from local restaurants and peruse these alone at night. Kate said: 'I could almost imagine I could taste what I was reading … Once I'd read it, it almost felt like I'd eaten it. I'd want it so badly but I couldn't let myself have it. And for me also, I'd look at it and think God how disgusting, eurgh'. A causal link between the neurological effects of starvation and fantasizing about food is clinically documented (see Keys et al. 1950) and famously discussed by George Orwell (2001) and Primo Levi (2004). Yet, aware of hunger's 'supposed brain effect', as she put it, Kate was emphatic that her menu collecting was not biologically-driven but was, rather, a modality of both 'making sure' that she was, and 'making [her]self', anorexic. To support this dual claim, Kate showed me the menus she had amassed in the EDU where, she pointed out, she was eating over 3,000 calories a day. At the time of that interview a pizza delivery menu hung on the board outside the Nursing Office. One of the healthcare assistants who ran DVD nights had suggested that the coming Saturday's DVD night could expand to incorporate pizza. Pizza night never materialized due to an unsurprising lack of enthusiasm but the menu hung on limply for several weeks. Just after our interview Kate and I happened to be standing in the corridor together. Kate, rapping her knuckles on the menu, said: 'You see, that's what I mean, I've read that menu a thousand times; every time I walk past I have to stop … I make sure I don't want it'.

Although Kate would seem to enact 'a simultaneous refusal to eat and incessant preoccupation with food' (Heywood 1996: 17), which has been widely discussed in relation to anorexia, by evoking yet refuting clinical paradigms of cause and effect she asserts the active and agential nature of her food refusal. That food embodies contamination and danger, Kate's account suggests, may certainly enable informants to frame a refusal to eat as a logical response to threat. But her emphasis on agency illustrates that we cannot, in turn, posit this threat as a cause of food refusal, or simply attribute it to anorexia's agency. Kate's sense of herself as anorexic depends on a notion of food as dangerous and 'wrong' but she also continually re-performs food's 'wrongness' by refusing it. By perusing menus, Kate draws close to food, albeit virtually, 'eating through [her] eyes' precisely in order to reject and resist it. She thereby manufactures distance from eating precisely by rupturing a self-created proximity. In a 'surveillance over, and against [herself]' (Foucault 1980: 155), Kate's 'hypperreal' (Baudrillard 1983) eating produces an absence made not only of lack but, importantly, of enforced lack. Whether or not Kate is restricting her calorie intake outside treatment or eating within it, the enforced lack that forms the substance of Kate's relationship of absence with food demarcates eating as 'other'. In processually affirming herself as someone who both *does not* and *should not* eat, she maintains herself as anorexic, importantly even in treatment where, after months on a re-feeding regimen, clinical markers such as BMI problematize this claim diagnostically (see APA 1994). 'Eating's just not me', explained Kate and, given that it is against eating that this 'me' – this anorexic personhood – is processually (re)produced day-by-day, it is beginning to emerge why so many informants echoed Kate in suggesting that they 'should not eat'.

In her interview Miriam said: 'If you eat you've given in, you've stuffed yourself silly and that's not right. You just shouldn't do that'. In hers, likewise, Libby said of improving, which meant in the language of the clinic eating with more ease, 'I find it really hard to improve because part of me feels like I shouldn't be improving cos it's wrong to improve, it's not anorexic to improve'. Yet, the word 'should' is, importantly, suggestive not only of horror at food, but also of desire for it. Desire resonates through pro-anorexia websites, which advise participants to: 'Spoil your food. As soon as you've cooked your meal, put too much salt, pepper, vinegar, detergent or perfume on it. That way you won't want to eat it'. A desire to eat was also present in Kate's narrative when she 'made sure' that she did not '*want*' the items on the take-away menu. In her interview, another informant, Abigail said: 'I don't hate food. I am terrified of it, absolutely terrified of it. But there are things I enjoy. And I hate to admit it but I do enjoy some foods. It just feels shit to admit that'. I suggest that underpinning informants' processual and active relationships of absence with food is an entanglement of horror and desire, with the latter as important as the former. The desire to eat and the visceral pleasure it might incite both emerge in narratives as aspects of eating that allow its enforced absence to become an achievement. Through these, we can see how food is experienced as threatening and horrifying and yet also that, as one informant Elle put it, 'starving feels like an achievement, it feels like you've done something right'.

'Achievement' and 'discipline' are words common to scholarly discussions of anorexia (see Eckermann 1997; Heywood 1996) but they are often used to argue that 'slenderness is the measure of one's moral calibre' (Ellmann 1993: 5), which locates achievement, and thus any measurement of it, in a corporeal 'end result' of anorexia – thinness. In contrast, in an anthropological exploration of anorexia, Sigal Gooldin argues that to understand achievement in anorexia, we need to pay attention to hunger; 'the experience of hunger, which involves physical pain and suffering, is transformed into a feeling of self-efficiency, power, and achievement that constitutes a sense of heroic selfhood' (Gooldin 2008: 281). On pro-anorexia websites this intermeshing of hunger and achievement is demonstrated by frequent advice to participants to 'feel your hunger, don't try and suppress it. You want to be hungry cos if you aren't then you're not doing it right. Hunger is not your enemy!' Yet, explaining 'how to get through it', as she put it, Milla said: 'Keep yourself really busy, almost to the point of being completely stressed out and then you can go for eight hours without feeling hungry'. In her interview Jumela said about emails she exchanged with other participants to pro-anorexia websites that, 'common goodbyes at the end of pro-ana emails are "stay strong" and "starve on"'. Jumela's statement suggests what many other informants also noted, that the physical discomfort of resisting hunger renders anorexia a processual rather than teleological achievement. It permits informants' measuring – or, following Foucault, 'surveillance' (1980) – of themselves against eating to be viscerally felt; through the embodied experience of hunger, eating's enforced absence is not only lived moment-by-moment but is also produced within the tangibly-lacking stomach. Yet, paying attention to informants' actively-constructed relationships of absence also highlights that the achievement of corporeally as well as conceptually resisting eating goes beyond an engagement on the part of informants with their hunger. Many also utilize the multi-dimensionality and materiality of the act of eating (against) itself to produce its own absence; hunger, taste, desire, pleasure and horror *all* form part of informants' othering of eating through 'chewing and spitting'.

In interviews some informants described a daily practice of chewing and then spitting out food. They recounted carefully, sometimes lavishly, cooking meals and then taking mouthfuls, chewing them, holding the food momentarily within the cheeks and on the taste buds before spitting it out. By permitting the experience of eating, or rather of tasting but not digesting, chewing and spitting constitutes a viscerally aesthetic, but subversive, engagement with food. It refuses it through, and indeed within, the body, heightening the achievement of resistance by locating it within the mouth. In this intimate encounter between food and mouth, chewing and spitting is dangerous. It navigates closely – too closely perhaps – the desire to swallow incited by bodily hunger, the pleasure of taste and, also, the swallowing reflex. I suggested earlier that Kate's perusing of menus was, following Baudrillard (1983), a 'hyperreal' eating. Chewing and spitting is also hyperreal but because of this increased danger, in its mimesis of eating there is not solely 'replication', but also 'simulation' (and indeed, stimulation). Simulation, argues Baudrillard, 'threatens the difference between "true" and "false", "real" and "imaginary"' (ibid.: 5). This blurring is illustrated by the fact that, as a number of informants argued and pro-anorexia websites also document, chewing and spitting is flawed; it is an 'imperfect' way of affirming oneself to not eat because, within the body, boundaries are not easily maintained; as one pro-anorexia website participant put it, 'the fat slides down your throat anyway'. Yet it is also clear from this statement that in the use of chewing and spitting to affirm eating as other, the throat is transformed into an important conceptual boundary, however imperfectly leaky. It becomes the temporal and spatial moment at which eating, and thus the achievement of resisting it, are located. As such, a contrasting 'real' eating coalesces around swallowing, where that signifies a *voluntary* ingestion of food, which many informants described simply as 'failure'. Chewing and spitting whilst in the EDU allows informants to affirm the otherness of eating whilst also *involuntarily* swallowing food that is both enforced by treatment and which, as many informants acknowledged in interviews, keeps them alive. This practice therefore enables informants to assert that, as Holly's put it, 'I just don't *want* [food] in me', even as it makes its material way through the body.

In its close engagement with bodily boundaries and pleasures, thus, the practice of chewing and spitting has once again brought to the fore the sense of precariousness and danger that pervades informants' discussions of eating. We have seen that relationships of absence with food must be continually re-performed to maintain eating at a distance; its threatening proximity and presence not only arise from the necessity of eating to staying alive, as Gillian reminded us at the outset, but also from how, especially in treatment, informants' affirmations that eating is 'not me' are concurrent with the daily swallowing of food. Taking up the entwined threads of threat, personhood and the sense of relentless 'liminality' (Turner 1967) to conceptually and actually keeping eating at bay, which have emerged in this half, the chapter now turns to consider why it is so important to informants to position themselves against eating and, thus, why eating is imbued with horror, contamination and danger. As I suggested earlier, it is insufficient to argue simply that this is *because* of anorexia. Rather, having established that eating is not merely a straightforward absence in anorexia, it is necessary to contextualize productions of its lack by reflecting on informants' relationships with their illness. This engenders recognition of how the absence of eating maintains and also mediates anorexia's presence.

Anorexia: Maintaining and Mediating a Presence

Cooking Group on the EDU is a therapeutic activity in which informants nearing the end of their stay are helped to buy and cook their own food in preparation for life beyond the unit. Seeing another patient, Milla, struggling to eat what she had cooked during one particularly fraught Cooking Group during my fieldwork, Hadia reassured her, 'the more you do it, the easier it'll get'. To this Milla replied emphatically, 'I don't want it to get any easier!' To Milla, as she explained later, there was a frightening temporal duality to this encounter between eating and anorexia. Whilst being able to eat with ease – without fear or horror – signified that anorexia had already begun to slip away, eating also constituted the production of that loss. As such, Milla demonstrates the danger posed by eating to anorexia and, in so doing, illustrates why eating *should not* be done. Yet to therefore suggest that relationships of absence with food maintain anorexia is not to deal in empty tautologies; it is not to claim, simply, that an absence of eating maintains starving. Rather, to understand how not eating 'matters' (in) anorexia, we need not only to trace how eating threatens anorexia, but also to reflect on what anorexia *is* and ask, thus, what it is that is threatened. This comprises listening to informants' narratives of anorexia as 'something you feel quite protective of; you don't want anybody to sort of rip it away from you because it stands for a set of things that you do', as Milla put it. Such statements suggest that ways of othering eating explored in the first half of this chapter go beyond embodied practices of self-production to be inter- as well as intra-subjective. As such, they offer insight into how anorexia both *is* not-eating practices and is also maintained *by* not-eating practices; its presence in informants' lives goes beyond the sum of its parts and that presence, to many informants, is that of a 'friend'.

Informants in both field sites offered friendship with anorexia as their reason for not eating. In his interview Laurie said: 'It's a friend, definitely a friend. It keeps me company … and it helps me … you know? It does help me'. And, Tara said of anorexia: 'It's been a friend to me for a long time'. Likewise, in her interview Shanice offered this explanation for not eating:

> The first thing that comes to my mind about anorexia is it's like a friend. It's like sort of … I dunno … yeah, more of a person rather than an illness. It's more of a friend, a close friend, rather than having an illness. Because, when it came along I was going through a really bad time and I was feeling quite depressed and it just sort of came along and rescued me in a way. It's always there for me. It doesn't abandon me, when the going gets tough. It'll always step in for me in a fight.

Another informant, Cally, said of a difficult marriage counselling session: 'It was there so I didn't have to be … like a friend would be. It just took over'. And, Jumela described why she wanted to maintain her anorexia by saying simply, 'it holds my hand'.

Descriptions of the illness as a friend are present in literatures from the anthropological (Warin 2006) to the psychological (Colton and Pistrang 2004; Serpell et al. 1999) and frequently occur in memoirs of anorexia (Grahame 2009; Hornbacher 1998, 2008). During my fieldwork, many reasons were offered as to why anorexia was a friend and all of these related to informants' experiences of anorexia's 'help'. This help was often discussed in terms of starvation's dampening of emotion. Elle, for example, said: 'It sort of helps to relieve feelings and stuff. It does its stuff so I don't have to feel them'. Claudine, another informant, explained that not eating is 'like you're telling yourself to shut up' and many informants described anorexia's protective 'numbness'. Informants' conceptualizations of anorexia's friendship therefore intersect with a wider recognition of anorexia as 'a functional coping strategy in which control of eating serve[s] as a means of coping with ongoing stress and exerting control' (Eivors et al. 2003: 96); they demonstrate why it has been termed 'an illness of the emotions' (Treasure, Smith and Crane 2007: 73) and is seen as a way to 'avoid dealing with other problems' (Cockell, Geller and Linden 2002: 77). Yet, informants' discussions of how anorexia 'steps in' or 'comes out' to take control in particular situations when they need it to offer a more nuanced insight into this 'friendship'.

From these narratives of 'help' the illness emerges as something that mediates and makes informants' everyday lives by offering them a way to contextually move through, interact with, or retreat from, their surrounding worlds and maintain a sense of control. By (re-)making anorexia day-by-day, not-eating practices maintain a personhood that not only 'unfolds' (Curtin and Heldke 1992) with the illness but that is also 'held' by it and that, therefore, depends on it. This imbrication of anorexia and personhood – or rather, perhaps, this anorexic personhood – was glimpsed earlier in Kate's assertion that eating was 'not [her]'. Through it we now see how maintaining anorexia might be underscored by a mingling of desire and necessity and also how recovery 'may appear terrifying' (Treasure 1997: 45) as it comprises the dual loss of anorexia and self. Relationships of absence with food therefore maintain one's personhood and also protect anorexia from being harmed, ruptured or effaced by the ingesting and digesting of food. Moreover, given the value placed by informants on anorexia's dampening of emotions, eating is not only distressing for what it signifies. It also actively incites distress; it is experienced by informants as violently rupturing the 'safety', as Eva put it, of anorexia, thereby engendering 'emotional fireworks' (Treasure and Ward 1997: 107) after months or years of self-starving. As such, eating *should* – indeed, *must* – doubly be kept at bay. Furthermore, not only does a lack of eating constitute the substance of anorexia's 'friendship', but the practices that produce this lack are themselves also a part of the relationship. In her interview Leila said about anorexia: 'Well, it helps you so you help it'. And this, she said, means 'you mustn't eat'. For Leila, it is the reciprocity of protecting her illness that makes her relationship with anorexia a friendship, which illustrates how belonging, such as that between Leila and her anorexia, comprises 'involvement and investment' (Grossberg 2000: 154) and is 'a production' (ibid.: 154; see also Probyn 1996).

Thus, informants' accounts of anorexia's friendship elucidate how practices that manufacture distance from eating keep anorexia close, and how both this distance and proximity are processual rather than static. Yet, there is a further dimension to these doubled processes, which disrupts but also reinforces their mirroring of each other. The liminality (Turner 1967), which I noted earlier, to these productions of distance and proximity lies not only in their relentlessness but also in their precariousness. In their narratives of eating and anorexia informants described how there are times when, as one informant put it, 'anorexia holds onto you too hard' – when it is *too close*; in her interview Shanice said:

> I suppose it's not a friend in a way because I suppose it's not really a healthy way of dealing with things and it makes you quite ill physically. It probably makes you mentally worse. It's sort of like it wants to help you but at the same time it'll kill you whilst it's doing it.

There is a clear, and indeed causative, parallel to being 'held' in a protective numbness by one's friend anorexia and being 'held onto' meaning trapped. These, moreover, are often juxtaposed in informants' narratives. In her interview Tanya said 'I just don't want to be without anorexia' and yet she also described the illness as 'hell'. Jumela, likewise, described herself as 'trapped in something I love and hate'. Such accounts illustrate how the balance of informants' relationships with anorexia may shift as the illness stops, at least momentarily, producing or helping to control their everyday lives and, instead, constrains or even destroys these. Descriptions of what some informants termed anorexia's 'betrayal' resonate with a sense of how its previously-helpful agency may painfully tighten the threads that bind personhood and illness. As such, it is clear why the clinical model of anorexia regards 'the patient as constrained and trapped' (Palmer 2005: 2) by anorexia. Yet, in interviews, many informants described this entrapment whilst also continuing to value, and even actively maintain, their illness. Moreover, just as being 'held' and 'held onto' mirror one another, so too do the encounters between food and anorexia that engender these dual subjectivities. In Kate's menu collecting and in chewing and spitting, above, engaging with food brought anorexia closer. Now, an attention to eating within this context of relationality not only illuminates how accounts of being 'engulfed' or 'trapped' by anorexia resonate with a sense of *excess* proximity but also how this is produced by unagential encounters with food. Recognition that being 'triggered' by food, as informants put it, is entangled with their wider subjectivities of anorexia as a friend offers insights into how informants' relationships with their illness *continually* change, both agentially and unagentially. It also suggests that the day-to-day complexities of these shifting relationships cannot be understood without taking account of both subjectivities and practices of food and (not) eating.

In her interview Miriam said: 'Sometimes I could sit and talk to someone perfectly rationally about food … but put me in front of it …. phewf, that's bad!' Like other informants, Miriam attributed this profound fear of, and horror at, encountering food to how it 'triggered' her anorexia. By this she meant that being near food gave rise to an unagential upsurge of anorexia within her. Miriam described this as having suddenly 'too much anorexia', about which she said: 'there's too much of it and then there's less me'. Likewise, one afternoon as we sat together in the Occupational Therapist's Office on the EDU planning what she would prepare later that day in Cooking Group, Leila also felt herself to be 'triggered'. Staring bleakly at a computer screen replete with chicken recipes, Leila suddenly said: 'Somehow everything just seems to get ruined; anorexia always gets in the way'. She went on to explain that whilst browsing the recipes on the Internet she had suddenly felt her anorexia to rise up, take over and push out not only herself, but also her 'friendly' relationship with anorexia. Clearly echoing informants' agential engagements with food in the first half of the chapter, in these unagential encounters with food – both actual and, in Leila's case like Kate's earlier, virtual – *too much* anorexia is produced. Narratives of being 'triggered' draw our attention to informants' subjectivities of the many and continual – even daily – shifts in the relationship between personhood and anorexia. Informants feel themselves to be more or less anorexic, more or less agential, in certain contexts and at particular times, and food produces these shifts both *with* informants and *against them*. Yet, despite offering us their mirror image, I suggest that 'triggering' does not undermine informants' relationships of absence. Rather, it underscores the importance of these to informants.

We cannot assume that feeling 'trapped', 'engulfed' or, as Leila put it, 'pushed out' by anorexia necessarily ruptures friendship with the illness. Rather, it became clear during fieldwork that it may also incite a desire to re-establish friendship and regain a relationship with a more 'friendly' anorexia. It is therefore here in an 'agency play' (Battaglia 1997: 506) enacted both against and also *with* eating that relationships of absence take on a further dimension of importance to informants. Positioned against the experience of being 'unagenetially triggered' as well as against eating, these both re-produce distance from the triggering food and also reclaim agency from eating *and* anorexia. To trace the threads of this argument, the chapter will now end where its discussion began, with Raja's weekend leave. Illustrating how 'agency arises out of a specific set of activities' (Desjarlais 1997: 202) Raja's narrative elucidates how both eating and not eating may be simultaneous and also mutually productive, as lack and substance fold into one another in the maintenance of both anorexia and (an anorexic) personhood.

Despite adamantly writing food out of her leave plans, almost every Sunday night Raja did cook, always with the same recipe and all the food in the house. This recipe was a dish of sausages and lentils from Nigel Slater's *Kitchen Diaries* (2005), which she had originally chosen, she said, 'because it looked like it'd taste nice'. In her interview, Raja described how during the weekends when 'on her own with [her] anorexia' and not having eaten anything at all for 48 hours, she would become scared that the illness was becoming too strong and engulfing. For this strengthening Raja offered many reasons, which included actively resisting eating to maintain anorexia and, also, having had 'too much triggering' accidental contact with food when out in public. Raja described feeling suddenly 'trapped' by a 'too close' anorexia and finding it difficult to 'get back', as she put it, to her valued 'friendly' relationship with her illness. To re-align this friendship, therefore, Raja meticulously cooked this meal of sausages to, she said, 'prove' to anorexia that she was 'in charge of it'. Then, fearing that all the food would itself be either too triggering or too desirable, where we have seen these to be entangled, Raja would always take two small mouthfuls and throw the rest away, taking the food to a dustbin outside the house and leaving the fridge once again completely empty. By voluntarily swallowing some of this meal, Raja arguably temporarily ruptured her relationship of absence with food and, thus, also her relationship with anorexia. By then throwing away – actively resisting – the barely-touched meal, however, she (re)produced eating's lack and anorexia's proximity, thereby reinstating both relationships discussed in this chapter. In so doing, Raja reclaimed anorexia as part of herself and eating as other, employing the threatening materiality of food to restore her desired balance of presence and absence.

Conclusion

It has been argued that in anthropological analyses, 'it is time to attribute to the people we study the kinds of complexities we acknowledge in ourselves' (Biehl and Locke 2010: 317). By turning its attention to eating, and thereby thinking beyond a more prevalent analytical focus on thinness, this chapter has sought to explore everyday complexities of being, doing and feeling anorexic. Practices to 'other' eating and processually maintain their illness elucidate the many ambivalent ways in which individuals live with, through and even inside anorexia. This is because attending to intricacies and intimacies of informants' (not) eating practices, rather than imputing from the visualilty of an already-emaciated body, alters the temporality of analysis. It engages with anorexia at the many daily moments of informants' food encounters and enables us to take account of how thinness is made without assuming this to be a goal of self-starvation.

Arguing that eating is not a simple lack in anorexia, this chapter has therefore explored how its absence, like its presence as this volume as a whole elucidates, is generative. (Not) eating practices cannot be dislocated from informants' subjectivities of relationality, and the illness has emerged from the chapter's discussions as both an intra- and inter-subjective presence. Listening to accounts of friendship with anorexia has revealed how the illness is experienced by informants as containing and mediating the world around them, and as offering a protective numbness. Anorexia's agency underpins friendship by 'coming out' and 'stepping in' for informants but it also renders the balance between anorexia and personhood precarious. Tracing the ways in which agency perpetually slips and slides along the threads between informants and their illness – as anorexia continually shifts between helping and engulfing – has illuminated the entanglement of these subjectivities with food and eating. Accounts of being 'triggered' by food have illustrated the many day-to-day shifts in anorexia's friendship, as well as how these intersect with the swallowing and not swallowing of food. Paying attention to triggering has also disallowed any assumption that anorexia's 'entrapment' is the end of the story – that it occurs temporally later than, and thereby undermines, practices to keep eating at bay. Rather, holding triggering and relationships of absence in one analytical space has elucidated their simultaneity. This has highlighted the dialectical engagement between – and multiple temporalities of – processually reproducing something that you already have, whilst knowing that it too produces and constrains you in ways both valued and unwanted. Such nuances in informants' narratives have illuminated compromised conditions of possibility in anorexia and also the creativity of individuals' ways of 'making do' (de Certeau 1984) with(in) these.

Thus, this exploration of what we might term *the contours of not eating* has reminded us that 'the topography of subjectivity is multi-dimensional' (Haraway 1991: 193); since we cannot assume illness to entirely flatten this landscape, we must ensure that our analyses do not do so either. By listening to anorexic informants' narratives, this chapter has sought to underline the importance of attending not only to *how* not eating matters (in) anorexia, but also to *why* anorexia matters to informants. Exploring their day-to-day experiences has suggested that such an understanding can perhaps only be forged by engaging with individuals' own intricate, complex and embodied conceptualizations of what *both* eating and anorexia *are* and *do*.

Acknowledgements

The PhD research on which this chapter draws was undertaken in the Anthropology Department at Goldsmiths, University of London. I should like to thank my supervisors there, Dr Simon Cohn and Professor Catherine Alexander. The PhD was funded by an Economic and Social Research Council studentship and a Goldsmiths Bursary, and it won the 2010 Radcliffe-Brown/Sutasoma Award from the Royal Anthropological Institute. My fieldwork on the inpatient eating disorders unit was granted NHS ethical approval (CoRec Number: 06/Q0706/128). I should like to express my gratitude to the many individuals in the EDU and on pro-anorexia websites who shared their stories with me.

References

Allen, J.T. 2008. The spectacularization of the anorexic subject position. *Current Sociology* 56(4), 587–603.

Allison, M. 2009. *Dying to Live: A Journey Through and Beyond Anorexia.* Brentwood, Essex: Chipmunkapublishing.

APA 1994. *Diagnostic and Statistical Manual of Mental Disorders IV*. Washington, DC: American Psychiatric Association.

Battaglia, D. 1997. Ambiguating agency: The case of Malinowski's ghost. *American Anthropologist* 99(3), 505–10.

Baudrillard, J. 1983. *Simulations.* New York: Semiotext(e).

Becker, A.E., Thomas, J.J. and Pike, K.M. 2009. Should non-fat-phobic anorexia nervosa be included in DSM-V? *International Journal of Eating Disorders* 42(7), 620–35.

Biehl, J.O. and Locke, P. 2010. Deleuze and the anthropology of becoming. *Current Anthropology* 51(3), 317–51.

Butler, J. 1990. *Gender Trouble: Feminism and the Subversion of Identity.* New York and London: Routledge.

Cockell, S.J., Geller, J. and Linden, W. 2002. Decisional balance in Anorexia Nervosa: Capitalizing on ambivalence. *European Eating Disorders Review* 11, 75–89.

Colton, A. and Pistrang, N. 2004. Adolescents' experiences of inpatient treatment for anorexia nervosa. *European Eating Disorders Review* 12(5), 307–16.

Curtin, D.W. and Heldke, L.M. 1992. Introduction, in *Cooking, Eating, Thinking: Transformative Philosophies of Food*, edited by D.W. Curtin and L.M. Heldke. Bloomington: Indiana University Press, xiii–xviii.

de Certeau, M. 1984. *The Practice of Everyday Life.* Translated by S. Rendall. Berkeley, Los Angeles and London: University of California Press.

Deleuze, G. 2007. *Two Regimes of Madness: Texts and Interviews 1975–1995.* New York: Semiotext(e).

Desjarlais, R. 1997. *Shelter Blues: Sanity and Selfhood among the Homeless.* Philadelphia: University of Pennsylvania Press.

Eckermann, L. 1997. Foucault, embodiment and gendered subjectivities: The case of voluntary self-starvation, in *Foucault, Health and Medicine*, edited by A. Petersen and R. Bunton. New York: Routledge, 151–69.

Eivors, A., Button, E., Warner, S. and Turner, K. 2003. Understanding the experience of drop-out from treatment for anorexia nervosa. *European Eating Disorders Review* 11(2), 90–107.

Ellmann, M. 1993. *The Hunger Artists: Starving, Writing and Imprisonment.* Cambridge, Massachusetts: Harvard University Press.

Foucault, M. 1980. *Power/Knowledge: Selected Interviews and Other Writings 1972–1977.* New York: Pantheon Books.

Giordano, S. 2005. *Understanding Eating Disorders: Conceptual and Ethical Issues in the Treatment of Anorexia and Bulimia Nervosa.* Oxford: Clarendon Press.

Gooldin, S. 2008. Being anorexic: Hunger, subjectivity and embodied morality. *Medical Anthropology Quarterly* 22(3), 274–96.

Grahame, N. 2009. *Dying to be Thin: The True Story of My Lifelong Battle Against Anorexia.* London: John Blake.

Grossberg, L. 2000. History, imagination and the politics of belonging: Between the death and the fear of history, in *Without Guarantees: In Honour of Stuart Hall*, edited by P. Gilroy, L. Grossberg and A. Robbie. London and New York: Verso, 148–64.

Haraway, D.J. 1991. *Simians, Cyborgs, and Women: The Reinvention of Nature.* London: Free Association Books.

Hayes-Conroy, A. and Hayes-Conroy, J. 2008. Taking back taste: Feminism, food and visceral politics. *Gender, Place & Culture* 15(5), 461–73.

Heywood, L. 1996. *Dedication to Hunger: The Anorexic Aesthetic in Modern Culture.* Berkeley, Los Angeles and London: University of California Press.

Hornbacher, M. 1998. *Wasted: Coming Back from an Addiction to Starvation.* London: Flamingo.

Hornbacher, M. 2008. *Madness: A Bipolar Life.* London: Harper Perennial.

Jackson, M. 2002. Familiar and foreign bodies: A phenomenological exploration of the human-technology interface. *Journal of the Royal Anthropological Institute (New Series)* 8, 333–46.

Keys, A., Brozel, J., Henschel, A., Mickelsen, O. and Taylor, H.L. 1950. *The Biology of Human Starvation.* Minneapolis: University of Minnesota Press.

Lavis, A. 2011. *The Boundaries of a Good Anorexic: Exploring Pro-Anorexia on the Internet and in the Clinic.* Doctoral thesis. Goldsmiths, University of London. [Thesis]: Goldsmiths Research. [Online]. Available at: http://eprints.gold.ac.uk/6507/ [accessed: 11 June 2012].

Levi, P. 2004. *If This Is a Man and The Truce.* London: Abacus.

Mental Health Act 1983. London: Department of Health.

Mental Health Act 2007. London: Department of Health.

Mol, A. 2002. *The Body Multiple: Ontology in Medical Practice.* Durham and London: Duke University Press.

O'Connor, R.A. and Van Esterik, P. 2008. De-medicalizing anorexia: A new cultural brokering. *Anthropology Today* 24(5), 6–9.

Orwell, G. 2001. *Down and Out in Paris and London.* London: Penguin Classics.

Palmer, B. 2005. Concepts of eating disorders, in *The Essential Handbook of Eating Disorders*, edited by J. Treasure, U. Schmidt and E. Van Furth. Chichester: Wiley.

Probyn, E. 1996. *Outside Belongings: Disciplines, Nations and the Place of Sex.* New York and London: Routledge.

Probyn, E. 2000. *Carnal Appetites: FoodSexIdentities.* London and New York: Routledge.

Serpell, L., Treasure, J., Teasdale, J. and Sullivan, V. 1999. Anorexia Nervosa: Friend or foe? *International Journal of Eating Disorders* 25(2), 177–86.

Slater, N. 2005. *The Kitchen Diaries.* London: Fourth Estate.

Tan, J. 2003. The anorexia talking? *The Lancet* 362, 1246.

Tan, J. 2005. Bridging the gap between facts and values. *World Psychiatry* 4(2), 92–3.

Tan, J.O.A., Stewart, A. and Hope, T. 2009. Decision-making as a broader concept. *Philosophy, Psychiatry, & Psychology* 16(4), 345–9.

Treasure, J. 1997. *Anorexia Nervosa: A Survival Guide for Families, Friends and Sufferers.* Hove, East Sussex: Psychology Press.

Treasure, J. and Schmidt, U. 2005. Treatment overview, in *The Essential Handbook of Eating Disorders*, edited by J. Treasure, U. Schmidt and E. Van Firth. Chichester: Wiley.

Treasure, J., Smith, G. and Crane, A. 2007. *Skills-Based Learning for Caring for a Loved One with an Eating Disorder: The New Maudsley Method.* London: Routledge.

Treasure, J. and Ward, A. 1997. A practical guide to the use of motivational interviewing in anorexia nervosa. *European Eating Disorders Review* 5, 102–14.

Tucker, I. 2010. Everyday spaces of mental distress: The spatial habituation of home. *Environment and Planning D* 28, 526–38.

Turner, V. 1967. *The Forest of Symbols: Aspects of Ndembu Ritual.* New York: Cornell University Press.

Warin, M. 2003. Miasmatic calories and saturating fats: Fear of contamination in anorexia. *Culture, Medicine and Psychiatry* 27(1), 77–93.

Warin, M. 2006. Reconfiguring relatedness in anorexia. *Anthropology and Medicine* 13(1), 41–54.

Warin, M. 2010. *Abject Relations: Everyday Worlds of Anorexia.* New Brunswick, New Jersey and London: Rutgers University Press.

Chapter 3

Home and Heart, Hand and Eye: Unseen Links between Pigmen and Pigs in Industrial Farming

Kim Baker

Introduction

Bangers and mash, pork pies, sausage rolls: all everyday food staples and typically traditional British dishes. But the amorphous goo of sausage meat retains no semblance of the pig it originated from, nor does it divulge evidences of the inter-related human and animal lives that preceded its making. All this is lost during the journey from pig-pen to plate, left undeclared and imperceptible to the consumer at point of sale. So what goes into a sausage: what underlies its skin?

In reframing Haraway's (2003) question, 'what do I touch when I touch [a] dog?' as 'what is the totality that I consume when I eat pork?' this chapter reveals aspects of human-animal relationships that are routinely withheld or glossed over as meat is consumed during an era when the actualities of farm life have receded far from public view. The popular metaphor of the 'food chain' frames contemporary farming as a mechanized system, an idea that bypasses the physicality inherent in production practices involving living bodies, both animal and human. Yet in answer to a question posed during my anthropological fieldwork, 'what do you do?' pigmen's accounts of their work reveal both connections and disparities between pigmen, pigs and consumers. This chapter therefore documents how both physical and metaphorical articulations between human and pig bodies are generated during the complex inter-species labour of meat-making. It discusses prevalent stereotypical images of pigmen and their work and considers how these representations are socially reproduced. This discussion draws on long-term ethnographic fieldwork conducted at an indoor intensive pig unit where 400 breeding sows, seven boars, and 3,500 progeny[1] (destined for meat production) were cared for by three pigmen. The classic anthropological theme of domestication underpins my reflections on relationships between the pigmen and pigs confined, or homed, together on pig farms. Like Cassidy (2007: 13), who emphasizes 'mutuality and the uncontrolled slippages that occur when people,

1 Artificial insemination was used to breed 95% of these progeny, or 'slaughter pigs'.

animals and things are brought into close proximity', in this chapter I show what is produced when people and pigs live and work side-by-side.

An industrial technocracy exerts immense power within contemporary animal husbandry, but this is not the whole picture. Although intensive pig systems are heavily mediated by technological means, the pig unit also constitutes the 'home' of the animals, who are after all referred to as 'domestic', or 'domesticated'. Much of the work of caring for them involves activities and approaches paralleled by work in the human domestic sphere – feeding, cleaning, nurturing young – all on a 24 hours a day, seven days a week basis. In noting that domestic work is not completely attuned to the measurement of the clock, Ingold (2000: 331) suggests that in the 'interplay between the task orientated time of the home and the clock time activities of the workplace' distinctions have been made 'along lines of gender, […] with women […] more committed to task oriented time and men more committed to clock time'. However, in the context of pig production, distinctions of the kind Ingold proposes, which link feminine and domestic time, are not clear cut. It is men who undertake the 'housekeeping' tasks, immersing themselves in routines necessary to support the exclusively feminine events of pregnancy, labour, and birth. In a workplace governed by production timetables, pigmen arrange their 'housekeeping' tasks so as to fully accommodate the female porcine biological rhythms that pervade the place: clock-time, production-time and body-time run concurrently. Pig business is big business, demanding optimal economic productivity achieved via the capitalization of human and animal bodies, a precondition which enmeshes pigmen in more overtly masculine labour objectives. As a site where humans and animals, bodies and technologies, nature and culture, domesticity and industry converge, the 'factory farm' epitomizes precisely the mix of imperatives that promote the open-endedness and slippage Cassidy proposes.

Conspicuous Absences: Stockmen and Stereotypes

> The social invisibility of the farm worker has made him an easy prey to caricature […] and even today he finds it difficult to shake off the derogatory term 'labourer', with its implication of unskilled toil and low status. […] The agricultural worker is even denied his own identity: it is the farmer – not the farm worker – who produces Britain's food and to whom we owe our gratitude, argues Newby (1977: 11).

Aspects of the situation Newby described three decades ago still persist, and are perhaps more acutely experienced by stockmen. Popular representations of 'pigmen' evoke diverse stereotypes ranging from the unlettered swineherd, to straw chewing yokel, to the blundering Grundy family of BBC *Archers* fame. Historical perspectives contribute to these prevalent misperceptions about farm workers. The dispossessed, unskilled Victorian 'labourer' who forms the subject of Jefferies's

(1880) *Hodge and His Masters* also haunts Newby's (1977) study of 'deferential' farm workers in twentieth-century East Anglia. Rossabi (1979: xiii) explains that Hodge was a derisory nickname for farm workers, signifying all that was 'slow, solid, sure and stupid'. The late eighteenth century saw the emergence of a clear tripartite class structure: landowners, tenant farmers, and landless labourers, understood collectively as the 'natural hierarchy of rural society' (Newby 1987: 56). Agricultural labourers were ascribed the lowest position in the 'natural order' both on the farm itself and in wider social contexts. Hodge's legacy continues in the twenty-first century as agricultural work continues to be socially divisive. The 'them and us' mentality persists: in terms of social power, agricultural workers perceive themselves to be disadvantaged, and socially weak in comparison to both the farmers *and* the 'well off' consumers they serve.[2]

During my fieldwork, pigmen's sensitivity about low status was evident when they alluded to being 'pissed off with', or 'tired of hearing about', stereotypes. While their occupation enables pigmen to develop strong intra-industry pride and reputation, in the wider world they contend with equally powerful currents of social disparagement. One man, married to a school teacher, explained further:

> I'm embarrassed to meet new people because one of the first questions is 'what do you do?' When I say 'I'm a pigman', they tend to sort of .... It's a joke to a lot of people. I don't see why, because I've trained to do this. Not anyone can just come and do it. It takes five years before you know the job. I mean I could empty bins, and could pick it up in a morning. But we seem to be thought of as lowly as bin men. We should get more credit than we do. [...] People have got this perception of you in smocks and straw hats with a bit of straw hanging out your mouth, just forking muck all day.

Here, being 'embarrassed' arises not from the work itself, but from frequent encounters with derogatory or erroneous preconceptions projected onto this pigman by 'outsiders'. In making explicit the frustration that is experienced in the face of dismissive assumptions concerning intelligence, skill, credentials and credibility, pigmen show that elements of 'spoiled identity', as described by Goffman (1990) figure strongly in the work.

Negative characterizations of pigmen are complicated by animal welfarist based perceptions. Wilkie (2010: 34–9) identifies a crucial ambiguity involved in the status of stockmen who are 'absolutely central to the production process' yet whose 'contributions have not always been formally acknowledged or appreciated'. She notes that whilst the activities of animal welfarist and rights groups have exposed problems, they have provided few examples of good husbandry practice: negative accounts of stockmen, deriving substantially from animal rights sources, tend to be both 'partial and partisan' (ibid.: 184). The imbalance in coverage has

2 The fact that there has been no agricultural strike in Britain since 1923 is indicative of a 'deference' which still lingers.

bolstered negative perceptions of livestock workers who are popularly believed to intentionally or unintentionally mistreat animals. These perceptions assume both that 'real' stockmanship and 'factory farming' are mutually and irretrievably incompatible and that factory farming is made possible by a particular kind of vilified, yet invisible workforce.

The problems generated by poor press were illustrated by a pigman who told me:

> It's not really to do with what we do. It's who's perceiving, and who's saying what's right and wrong. They're not saying that from a fair viewpoint. They're just saying what we do is wrong. They've got no perception of what we do, or the difference between right and wrong.

This remark highlights the difficulty: who do critics think they are talking about? Importantly it also implies the mutual absences which are created as meat passes along the production/supply chain. As a result of the series of catastrophic livestock disease epidemics which occurred between the mid-1990s and 2004, pig farms have become highly sequestered places as biosecurity precautions have inevitably contributed to screening stockmen and the detail of their work from public view. And, just as the farm is hidden from consumers' view, consumers too are hidden from pigmen's view. In the space between them, misconceptions proliferate: pigmen emerge as indifferent, disengaged from pigs, or downright cruel, while consumers figure as uncertain, inconsistent, critical and volatilely judgmental and these misperceptions are compounded by pigmen's cultural invisibility.

The recent retailing trend for showing farmers' photographs on meat packaging encourages consumers to identify with individual producers and specific farms, a tactic enabling retailers to 'personalize' meat, a quintessentially generic food. This visual marketing strategy reflects the cautious approach of most retail chains to participating in the agenda of 'reconnecting with food'.[3] Such 're-connection' may present a challenge to some consumers when the food in question originates from animal rather than vegetal bodies. Retailers' celebration of farmers produces an unintended side-effect of concealing stockmen,[4] figuring them as a distant, faceless labour force invisible to the consumer. In the hierarchy of visibility that retailers construct by applying such farming imagery to meat packaging, farmers are at the top, portrayed relatively frequently. Pigs make very occasional appearances, although during fieldwork one farmer recounted how meat-purchasers at a major supermarket chain had informed him that, 'no one will eat it if it's got live pigs on it'. Pigmen are at the bottom of this hierarchy, never appearing. This model denies consumers the chance to see those most closely involved with pigs and, by extension, obscures the actuality of relations of production. Yet, it is also arguably

3 See Holt (2005), Tregear (2007), Holloway and Kneafsey (2000), for accounts of re-connective food marketing strategies and Tovey (2003) on animal 'invisibility'.

4 'Stockman/stockperson' and 'pigman' are used interchangeably.

such relations that give rise to the sense of 'spoiled identity' (Goffman 1990) both culturally attributed to, and felt by, pigmen.

If 'good' qualities can reciprocally be transposed between humans and animals during specialized livestock husbandry, as Ritvo (1987) shows in her account of aristocratic 'improvers',[5] then it follows that so too can 'bad' ones. Arluke and Sanders (1996) outline the 'sociozoologic scale' in which humans and animals are ranked according to the socio-moral roles expected of them. Farm animals appear high in the rankings, as 'good animals', equal to pets, and they also appear in the second positive ranking as 'tools'. The lowest ranks include 'vermin', and 'killers', classified as 'bad animals'. In terms of this scale, the position of pigs is ambivalent; although good and useful, their natural predisposition to scavenge and fight qualifies them concurrently for a place in the 'bad' category. In detailing secular and religious medieval iconographies, which consistently portrayed pigs as contemptible beings capable of transmitting shame to their human associates, Phillips (2007: 376) shows that in the Western imagination pigs have long been regarded as despicable. The iconographies she discusses depict pigs' least attractive qualities, which include being filthy, coprophagic, stinky, greedy, gross, lustful and dangerous – negative attributes too readily reassigned to their handlers.

As Douglas (2002: 66) observes, close co-existence between humans and animals can be problematic when it results in perceived 'mixing, or confusion', of categories. Pigmen, more than any other category of agricultural worker, are susceptible to harmful analogies being made between themselves and the polluting 'dirt' that their charges both emanate and are. While the practice of pigmen and pigs living in close proximity may engender affinity and closeness, as explored later, there is an irony in the fact that the arrangement may also contribute directly to negative caricaturing of pigmen. The concepts of living *near* pigs and living *as* pigs are too close for comfort, too easily mixed up. Human co-habitation with pigs also fuels the idea of pork as highly threatening – a substance capable of symbolically conflating or holding together categories that meat-eaters might prefer to keep apart. For the consumer, such confused or confusing relationships are indigestible. On the journey to the plate it is precisely the need to make distance from 'dirt' that initiates literal and metaphorical sanitization of pig bodies. The transformation from polluting pig to pristine pork involves the removal of impure, defiling bodily wastes (hair, claws, guts, blood) and, equally importantly, the editing out of evidence of 'contaminatory' relationships forged in interspecies domestic co-existence. Without the latter the total objectification of the pig's body as meat would be unattainable. By exploring alternative ways of looking at the resonances between the bodies of pigs and pigmen, the next section offers a counter-argument to the stereotypes of pigmen discussed so far.

---

5 Ritvo cites connections between selective livestock production and the governing classes in eighteenth-century England, with dukes, earls and the King, 'Farmer George', contributing to national and class based prestige: 'For both animals and people a distinguished lineage divided those with hereditary claims to high status from arrivistes' (1987: 61).

Making Pigs, Making Pigmen: Embodied Productions

So, what are pigmen like, and how do they explain their working world? Many of the older men I interviewed referred to leaving school early having 'not done very well'. A successful manager recalled how he had been a 'layabout and a twit' who had initially 'drifted' into stockwork. Despite some introducing their job choice like this, the pigmen I met were almost unanimous in claiming a family background in stockmanship, with most starting the job on leaving school between 14 and 16.[6] When describing what they habitually referred to as 'the culture of stockmanship', most insisted that working with pigs requires 'a feeling for the animals' which one must be 'born to'. Many were adamant that the 'feeling' must come 'from the heart', and if a man does not possess this vital inherent trait, then he cannot be taught it.

These conceptualizations indicate the faith that stockmen place in perceived linkages between ability and heredity, an idiom which they commonly apply to both people and pigs.[7] The strength of belief in the relationship between aptitude and heredity was graphically illustrated when pigmen drew out diagrams of their own family trees which they annotated with indications of familial attributes and skills, understood as their professional 'legacy'. Pigmen often emphasized their understanding that the cultivation of sensate skills, and empathetic relationships with animals, were attributable to long- standing family predisposition 'handed down' to them via their antecedents. These were ideas that were frequently transposed onto the way they viewed pigs for whom they understood the same principles to apply. One man drew out a 'family tree' for his herd, and the sketch illustrated his conviction that the desirable 'professional' traits required of commercial pigs also originate with ancestors and are transferable across generations. This way of thinking provided an insight into how pigmen regard themselves and their pigs as similarly constituted; both derive from 'working' backgrounds where qualities of physical robustness, vigour, and fitness for purpose are prime transmissible values. Voicing the idea that 'good places' produce good workers, who in turn propagate more good workers, several middle-aged interlocutors drew attention to the cross-generational 'service' their family members gave when employed on prestigious local agricultural estates.[8] This interpretation suggests that professional reputation, expressed through 'good workmanship', is acquired partly by 'breeding', or genetic inheritance, *and* partly via osmotic transmission-by-contact with elite agricultural enterprise.

6 Many had also worked sporadically in other local jobs which had initially promised better pay and conditions.

7 Gray (2000) and Cassidy (2002) respectively document how Scottish sheep farmers and racehorse breeders use the metaphor of genealogy similarly.

8 In the nineteenth century these estates were owned by an influential family who were instrumental in developing the 'Suffolk Trinity': Red Poll cattle, Suffolk sheep, Suffolk Punch horses.

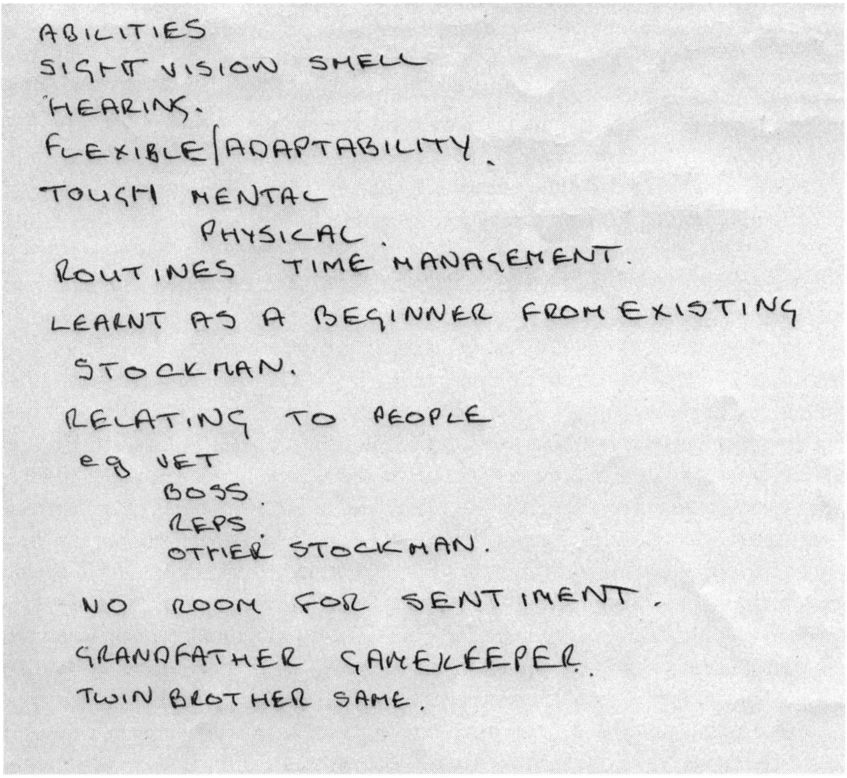

ABILITIES
SIGHT VISION SMELL
HEARING.
FLEXIBLE/ADAPTABILITY.
TOUGH MENTAL
PHYSICAL.
ROUTINES TIME MANAGEMENT.
LEARNT AS A BEGINNER FROM EXISTING
STOCKMAN.
RELATING TO PEOPLE
eg VET
BOSS.
REPS.
OTHER STOCKMAN.
NO ROOM FOR SENTIMENT.
GRANDFATHER GAMEKEEPER.
TWIN BROTHER SAME.

Figure 3.1 A unit manager's ranking of the abilities and aptitudes he considers essential to good stockmanship. Vision and sensate skills are paramount and he draws attention to his family history of working with livestock

Yet, although a predisposition for stockmanship may be 'in the genes', this legacy does not mean that any stockperson is born replete with the skills for the job. Expertise is hard-won and only acquired through protracted or indefinite embodied learning. So how are stockmen 'made' or cultivated? How does their learning contribute to a sense of identity? Although many of the stockmen I met stressed hereditary predisposition, they also acknowledged that husbandry skills could be acquired. Having emphasized his own inherited credentials, and knowing that I had none, the unit manager hosting my fieldwork confidently predicted 'We'll make a stockman out of you!', an assertion which revealed his own ambivalence about the 'born' versus 'made' conundrum. Brian, a farrowing manager[9] told me:

9 This man specialized in the care of pregnant sows and newborn piglets.

> You can teach people to look after stock, but you can't teach people to be a stockman. Stockman comes from your heart. You do learn it as you go along but there has to be an initial part of you which is stockman. Not everybody can come and do it. […] Being born and bred on a farm is most of it. Yes, I did pick up a lot of things from college. […] but I learnt from young. When I went to college I had a lot of physical knowledge I took with me. Then I did learn a lot about academic stuff. I still learn now, see different things every day.

The stockmen with whom I worked used a confusing array of terms with reference to their 'training' or 'apprenticeship', their accounts bearing out Simpson's (2006: 153) idea that apprenticeship assumes many forms. Some stockmen had followed a 'traditional' training, learning as boys from fully-experienced men. Others had attended courses at the local agricultural college, and some combined both routes. This diversity is partly responsible for further contradictions that appear in stockmen's self-perception of their role and status. Several middle-aged men recalled how, after poor achievement at school, they were warned of 'ending up with the pigs', being 'only a pigman' as a last resort. By contrast, college initiatives of the 1960s and 1970s attempted to improve the job image by rebranding stockmen as 'pig technicians', although this term was not used by stockmen I met. Some linked the term 'labourer' unfavourably with unskilled migrant European workers,[10] and 'farm operative' and 'farm hand' were also disliked since they implied the sterility of 'mindless' factory work.

Since he had alluded to 'apprenticeship', I asked Brian whether he saw himself as a craftsman. 'No! Not officially. I don't look on it as a craft. If I was a carpenter I'd be a craftsman' was his initial response. But on reflection he conceded 'I must be one', qualifying this by saying 'I look on what I do as a skill. That's the way I perceive it'. In distinguishing differences between his work and that of other agricultural craftsmen, such as tractor drivers and mechanics, Brian considered these people to be less skilled, since their work was limited to knowledge of machines alone. Sennett's (2008: 8, 9) idea that craftsmanship consists of the 'skill of making things well' and is an 'enduring, basic human impulse, the desire to do a job well for its own sake', may be more compatible with the way that stockmen perceive their skills within an overarching hierarchy of knowledge. The fact that they attribute different values to college learning and on-the-job learning was illustrated by a pigman who said that 'having a driving licence doesn't make you a good driver', implying that participation in formal training schemes and possession of qualifications do not automatically predetermine a good reputation or any ultimate guarantee of competency. In fully recognizing the difference between knowledge *generated by and through the body as* opposed to knowledge abstracted *from* the body, many stockmen respond sceptically to educational agendas that attempt to reproduce practical and intuitive skills through detached, reductive 'book learning'. For them 'ability' is what counts. When pressed to

10 That said, others praised the innate skills of European migrant stockworkers.

explain ability, a propensity to deal effectively with variability emerged as a cultivable attribute rather than an imaginary resource. Emphasizing the variables he encountered in his work, Brian said:

> It's not like making a door, painting it, polishing it because you can do fifty of those and they'll all be the same. With livestock it's always different, always something going on; weather, temperature, hard pigs to move, problems with water .... It's never going to be a science. There's no fixed formula to what we do. It's all individual stuff, treating them individually. [...] There are standard rates you should be working at, but within that there's a lot of individualism – what I do with *each* animal.

Brian's words show how he sees direct relationships between ability and adaptability. Experience has taught him that with pigs, results and effects are never exactly reproducible because variability can never be eliminated from people, animals, their interactions, or the working conditions they share. Ability is essentially the skill of responding appropriately to *individual* animals within a highly choreographed set of production procedures. As I spent time with pigmen, the caricature of the uncaring, bullying, deskilled dolt simply failed to fit with the actuality of the energetic, knowledgeable and quick-witted individuals I encountered. This account, given by the manager of an 800-sow unit employing four other stockmen, indicates how he manages close relations between pigs and pigmen themselves:

> A lot of my time is spent keeping people busy, efficient. Pigs are easy. It's people that are difficult. Stockmen are funny. A decent one is very passionate about what he does. They take pride in what they do. It's easy to be a manager and say 'do that!' but I have to show good standards, take the lead, and control the sort of uncertainty that happens with stockmen. I prioritize the jobs [...] manage time. You have to be on the ball and keep at it. [...] I know the gilts[11] are really well [...] and that gives us confidence. It wouldn't be good if we ended up in the press with a welfare problem. We have a reputation so that's where the pride is. We're a close-knit crowd.

The idea of 'close-knit' relations also extends to pigmen's living arrangements. Now, as in the past, the practice of 'payment in kind' in the shape of accommodation and other perks compensates for low pay and keeps the employment contract semi-informal.[12] For pigmen, good domestic practice at home and at work is crucially interwoven. While human welfare may be irrelevant to the pigs, pig welfare is

11 Young breeding sows.

12 The Farm Animal Welfare Council's (FAWC) (2007: 8) report on stockmanship states that 50% of stockmen are aged under 35, a demographic suggesting that family housing is an important requirement.

critical for the security of pigmen's families; stockmen have more than just a job to lose if things go wrong. The tied cottage represents both incentive and risk.[13] Brian emphasized the house as the lynchpin which made a whole way of life possible, connecting job, family, locality and self-worth in relations of continuity:

> I like living in the countryside. I'd have never been able to buy a four bedroom house. I wanted my family to grow up in this environment, to have the chance to do fishing and shooting and bike riding without the constraints of living in the town. I couldn't do it without family. My wife accepts there's evenings when I'm backwards and forwards. When you marry a stockman, that's part and parcel of the job. It's not a nine-to-five job. It's a career where you're involved all the time in lives, and birth and death. It's a way of life rather than just a job. I don't consider the house to be a perk but I consider the fact of where I'm living to be a perk. The house is really a part of my wages, part of my deal.

Tied accommodation is often located adjacent to, if not actually within, the confines of the pig unit itself, meaning that pigmen and pigs literally live and work side-by-side. Examples of mutuality between the domestic spheres of pigmen and pigs were inescapable during fieldwork as pigmen multi-tasked between home and pig unit. They alternated freely between caring for human and pig infants, shifted juveniles of both species to the right places at the right times, kept track of who was 'off colour' and who ate what, fetched and carried, cleaned and tidied, made beds, repainted and mended, turned out refuse, adjusted the central heating and kept the plumbing flowing. The overlaps are too numerous to list fully, but collectively they suggest farming and stockmanship are closely related versions of domesticity. In the relationships I observed between pigmen and pigs, domestic care and imperatives of production were so closely integrated as to be inseparable.

Thus, what constitutes the 'culture of stockmanship' that is produced through this entangled domesticity? The next section suggests that for pigmen the answer embraces empathy, as well as particular sensate and physical skills. Collectively these are regarded as cultivable assets but, importantly, they are also part of a network of relationships between stockmen and pigs. Such relations are both central to the making of, but absent from, the meat eaten by the consumer. They also do not fit with culturally-widespread negative images of factory farming.

13 In a different human-animal context Cassidy (2002: 121) notes a comparable situation amongst Newmarket stable staff.

Networks of Hidden Relations: Welfare, Empathy and Vision

The negative aspects of industrialized livestock production methods established in the mid-twentieth century are well documented.[14] Yet, the National Farmers Union president claims that British pig producers are acknowledged by their European counterparts as 'high welfare achievers' (Jones 2010) and this is enshrined in policy. Brambell's (1965) seminal report has informed much of the animal welfare legislation that has appeared since the 1960s in Britain and Europe, and recent policy makes increasingly explicit reference to stockmanship and the values it enshrines. In Britain the Animal Welfare Act (2006) prohibits unnecessary suffering and endorses a 'duty of care'. The Department for Food, Environment and Rural Affairs (DEFRA) Code of Recommendations for the welfare of pigs (2003: 6) suggests that 'no matter how acceptable a system may be in principle, without competent diligent stockmanship, the welfare of the animals cannot be adequately catered for'. The Farm Animal Welfare Council's (2007: 7) 'three essentials' of stockmanship consist in: 'knowledge of animal husbandry' (animal biology and how animal needs may best be met); 'skills in animal husbandry' (observation, handling, care, treatment, problem detection and resolution); and 'personal qualities' (affinity, empathy, dedication and patience).

When I asked Simon, a finishing manager,[15] about professional pride in the welfare of pigs, his reply showed that the pride he felt had both a private dimension and one that was projected beyond the farm:

> I always try and make their life as nice as possible. There's nothing else I can do. So I don't feel guilt when they go. I get job satisfaction when everything is healthy, and when I'm actually putting them on the lorry, that's when I can be proud of them. [...] There are weeks when you don't get such a good batch ... the people at the slaughterhouse don't know me ... but I imagine the pigs coming off the lorry, and the vet and the other people standing there looking, and you feel you're being judged. [...] You look at it and you realize the enormity of it all; my God, that's a huge lorry packed full of pigs.

Simon's words highlight the contradictions implicit in work where every ounce of care a pigman invests moves the pigs closer to a single inescapable destiny – pre-arranged premature death. Every week the 200 six-month-old pigs who go to slaughter are instantly replaced by 200 newborn piglets. Birth and death are continuous, held in an unremitting equilibrium. Thus, when stockmen say 'got livestock – got deadstock' they refer to the fact that both predetermined animal death and accidental 'wastage' (through fatal sickness) are inherent parts of a production system predicated on the dichotomy between creation and destruction. Yet, the entanglement of this dichotomy with empathetic ways of relating to the

---

14 See Harrison 1964.

15 Finishing is the final stage of 'growing' or fattening prior to slaughter.

pigs in their charge was clear in pigmen's accounts of their work; in response to being asked 'what is it that you think you do in this work?' Simon drew production efficiency into a direct relationship with ethical treatment, saying, 'I'm trying to get them to finishing weight as quickly as possible, with as less stress on them as possible … that's it really. Trying to keep them healthy. Trying to keep them happy'.

For Hemsworth (2007, and n.d) livestock ethics encompass economic, philosophical, social, cultural and religious aspects. In the 'asymmetrical nature' of human-livestock relations, the ethical dilemma involves making choices between profitable animal production with higher moral costs rather than non-profitable animal production with lower moral costs. Hemsworth emphasizes the explicit responsibility of stockpeople to care for and manage their livestock to achieve efficient economic performance: productivity and welfare are given in relations of reciprocity.[16] Pig production manuals, such as Gadd (2005: 114–16), suggest that deployment of sensate and organizational skills alone would not constitute stockmanship, since the quality of empathy is essential. This quality is defined by English et al. (1992: 29–36) as part of a reciprocal relationship involving 'emotional attachment of man and animal' and 'careful and gentle handling', in which empathy transcends 'solely using technically correct methods'. When English et al. suggest that 'the art of stockmanship' consists in the special ability of 'some individuals to achieve a response from animals which cannot be achieved by others using the same resources and techniques' (ibid.: 36), they imply that qualities of mystique or enigma are constituent elements of some stockmen's work.

The kind of 'emotional' engagement that has emerged here is not to be confused with mawkish sentimentality, or inane petting. As English et al. (ibid.: 36) suggest instead, in cultivating such empathetic relations stockmen may, at different moments in the production process, assume the roles of boss animal, substitute peer animal, mother, leader or friend, all terms implying different varieties of mutual emotional attachment. This is an idea borne out by the pigmen I worked with. They extended the repertoire of roles and relationships, often alluding to themselves as the slaves, mothers, chaperones, bosses and workmates of the pigs, and they routinely and unselfconsciously characterized the breeding sows as their 'sweethearts', 'ladies', 'girlies', 'darlings'. Their usage of such terminologies is indicative of the sense of affinity and affection that many stockmen extend to their animal co-workers.

Thus, as both Wilkie (2010) and English et al. (1992) note, extensive research into animal handling (conducted from agricultural and animal science perspectives) has hitherto been too narrow in its understanding of the role of stockmanship, and has discounted the importance of cognitive, ethical, and emotional dimensions. These omissions may have unintentionally contributed to formulations of stockmanship as an ineffable 'gift' which cannot be taught, as noted above.

16 Hanna et al. (2009) and Ravel et al. (2006) advance similar ideas.

Figure 3.2 Piglets being barrowed to the nursery. Here the piglets are dependent on the pigman's 'mothering' skills, which are mediated and augmented by technological means

However, the alternative strategy of devising formalized ways to identify and mobilize empathy or 'emotion work' (Wilkie 2010: 176) within competitive commercial settings may risk producing negative side-effects. Wilkie warns that 'adopting such a calculative perspective [...] diminishes the stockperson-animal interface to [...] another productive factor that needs to be understood so that it can be fully exploited. Not only are animals increasingly controlled in productive contexts; so are those who handle them' (ibid.: 36). These observations are important. In suggesting that pigmen and pigs are both subject to control, Wilkie's thinking contradicts the popularist idea that 'anything goes' in the behaviours that stockmen extend towards livestock.

Just as the farm creates a social order, pigmen too rank the bodily skills they possess and these cannot be dislocated from the ways in which pigmen relate to pigs. Reliance on visual acuity in animal husbandry contexts is well-documented; Hudson (1910), for example, shows how shepherding and 'looking' are synonymous, a theme taken up in Grasseni's (2005, 2007) recent accounts of 'skilled vision' and 'good looking' amongst cattle breeders. For Grasseni's interlocutors, being able to 'see' constitutes a shorthand for 'being integrated into the environment of one's practice and for having developed a thorough sensory relationship' (2009: 86) with one's livestock. The pervasive topic of faith in vision was raised by pigmen in many different work situations, and they were explicit about the attitude of watchfulness they used to detect any and every sign (of health, sickness, hunger, satiety, discomfort, sexual receptivity) which they could read directly from the bodies and behaviours of the pigs. Against the background of the above discussion of ethics, vision emerges as an embodied ethics, a practical enactment of pigmen's affinities with their animals. Simon explained:

> You're concentrating all the time. I often feel like a sheepdog that's watching the whole lot and he never takes his eyes off them, watching for every single little tiny movement, any indication, he adjusts himself to every tiny thing. I'm a bit like that. I can't just watch one pig or two pigs. I'm also watching … [He spread his arms as if to span all the pigs in the building].

The emphasis given here to an all-encompassing 'watching' suggests that Simon's use of sight is neither passive nor reducible to a merely coincidental 'seeing-in-passing', and his vocabulary shows that the visual aspects of stockwork are concentrated around the combined activities of well-practised watching and observing. Marvin (2005: 4, 5) distinguishes usefully between seeing (as basic and least intense engagement), looking (as active and directed application of sight), watching (as more attentive, time consuming, continuous alertness), and observing (consisting of concentrated investigative viewing that is both recorded and guided by intention). The actuality of Simon's behaviour contradicts the common perception that modern stockmen do not 'know' their animals. Elaborating further, he continued:

> Someone might think I walk through my buildings quite quickly, and think I haven't given the pigs a good look. But I can guarantee if there's a lame pig in there I would spot it. Even if it's laying down. I can just tell by the angle. It's something that comes from working with them for years and years. I don't look at the pig and think I'll check this and I'll check that. It's just one complete thing.

> I saw a programme – Paul Mackenna who does all the mind stuff – he did this trick where he knew, without counting, there were sixty-six little things on a tray. If you do that often enough you know what sixty-six looks like. It's just the same when I walk through. I know what thirteen looks like. I haven't counted them but I can *feel* it.

Another man agreed: 'You can tell when you walk into the room, she's not happy, not going to get up. You can see as soon as you look at her. […] You're thinking what they're thinking, thinking to yourself about how they're behaving'.

These accounts suggest that knowledge acquired by eye, and accumulated over time during repetitious acts of watching, is gradually internalized so that an established, cross-referencing fund of information informs every new encounter so powerfully that the pigs are intensely 'felt' rather than superficially 'seen'. It is as though the un-intrusive look or alert glimpse of brief duration activates an intricately detailed precognition; layers of hindsight inform the insight of the moment. The emphasis placed on vision by pigmen indicates that now, just as much as in the past, livestock are 'known' (and by extension, cared for) through, and because of, their keepers' highly developed visual capabilities. The examples above show that the education of the eye, and other sensory skills, are well-recognized and highly valued amongst stockmen. Possession of such aptitude contributes to the positive reputation of experienced stockmen who come to be known for their capabilities not just on the farm, but more widely within the pig producing community at local and regional levels.

Pigmen's ability to work in a patient, systematic way is both essential and deliberately developed. Because pigs are highly stress-prone and can neither speak nor comprehend human speech, pigmen rarely shout, and their movements are carefully premeditated rather than ostentatiously animated. Displays of temper are very rare and pigmen consistently exercise careful control over their physical actions around the pigs. Because aggression is understood as counterproductive, threatening or violent actions are discouraged and trainees are taught to exclude them from their body language repertoire. One man referred to his carefully premeditated movements when handling pigs as a deliberately structured 'dance'. Inasmuch as the system imposes a requirement for pigmen to manage and control pigs, it also demands that they exercise constant *self*-discipline and control of their own physical and mental demeanour.

Pigmen also draw on another kind of aptitude involving self-control, one seldom discussed amongst themselves, but sometimes obliquely referred to as 'sixth sense'. The existence of such rarely-acknowledged expertise was hinted at during routine tasks when I worked with a finishing manager. This highly-experienced pigman was reserved, and initially under-confident about talking with me, but as we grew to know one another he repeatedly mentioned 'self-hypnosis' and the 'state of mind' he used in his work. In a lengthy interview he haltingly revealed his experience of the special psychological state he regularly entered, and which he believed his colleagues experienced too:

> It just comes over you as you're walking through. I'm almost not aware of me walking along sometimes. I just find myself gliding along. […] I'm not thinking about time. Time can go … It's usually when I'm alone. […] You switch off everything else in your life. I can feel it coming over me. My breathing slows slightly. I just feel at one with the pigs and everything around […] totally relaxed. […] I know that when I've got that sort of feeling things are going to go fairly well. […] I believe when I'm like that, the pigs are more relaxed. They feel relaxed with me as much as I do with them. […] I feel totally alone, but totally at home. That's totally comfortable and I know the pigs are comfortable as well. I want to disturb them as little as possible. If I keep calm, they keep calm. I want to let them rest. When I'm in that … I know that everything's under control.

This pigman emphasized his sense of seamless connection between himself, the pigs, and the place, a sensation that was 'extra real, beyond words', and which induced a rush of confidence. Several times he spoke of an extreme reluctance to let 'other people' near the pigs. Allowing intrusion would risk 'chaos', a loss of the protective control he maintained. At face-value, his description of this 'state' might be understood as a simple clarification of the empathetic behaviour that stockwork requires. However, the features of this experience go beyond the prescribed attributes of 'conscientiousness' or 'agreeableness' mentioned previously, and are instead manifestations of the enigmatic 'extra something' that English et al. (1992) point towards.

Changes in perception and consciousness, increased perceptual focus, loss of ability to gauge time, sensations of merging with material forms or technologies, and hyper-real experience beyond verbal description are cited in Lyng's (1990) account of the 'edgework' performed by people engaged in risk-laden activity, skydivers, police, firefighters for example. Edgework typically involves a threat posed to participants' physical or mental well-being, or to their own sense of an ordered existence. Taking place on the boundaries between the conscious and the unconscious self, the ordered or disordered self or environment (ibid.: 857) edgework offers opportunities for developing skills via 'planning, purposive […] action and concentration' within status hierarchies whose aim is to control situations verging on chaos (ibid.: 874). These features of edgework coincide closely with the issues raised in the pigman's account, which focuses on his total determination to avert chaos and thereby maintain calm and comfort for his pigs. The risk, or threat, he perceives centres on prospective breakdown of the whole system should stress or disease be introduced into the herd, and the force of this concern drives the intensity of the sensations he experiences. Langer (1975) suggests that complete mastery of an environment would imply, by extension, ability to control chance events; 'The greatest […] feeling of competence would therefore result from being able to control the seemingly uncontrollable' (ibid.: 323).

Figure 3.3 Much of the daily routine for pigmen involves 'walking through' the buildings visually checking the welfare of pigs. Here the pigman administers medication to a juvenile animal showing signs of lameness

The concept of edgework also resonates with the emphasis that many stockmen place on the value of creative resourcefulness, or willingness to 'try something else' to promote pig well-being and productivity. Lyng (1990: 868) also draws attention to the paradoxical situation in which free activity only arises with constrained or necessary activity as its basis. Against the backdrop of 'the realm of necessity' workers may 'organise the work process in a way that reflects their own human needs' (ibid.: 869). Within the rule-bound context of industrial livestock production, good stockmen devise ways to balance production-led demands with sensitive treatment of the animals they care for. In this process, the logic of industrialism fails to wholly preclude positive interpretations of animal 'needs' such as calm, comfort and wellbeing, each borrowed from the human domestic realm and transposed onto the 'shop floor' of the pig sheds. The production schedule, led by market forces, constitutes 'a realm of necessity' against which stockmen extemporize so as to devise highly-nuanced working knowledges. Given that the edgeworking model, 'sixth sense' and the 'factory farm' are not completely incompatible, as stockmen's personal accounts conclusively show, industrial farming involves experiential intensity of a kind that is seldom acknowledged.

Conclusions

This chapter has suggested that inasmuch as making pigs produces meat, it also contributes to the production of knowledgeable selves and knowing bodies. Although space does not permit a comprehensive account of stockmanship, I have indicated rarely-acknowledged articulations between hereditary skill, intuitive behaviour, embodied knowledge and technical expertise. Pigmen's own accounts of their work coincide closely with Marchand's (2008) idea that the deployment of manual craft-based skills involves far more than 'muscle' or brute strength alone, since morals and mind are in play too. In detailing pigmen's professional self-perceptions, I have suggested that a series of unseen and widely-differing logics operate simultaneously. The picture of production popularly imagined from outside the farm is both limited and misleading in constructing a single logic of undiluted exploitation. Looking at it as pigmen do – as a living, a way of life, a domestication practice, a craft, and a series of technological production procedures – reinforces the idea that each of these 'versions' of the work are held in tension: collectively they constitute a well-defined, albeit hidden and under-acknowledged, culture. In arguing for the positive aspects of the 'culture of stockmanship', my account departs from prevalent notions of contemporary capitalist 'agri-business' as a dystopia in which animals are routinely subjected to calculative discipline, dominance, indifference or sheer brutality. Of course the pig farm is not a utopia for humans or animals, but the alternative picture offered here suggests that care and commitment involving intimacy, affection, professional pride, and ethical treatment are not automatically precluded from so-called 'factory farming'. Moreover, these qualities are 'consumables' too, invisibly embedded in the meat we eat; they are present precisely because of, rather than in spite of, the speed and magnitude of production. As the twenty-first century unfolds there are signs that retailers have spotted the potential of marketing the consumables mentioned above, but together with farmers and pigmen they face the 'damned if you do and damned if you don't' dilemma attaching to the objective of revealing the details of meat production. They are well-aware that amongst consumers the appetite for such knowledge is as yet uneven. In the meantime, the facts of human and animal lives at the farm end of the production line hover half-in and half-out of sight for the very reason that they undermine clear-cut distinctions between species, bodies, commodities and the mouths of meat-eaters.

Acknowledgements

The PhD research that forms the basis of this chapter was supported by a grant from the Arts and Humanities Research Council and by a Goldsmiths Anthropology Departmental Bursary. I should also like to express my gratitude to Professor Rebecca Cassidy and Dr Emma Tarlo.

References

Arluke, A. and Sanders, S. 1996. *Regarding Animals.* Philadelphia: Temple University Press.

Bock, B., Van Huick, M., Prutzer, M., Kling Eveillard, F. and Dockes, A. 2007. Farmers' relationship with different animals: The importance of getting close to the animals. Case studies of French, Swedish and Dutch cattle, pig and poultry farmers. *International Journal of Sociology of Food and Agriculture* 15(3), 108–25.

Brambell, F.W. 1965. *Report of the Technical Committee to Enquire into the Welfare of Animals Kept under Intensive Livestock Husbandry Systems.* London: Her Majesty's Stationery Office.

Cassidy, R. 2002. *The Sport of Kings: Kinship, Class and Thoroughbred Breeding in Newmarket.* Cambridge: Cambridge University Press.

Cassidy, R. 2007. Introduction, in *Where the Wild Things are Now: Domestication Reconsidered*, edited by R. Cassidy and M. Mullin. Oxford: Berg, 1–27.

Cole, M. 2011. From 'animal machines' to 'happy meat'? Foucault's ideas of disciplinary and pastoral power applied to 'animal-centred' welfare discourse. *Animals* 1, 83–101.

Coppin, D. 2003. Foucauldian hog futures: The birth of mega-hog farms. *Sociology Quarterly* 44(4), 597–616.

Department of Food, Agriculture and Rural Affairs (DEFRA). 2003. *Code of Recommendations for the Welfare of Livestock; Pigs.* [Online]. Available at: http://www.defra.gov.uk/foodfarm/farmanimal/welfare/onfarm/pigs.htm [accessed: 10 October 2010].

Douglas, M. 2002. *Purity and Danger: An Analysis of the Concept of Pollution and Taboo.* London: Routledge.

English, P., Burgess, G., Segundo, R. and Dunne, J. 1992. *Stockmanship. Improving the Care of the Pig and Other Livestock.* Ipswich: Farming Press.

Farm Animal Welfare Council (FAWC). 2007. *FAWC Report on Stockmanship and Farm Animal Welfare.* London: Farm Animal Welfare Council.

Gadd, J. 2005. *Pig Production. What the Textbooks Don't Tell You.* Nottingham: Nottingham University Press.

Goffman, E. 1990. *Stigma. Notes on the Management of Spoiled Identity.* London: Penguin.

Grasseni, C. 2005. Designer cows: The practice of cattle breeding between skill and standardisation. *Society and Animals* 13(1), 33–49.

Grasseni, C. 2007. Good looking: Learning to be a cattle breeder, in *Skilled Vision: Between Apprenticeship and Standards*, edited by C. Grasseni. Oxford: Berghahn Books, 47–67.

Grasseni, C. 2009. *Developing Skill, Developing Vision. Practices of Locality at the Foot of the Alps.* Oxford: Berghahn Books.

Gray, N. 2000. *At Home in the Hills. Sense of Place in the Scottish Borders.* Oxford: Berghahn Books.

Hanna, D., Sneddon, I.A. and Beatie, V.E. 2009. The relationship between the stockperson's personality and the productivity of dairy cows. *Animal* 3, 737–43.

Haraway, D. 2003. *The Companion Species Manifesto. Dogs, People and Significant Otherness.* Chicago: Prickly Paradigm Press.

Harrison, R. 1964. *Animal Machines.* London: Vincent Stuart Ltd.

Hemsworth, P. 2007. Ethical stockmanship. *Australian Veterinary Journal* 85(5), 194–200.

Hemsworth, P. Not Dated. *Ethical Stockmanship and Management of Animals* [Online]. Available at: www.daff.gov.au [accessed: 30 September 2010].

Holloway, L. and Kneafsey, M. 2000. Reading the space of the farmers' market: A preliminary investigation from the UK. *Sociologia Ruralis* 40(3), 285–99.

Holt, G. 2005. Local foods and local markets: Strategies to grow the local sector in the UK. *Anthropology of Food* [Online]. Available at: http://aof.revues.org/index179.html [accessed: 4 June 2009].

Hudson, W. 1910. *A Shepherd's Life. Impressions of the South Wiltshire Downs.* London: Methuen and Co. Ltd.

Ingold, T. 2000. *Perceptions of the Environment: Essays in Livelihood, Dwelling and Skill.* London: Routledge.

Jefferies, R. 1880. *Hodge and His Masters.* London: Smith, Elder & Co.

Jones, G. 2010. *Driving Improvements in Sustainable, Safe Animal Health and Welfare*. Keynote address to Inside Government Conference: Improving the Provision of Animal Health and Welfare in England, London, 5th October 2010.

Langer, E.J. 1975. The illusion of control. *Journal of Personality and Social Psychology* 32(2), 311–28.

Lyng, S. 1990. Edgework: A social psychological analysis of voluntary risk taking. *The American Journal of Sociology* 95(4), 851–86.

Marchand, T. 2008. Muscles, morals and mind: Craft apprenticeship and the formation of person. *British Journal of Educational Studies* 56(3), 245–71.

Marvin, G. 2005. Guest editor's introduction: Seeing, looking, watching, observing nonhuman animals. *Animals and Society* 13(1), 1–11.

Newby, H. 1977. *The Deferential Worker. A Study of Farm Workers in East Anglia.* London: Allen Lane.

Newby, H. 1987. *Country Life. A Social History of Rural England.* London: Sphere Books.

Novek, J. 2005. Pigs and people: Sociological perspectives on the discipline of nonhuman animals in intensive confinement. *Sociology of Animals* (13), 221–44.

Phillips, S. 2007. The pig in medieval iconography, in *Pigs and Humans. 10,000 Years of Interaction*, edited by U. Albarella, K. Dobney, A. Ervynk and P. Rowley-Conwy. Oxford: Oxford University Press, 373–87.

Ravel, A., D'Allaire, S. and Bigras-Poulin, M. 1996. Influence of management, housing and personality of stockperson on preweaning performances on independent and integrated swine farms in Quebec. *Preventative Veterinary Medicine* 29(1), 37–57.

Ritvo, H. 1987. *The Animal Estate. The English and Other Creatures in the Victorian Age.* Massachusetts: Harvard University Press.

Rossabi, A. 1979. Introduction, in *Hodge and His Masters*, edited by R. Jefferies. London: Quartet Books, vi–xxiii.

Sennett, R. 2008. *The Craftsman.* London: Penguin.

Simpson, E. 2006. Apprenticeship in western India. *Journal of the Royal Anthropological Institute* 12, 151–71.

Skarstad, G., Terragni, L. and Torjusen, H. 2007. Animal welfare according to Norwegian consumers and producers: Definitions and implications. *International Journal of Sociology of Food and Agriculture* 15(3), 74–90.

Tovey, H. 2003. Theorising nature and society in sociology: The invisibility of animals. *Sociologia Ruralis* 43(3), 196–215.

Tregear, A. 2007. Proximity and typicity: A typology of local food identities in the marketplace. *Anthropology of Food.* Special issue on local food products [Online]. Available at: http://aof.revues.org/ index438html [accessed: 26 April 2009].

Wilkie, R. 2010. *Livestock / Deadstock. Working with Farm Animals from Birth to Slaughter.* Philadelphia: Temple University Press.

Interlude

Eating Practices and Health Behaviour

Simon Cohn

It is perhaps an obvious point to say that eating involves moments of not eating. The word 'eat' both refers to the moments when we actually put something in our mouths to consume it, and the more extended periods of time during which a succession of these moments takes place. In other words, we eat a meal but always through a succession of eating mouthfuls. The in-between periods of not consuming, however, are key features of the act: savouring the taste and texture of current morsels, the interlude when we attend to what we see and smell, and the power of expectations of further consumption are surely all part of the same general activity. In fact, we sometimes mark out the pleasure of eating by ensuring that we pace out each individual forkful and each subsequent course, while parents may chastise their children for rushing their food and treating mealtimes merely as a means to get food into the belly. The point is that we all intuitively know that eating is a bundle of activities and experiences.

If there are rhythms to meals that accord with particular contexts – whether they be snacks, sandwiches at a computer desk or formal dinners at a high society event – then the articulation of absences is clearly as much a part of the experience of eating as the actual moments of consumption and internalization. The experience of consuming or not at any one time is determined by another simple pairing – that between the inside and the outside of the body. After all, an absence does not necessarily mean an absence of the availability of food, but rather of its crossing of the boundary from outside in. The logic does not necessarily only reside in the semiotic meanings associated with food and eating, but just as equally in the very everyday, embodied experiences of food practices themselves.

This centrality of conceptual boundaries determining the practices of eating is alluded to in Baker's chapter. It starts with a challenging, if unsavoury, question – what exactly is *in* a sausage? We all have our hesitations about such meat products of course, because we all have heard stories of them containing the scraps of rendered pork that cannot be called 'meat' in and of themselves. Our concern, then, is that a meat product may not so much be the product of meat but the production of meat from something else. And although the author goes on to give a fascinating insight into the role of those who look after pigs, this initial question of ambiguity nevertheless lingers. Two juxtapositions that Baker refers to, both of which are orientated around the question of production, resonate with this general concern: 'factory farm' and 'pig-men'. In each, it is not merely the 'slippage' that Baker describes, but the potential 'mixing' that is equally concerning and compelling.

How can a farm, with all its cultural connotations of simple goodness, be coupled to the word factory and the inherent alienation that it implies? And equally, how can a person be so closely associated to pigs that even their name reflects an intimate and blurred relationship? Both pairings remind us that it is probably the not-knowing that is sometimes essential if we have any chance of enjoying the products we eat. And though we tend just to think about the production of food like pork in material ways, worried only about how much fat, gristle, salt, or grain something might have, we are actually consuming the outcome of a particular system of production that involves not only pigs but people, technologies, skills and histories that are all thrown into the mix.

I want to draw further upon this observation, both that eating is a set of multiple practices of doing and not doing, and that what we eat is the product of other similarly varied practices, to reiterate a key theme of this volume; that is, by concentrating only on the 'ingredients' of food, whether in terms of nutritional composition or indeed symbolic significance, the inherent and active social dimensions of eating are frequently ignored. All too often, eating is equated solely with the actual consumption of food. As such, not only is it abstracted from the wide range of activities that are integral features of anything to do with food and eating, but the boundary between the external and the internal is regarded as a simple and singular matter.

In my own role as a medical anthropologist working within large disciplinary teams on public health issues, it is all too apparent that eating has inescapably become a medical issue. Whether one is referring to campaigns about addressing the rising issue of obesity, the current fashionable 'super food' said to contain some rare but essential trace element, or the implementation of accessible graphics on supermarket products to highlight nutritional composition, food is now invariably thought of, and possibly experienced, as a medical matter. It is not surprising, of course. We are constantly reminded that changes in demographics and the decline of communicable disease in Western populations have resulted in the rise of so-called chronic illnesses, in particular those labelled 'lifestyle conditions'. This has led health professionals to turn their gaze from eradicating external agents of disease to altering a wide range of interacting factors identified as causing ill-health. Consequently, growing attention paid to the ways in which the environment can interact with the body throughout our lives has led to social factors increasingly falling under an expanding medical gaze. Both aetiological models and strategies for therapy have had to find ways to embrace not merely what is inside the body, such as the continuous interactions between physiology, metabolism and genetics, but crucially what is conceived as crossing the boundary from outside in to affect our individual health status.

Core to all of this is the use of the word 'diet', which not only captures dominant association between food and health, but also a restricted conceptualization of eating solely in terms of what exactly is consumed by the individual; it is devoid of context, meaning or any sense that it is inescapably a complex practice comprised of absences and presences. As a result, both within the UK context and in many other countries, government policy itself is increasingly categorizing diet as a health behaviour, finding in this overarching term a powerful means to redefine individual-state responsibility, distribute resources and reframe health inequalities through a neoliberalist paradigm. This is succinctly captured in the following statement from a UK Department of Health report:

> Persuading people to adopt healthier behaviours has become a central theme of modern public health policy. It underpins the goal of making the NHS more of a health service rather than a sickness service. Yet it is rarely easy. Most of us do things that damage our health. And most of us have habits that we would like to change (PoST 2007).

The idea of health-related behaviours as being the key to so many health problems in the twenty-first century means they have also become valid targets for potential intervention. However, in order for them to be integrated into existing causal explanations of biomedical knowledge, and so that they can be evaluated according to standard criteria of medical evidence – in particular the randomized controlled trial, medical investigations into health behaviours such as diet have to construct them in very particular ways. The research process therefore tends to consist of a sequence of refinements to ensure there is commonality between the varied elements, and that a common ground can be established across very diverse factors. As a result, health behaviours have to be sufficiently delineated out of the inherently chaotic variation of human activity, including what we might call the social domain, in order for them to sufficiently resemble other clinical variables. In this way, recognition of the complexity of many contemporary health issues is in practice transformed into establishing more mechanistic approaches to incorporate new variables that can be measured, manipulated and combined with existing ones.

Armstrong describes how this overarching concept of health behaviour is surprisingly recent (2009), and reiterates the argument that these could now be said to operate as a target for medical intervention in the same way that pathogens have done in the past (Matarazzo 1983). Derived from psychology and behavioural science, health behaviours are consequently presented as obvious elements that can logically be integrated into the existing medical research paradigm. Although this work does not equate directly to classic behaviourism, what nevertheless constitutes 'behaviour' remains conceived of as definitive and observable and ultimately self-evident (see for example the text by Di Iorio 2005 in which health behaviour itself is never problematized). Although some have suggested that by seeking collaboration with health psychologists, medicine has gone beyond its traditional frame of reference (see Murray 2004), it seems clear that this productive relationship is perhaps more than merely a pragmatic convergence of interests. Rather, it is also because the disciplines share many of the same epistemological assumptions: the same notions of objectivity, causation and, ultimately, that the essential unit of analysis is the individualized body and person. Given this, if practices related to food and eating are rendered into a specific health behaviour, it is not surprising that they take on a very particular nature. It is in the context of this that the word 'diet' appears to seamlessly accord with the notion of health behaviours more generally. But many matters relating to food and eating do not necessarily align completely with this apparently-straightforward term. A characteristic of conceiving of 'diet' as a specific behaviour and a focus for health interventions is that it conceives of the practices of eating solely in terms of what is actually consumed, rather than as a culturally-meaningful patterning of eating and non-eating, whether that be over the duration of a single meal, or everyday routine.

O'Connor's chapter suggests just such a mismatch in her description of how an item of food from the sea – namely seaweed – has a largely secretive, surreptitious presence in the eating habits of many inhabitants of Wales. Not only is it so physically transformed and combined with other food products to ensure its origin is only clear to those who know, but its covert status is amplified by the fact it is rarely presented on the plates of outsiders. Does it really make much sense, therefore, to talk simplistically of the extent to which *laver* is a component of the Welsh diet? The author contrasts this to Japanese cuisine and the celebratory uptake of sushi globally, in which seaweed is not only prominently served but openly savoured. With growing concern about the ways current world food production is having enormous environmental impact, seaweed may well be a more significant source of food in the future, as it perhaps once was in the past. But perhaps more of interest to me is that it addresses how the term 'diet' unavoidably appears as a mundane summary description from an outsider's perspective, encapsulating very little of the cultural history, or current practices, of what people choose to eat or why.

To return to my own focus on how diet has become a 'health behaviour' and thereby conceived of as a valid object not only of study but for the design of medical interventions to improve people's health, initial research questions are concerned not only with what might be the specific medical concern to pursue, but equally with the practicalities regarding how diet can ever be reliably measured. Whatever the technique adopted, the aim is not merely to find an accurate and reliable method to measure diet but implicitly to thereby define what diet actually is. A clear distinction is frequently made between subjective and objective measures. The former consist of various methods of participant self-reporting – usually by questionnaire or diary – and tend to be regarded as unreliable. It is accepted that people are either deliberately, or inadvertently, not very accurate when it comes to reflecting and quantifying aspects of their life – especially when it comes to matters of eating since these are likely to have strong moral associations. Because of this, these techniques tend to be used when no other method presents itself. Such a sentiment reveals the assumption that the object of the research – the 'actual' diet behaviour – in some way has a meaning or existence prior to, and independent of, the participants' own awareness of how they live their lives. In contrast, therefore, 'objective measures' are those techniques that effectively short-cut having to rely on the person and are able to record diet independently, such as body fat scans or blood tests that identify nutritional markers that can be associated with an overall pattern or food consumption. In practice, they generally require greater input from the research team to provide the technical and expert knowledge to produce them, but curiously, they could be said to measure behaviour independent of the participants.

Of course, many have commented on how the act of measuring things inevitably plays a role in constructing them in certain ways (Cartwright 1989). This relates not only in a simple way to how methods of recording and quantifying can influence the activities of people, but also to the extent to which they actually shape and predetermine just how the objects of study are conceived and defined. Thus a 'measure' is never a straightforward entity, since it is both a property of a thing *and* a property that is conferred on to it by the observer; it has built within it what Hacking might call a 'looping effect' (Hacking 1995). In this way, although the 'measurement effect' can be interpreted as a flaw in scientific investigation, it can alternatively be recognized not only as an inevitable, but also as a highly productive and necessary, process to nominate and extract features from their complex everyday natures in order to constitute particular objects of enquiry. Such slippage is vital, since all scientific practice relies on models and representations of reality in order to make sense of it. By being constructed as a measurable object and, in tandem, constructed as a discrete element that can be targeted and altered, the resulting notion of diet as a specific health behaviour has much of the logic of an intervention embedded within it. In other words, describing it in terms of units of individualized action independent of the complexity that constitutes people in their everyday lives, allows for the quantitative analysis to proceed unhindered.

That eating encompasses much more than this restricted sense of behaviour and the behavioural choices that 'diet' has come to stand for is clearly emphasized by Lavis. Examining eating practices that could be said to be extreme, but thereby particularly revealing of broader underlying tensions, she directly challenges any simple depictions of anorexia or the people who live with anorexia, through a sensitive account of the inherent tensions and contradictions that are lived with and constantly negotiated. One of the most compelling things she points out is something that many have been made very aware of, that the experience of acute hunger can be all-encompassing, conceivably even violent. It therefore makes no sense to portray the act of not eating as passive, or a person who draws on such a wide variety of creative strategies to resist their urge to eat as someone who is merely the subject of external forces. Despite possibly having a slight and frail body, people with anorexia tragically demonstrate opposite characteristics: their emaciated forms suggest a degree of mastery over their bodies and their most basic visceral nature that cannot be simply dismissed through a psychiatric label that safely denies their agency. This, then, raises a further critique of the process of rendering the complexity of food and eating to be researchable in any simple or definitive way; not only is what tends to stand as diet a very restricted notion that focuses only on trying to measure what is consumed rather than also what is not, but also by being so wholly extracted from the context of people's everyday lives and experience, the resulting findings never need address the broader, overlapping and frequently-contradictory circumstances that shape what and why people eat and do not.

Through exploring themes drawn out from three contributing chapters in this volume, I have described how, in my own field of medical anthropology and public health 'diet' is frequently described unproblematically as a health behaviour. Whilst the general term might be a relatively recent construction, individual health behaviours now serve as core measurable objects that are able to convert some of the complexity of contemporary health problems into units for intervention design. The concept of health behaviour has consequently enabled classic clinical research and methods to expand into new territories. As a unit of individualized action, a health behaviour is independent of the complexity that constitutes people in their everyday lives and thereby allows for the quantitative analysis to proceed unhindered. Although the idea of 'context' is sometimes used in discussions of this kind of medical research into diet, the term indirectly serves not only to reinforce the assumption that the other things are of primary importance, but also signifies a resignation that perhaps broader and more diffuse factors cannot be reduced to measurable components and so must remain an undefined general category. For example, even though some medical researchers may make mention of the effect of socioeconomic circumstances as a backdrop against which individuals make their dietary choices, by nevertheless still attributing the problem and the potential intervention to the individual, it is impossible to reinsert any genuine engagement with the politics of health. In other words, by employing an apparently neutral set of categories, such as diet, the specifics of health behaviour are successfully purified or extracted – and are fore-grounded against a backdrop that falls outside the experimental design.

The principle of encouraging people to change their food-related behaviour resonates strongly with the general idea of libertarian paternalism that underpins much health research and policy, and which has been presented as a way of balancing old-style critiques of medical imperialism and control with an apparently liberal philosophy that values individual freedom and autonomy (Rizzo 2005). Proponents of the idea claim to be able to reconcile the potentially opposing values of freedom and constraint by devising strategies that only address those instances in which people need particular assistance in order to be able to think rationally, and thereby make the right choice. For example, Sustein and Thaler present a number of case studies that they consider to successfully tread a path between these two values by ensuring only those individuals who clearly require assistance are in the end targeted by an intervention (2003). The overall approach, however, relies on a remarkably uncritical belief in human rationality; or more correctly, a belief that rationality can be judged objectively. In addition, by presenting the potential influences of what are variously termed circumstance, context, framing, and everyday conventions, as either distracting or corrupting – in other words, by reformulating the social and physical context in this way – the division between libertarianism and paternalism is simply shifted onto the boundary between what is deemed individual choice and what are labelled as externally-derived 'automatic' or habitual processes. In this way the oxymoronic nature of 'libertarian paternalism' becomes naturalized. But perhaps an equally valid interpretation is that it is merely a conceptual device to mask the inevitable issue of power that is simply unavoidable whenever the issue concerns one person trying to influence what another person does. Unlike the construction of a specific behaviour such as diet that can then be the focus of a medical approach that claims it to be neutral and external, everyday practices are always composites of an infinite number of actions and activities. They could be said to be the result not only of internally-driven processes of the individual, or defined by the internalization of food, but of the relationality of a person within a specific context.

Conceiving of diet as a health behaviour, representing the individual as always having the capacity to be in control and therefore ultimately responsible for their actions, and by considering the social merely as the backdrop to individual action, mean that issues of inequality, whether they be economic or other, never need be considered as direct causes in and of themselves. In essence, the objectivity that is marshalled through the research endeavour serves not only to disengage the social but also to delete any political context of health and illness. In contrast, ethnographic accounts have the potential to warn us that not only is it impossible to conceive of the range of actions that might potentially be associated with eating as a single 'behaviour' that can be attributable solely to the individual, but also that by doing this any notion of power comes to be made not only invisible but irrelevant.

References

Armstrong, D. 2009. Origins of the problem of health-related behaviours: A genealogical study. *Social Studies of Science* 39(6), 909–26.

Cartwright, N. 1989. *Nature's Capacities and Their Measurement*. Oxford: Oxford University Press.

Di Iorio, C.K. 2005. *Measurement in Health Behavior: Methods for Research and Evaluation*. San Francisco: Jossey-Bass.

Hacking, I. 1995. The looping effect of human kinds, in *Causal Cognition: An Interdisciplinary Approach*, edited by D. Sperber. Oxford: Oxford University Press, 351–83.

Matarazzo, J.D. 1983. Behavioral immunogens and pathogens: Psychology's newest challenge. *Professional Psychology: Research and Practice* 14(3), 414–16.

Murray, M.P. (ed.) 2004. *Critical Health Psychology*. Basingstoke and New York: Palgrave Macmillan.

PoST [Parliamentary Office of Science and Technology]. 2007. Health behaviour. *Postnote* 283.

Rizzo, M. 2005. The problem of moral dirigisme: A new argument against moralistic legislation. *New York University Journal of Law & Liberty* 1(2), 789–843.

Sunstein, C.R. and Thaler, R. 2003. Libertarian paternalism is not an oxymoron. *University of Chicago Law Review* 70(4), 1159.

PART II
Intimacies, Estrangements and Ambivalences: How Eating Comforts and Disquiets

The chapters in this section explore the multiple relations that are made and unmade by eating, and in which both eaters, and the foods they eat, are enmeshed. As such, they highlight how the ingestion of food is a simultaneously sensuous and symbolic experience, and demonstrate how eating informs the ways in which bodies, relatedness and places are socially (re)produced. Accompanied by an interlude by Jon Holtzman, together they investigate how notions of, and boundaries between, self and others are triangulated by the act of eating as both proximity and distance are established across geographical, temporal and social spaces. They also examine how eating is framed across value systems that construe bodies, and their relationships with others and with foods, in different ways. In so doing, the contributors also explore the tensions and ambiguities – between foods and bodies, and consumers, producers, retailers and nutritionists – that are inherent in the act of eating. Each in their own way thus elucidates how food is laden with contradictions: they stress consumers' ambivalence over certain foods and their valuing of others. In so doing, these chapters further demonstrate the raw sensual power of food; a sweet aroma and taste of a freshly picked fruit may invoke kinship and reiterate a sense of home, but it can also provoke a feeling of alienation and estrangement, and an awareness, heightened through the visceral response of the stomach, that the eater's body is 'foreign'. Likewise, the taste and crunch of a salad leaf, which may not quite satiate hunger as a biscuit could, can serve to remind eaters of failed diets and a sense of being uncomfortable in their own bodies. The same leaf may be approached with trepidation – as a source of unknown germs – or with pleasure, the texture and freshness signalling its health-giving properties. Food, then, can be a source of nourishment, both emotional and physical, but it can also be a site of conflict, anxiety, guilt and risk. In engaging with such ambivalences, the chapters in this section highlight how intimacies and estrangements are created through the act of eating. They also examine the multiple mechanisms and frameworks – from 'organic' certification and relations of trust, kinship and community, to weight management programmes and the transmission of knowledges – through which actors and agencies attempt to minimize risk and (re)assert control not only over the sites of food production, procurement and consumption, but also their own, and others', bodies.

Chapter 4
Advancing Critical Dietetics: Theorizing Health at Every Size

Lucy Aphramor, Jennifer Brady and Jacqui Gingras

Introduction

In his opening remarks in the UK's latest anti-obesity policy, *Healthy Lives, Healthy People* (Department of Health 2011), the Secretary of State for health reassures the reader that government 'will ensure that local effort is supported by high quality data and evidence of 'what works' (ibid.: 4). 'What works' refers to targets to reduce population weight; the strategy proposes 'helping' people achieve calorie deficit through a combination of eating less and being more active. This 'body as bomb calorimeter' approach – where a bomb calorimeter is an instrument designed to measure caloric heat output from food by exploding it – underlies prevailing health recommendations preoccupied with eradicating obesity. This chapter takes a critical stance to such an approach. We demonstrate that it is aligned with a universalizing and reductionist perspective to health that strips food and eater from social and affective contexts and denies personal and local knowledges. In this way, we argue that 'mainstream' dietary health interventions eclipse the gendered, embodied realities of eating, and support a hierarchy that privileges some and disadvantages, and devalues those who are othered. We thus explore the politics of knowledge production within obesity policy and dietetics, and highlight how the mainstream approach to fatness is experienced: how structural inequalities, fear of eating and body hatred play out in people's lives. One of our central purposes is thereby to show how treating people as tightly regulated eating machines skews their relationship with food and their bodies and renders other dimensions of health and wellbeing invisible.

We write as three dieticians contributing to a movement committed to changing strategies for knowledge creation and professional socialization in dietetics. This movement, Critical Dietetics, does not seek to measure success by demonstrating improvements in existing outcomes or greater efficiencies in established practices but instead moves beyond dietetics' mapped out perimeter to seek new vantage points, examining connections between power, social justice, disadvantage and health. Bridging the mainstream and the edgelands, it asks disturbing questions about what and who counts in obesity discourse and dietetics and examines the consequences of these processes in how and why people eat. In this chapter then, we also explore the impact of an alternative approach to nutritional wellbeing, known as health at every size (HAES): We recount one author's professional route to HAES practice in a UK context, and explore the philosophies and strategies that characterize a HAES approach through the voices of those participating in a community HAES course. The advantages of dropping the metaphor of the 'body as bomb calorimeter' become apparent as we hear from people who have learnt an alternative, embodied, and relational approach to food and eating. Finally, we reflect on the advances in critical weight science, and the promise this holds for transforming people's relationships with food, paying particular attention to recent progress in dietetics.

The Problem with Anti-Obesity Policy

The approach advocated in *Healthy Lives, Healthy People* is founded on the taken-for-granted assumptions that: fatness *per se* is strongly associated with poor health; thinness *per se* is strongly associated with favourable health outcomes; fatness and thinness are largely determined by eating and exercise habits; weight loss behaviour is clinically safe; weight loss behaviour leads to long-term weight loss; weight loss improves health outcomes; a weight-centred agenda leads to improved health and has no adverse impact on society or individuals. There is nothing, however, that hints at the growing body of critical literature that challenges the rationale or the risks and inconsistencies of the weight-centred approach, of which we are part.[1] This literature problematizes the assumptions of health promotion campaigns predicated on thinness and foregrounds the shortfalls of an over-reliance on positivistic science more generally. For instance, critical weight scholars highlight ways in which the health dangers of being fat are exaggerated (Campos et al. 2006). In contrast, 'mainstream' authors typically focus on the links between fatness and poor health, often obscuring the health risks of thinness as they do so. For example, Pischon et al.'s discussion of fatness and fat distribution concluded, 'the findings of our study suggest that general and abdominal adiposity are both associated with the risk of death' (2008: 2118), despite the fact that the results (using adjusted relative risk) showed that, in men, best survival was in the overweight category (BMI 25–8) rather than the recommended 'normal' weight category (BMI 18.9–24.9) and the risks of being underweight (BMI < 18.5) exceed the risks of being obese (BMI >35). Moreover, they report that, in women, the risk of being underweight also exceeds the risk of being obese and that their best survival occurs at BMI 24.3. This confirms similar findings that question the link between 'overweight' and increased mortality and demonstrate that higher BMI measures can lead to improved prognosis in a number of common illnesses,

1 See, for example, contributions from across disciplines including medical sociology, human geography, activism (see Rich et al. 2010), education (Evans et al. 2002), populist public health writing (Campos 2004), psychology (Riley 2007) and sports and exercise physiology (Blair and Church 2004; Lyons and Miller 1999).

such as heart disease and hypertension (cf. Flegal et al. 2005). In short, arguments advanced in the critical literature call into question the causal relation between fatness and excess mortality, and the assumption that thinness carries survival advantage.

In the words of Helen,[2] a chronic dieter, the tone of the anti-obesity messages, such as those found in '*Healthy Lives, Healthy People*' has the effect of making people 'feel like a second class citizen ... they're panicking that the National Health System isn't going to be able to cope with all the complications of obesity and, therefore, it's blame a victim'. Rather than helping people develop a healthy relationship with food and promoting self-care, Helen pointed out that the strict enforcement of BMI as a measure of health was counter-productive; thereby indicating that when dieters failed to reach their goals they felt guilty, anxious for their health, and discouraged, turning to food 'for comfort or rebellion'. Helen had asked to be referred to the Eating Disorders Clinic. She explained how the psychologist there told her 'how much damage the diet-binge cycle does ... the so-called wisdom that I had been fed all my life was wrong and that it had damaged me'. The damage caused to Helen has been written about extensively in activist and feminist press (see, for example, Riley et al. 2008; Solovay and Rothblum 2010). Moreover, each of the three chapter authors has also highlighted these issues from a dietetic perspective (see Aphramor and Gingras 2009; Gingras and Brady 2011) and has experienced considerable resistance from dietetic organizations, health care colleagues and funders when raising these issues. This resistance is documented throughout the literature (Bacon 2010; CFCF 2010; Gingras 2009) and attributed to financial conflict of interest, pressure to conform, and fear of reprisal for dissent that perpetuates the status quo through silencing marginalized views. That policy makers' power to define what knowledge counts, when harm matters, and to set agendas has real life embodied consequences is illustrated through the testimonies we present.

Helen's difficulties with food also highlight the fallacy of the construct of the rationally motivated eater who exercises cognitive restraint over their food choices in direct response to professional advice. This practice paradigm informs mainstream public health nutrition policy and is characterized by what Gingras and Brady (2010) describe as 'control discourse'. They argue that policy written from this perspective does not pay significant enough attention to the role of poverty or structural access to food in influencing people's food 'choices' and further 'constitutes individuals' eating patterns as a series of reasoned, discrete, and quantifiable choices (i.e. weigh, measure, limit, and avoid) in direct contrast to views that eating is determined by emotion, hunger, appetite, and sociality. We will see later how Helen's rejection of these externally imposed injunctions finally enables her to start healing her relationship with food. Meanwhile, the reliance on BMI as a measure of health meant years of failed dieting attempts for Helen until she eventually conceded, 'I've been trying this dieting lark for 30 years and I

---

2 All participants have been given pseudonyms.

haven't got the hang of it yet, I'm going to be fat and happy. And I was. I stopped worrying about it. My weight stabilized ... and then I got diabetes'. She continued: 'the whole guilt trip ... was there in my head, when your toes drop off and you go blind and you get renal failure, it'll be your fault and you'll be causing no end of problems and expense to everybody else'.

Helen had been led to believe that weight is completely controllable ('everybody was telling me this was the right thing and I just couldn't do it and you inevitably feel guilty') and told that she had in effect given herself diabetes by failing to control her weight. Diabetes is a metabolic condition with a link to weight (Bacon and Aphramor 2011) and to the material substance of body fat (see Kendrick, this volume). Yet, this link with weight is contested, having not been proven to be one of simple cause and effect. Whereas, insulin resistance, which is a core feature of pathology in diabetes, is known to be raised by weight cycling and stress (Montani et al. 2006; Vitaliano et al. 2002), both of which featured prominently in Helen's struggles with food. As weight cycling reduces survival (BNF 1999), anti-obesity policy thus exacts a physiological and psychological cost from the everyday, determined, dieter.

Mainstream literature increasingly acknowledges the limitations of BMI, warning of errant readings in people with exceptional muscle mass, and suggesting lower cut-off points for some populations (Razak et al. 2007). Others urge we use waist circumference, and/or consider the location of fat in the body, as a more sensitive indicator of pathology than BMI (Chan et al. 1994). Helen has shown us how the message that people can, and should aim to, achieve a particular body composition is antithetical to the first principle of medical ethics, that we do no harm. Dieting can be seen not to work in this context and BMI can be a poor indicator of health, but in many respects that is not the point. Whether or not dieting works, or anthropometric readings measure health, vilifying fatness and valuing thinness is not improving Helen's health; and this indicates significant implications for social justice.[3] As Burgard writes, 'deeming a particular BMI pathological is a political rather than a scientific act' (2009: 44).

Healthy Lives, Healthy People does acknowledge that the 'evidence base in relation to tackling obesity is still emerging' (Department of Health (DoH): 26). But the document's overall tone is strongly in support of weight loss, tacitly suggesting that existing data point towards positive trends and implying that it is only a matter of time until evidence to support policy emerges. Anti-obesity policy needs to maintain an air of objective authority and to do so it downplays its profoundly moral and ideological dimensions. Thus, weight science findings are presented as neutral facts, contradictory findings are overlooked, and a research culture that routinely fails to report adverse effects is relied upon (Avenell et al. 2004; NICE 2010). This seamlessness is maintained at the cost of inattention to

3 When we talk of social justice we are referring to justice within and through a web of contingent and shifting relationships that differently influence individuals' and communities' access to material and non-material resources, processes and outcomes.

political drivers of health and the cumulative impact of structural disadvantage and discrimination. Consequently, an equality impact assessment (EIA) accompanying *Healthy Lives* reads: 'We expect that the new [sic] approach to obesity will not have an adverse effect in terms of discrimination, harassment or victimisation' (DoH: 14). In making this statement, decades of relevant research on discrimination, harassment and victimization, that is readily accessible via the major clinical databases, from activist writers and critical health researchers (Rothblum and Solovay 2009; NAAFA 2012; Puhl and Brownell 2001), becomes devalued and the impact of control discourse on people's relationship with food and self-care, as witnessed in Helen's story, is overlooked. As Helen demonstrates, scientific narrative has embodied consequences: she mentioned the 'Big Brother' effect several times, building a picture of feeling constantly judged, being seen as greedy, lazy, undisciplined and selfishly draining the National Health Service (NHS). Yet, the power vested in the establishment can also be experienced as positive. Helen described her relief at meeting the psychologist who validated her struggles saying she felt totally vindicated. Similarly, she spoke positively of the role of (someone in) authority in affirming her innate self-worth: '[The facilitator], as a professional, is saying I'm okay and no professional has ever done that before and she's representative of the powers that be and Big Brother … and so now it's not a case of I've got to do this, otherwise I'm a bad person'.

Trust a Dietitian …

… to know about nutrition: so reads the British Dietetic Association's car sticker strap line. How does the dietitian come to this knowing? It is via textbook and policy, or is it according to dietitians' knowledges of their own/clinical experience and that of their clients? Given that weight loss diets, as we have seen from Helen, can be counterproductive, dietitians can find themselves faced with experiential knowledge on weight management that contradicts that which has been learnt (and internalized) from the professional canon. This was my (LA) experience. I started work in the NHS as a part-time dietitian in Coventry in 2001. The ninth largest city in the UK, until recently Coventry was a centre of car manufacturing, but now 50% of its population live in the top 34% most deprived LSOAs[4] nationally (Coventry Partnership 2010). Early on, my post was that of a community dietitian and I worked from General Practitioners' (GP) surgeries, seeing mainly adult patients on a one-to-one basis. Most of these patients were seeking advice for weight management and presented with a diagnosis of diabetes, hypertension, arthritis, heart disease, 'obesity' and/or depression. As I cycled between GP surgeries there was no escaping the poverty – of opportunity, income, access to food and desirable housing – and these material realities were reflected in

4 Lower Super Output Area is a geographical area used for the collection and publication of small area neighbourhood statistics.

my patients' narratives. Many of the people with whom I was working had learnt to tolerate chronic poor health and it had been a long time since they had ever felt physically well. Still, in the context of the clinic, the presenting problem was fatness and this was where patients and I directed our energies. To little avail, as it happens, for despite all our efforts significant weight loss remained an elusive target. But we persevered. Many patients – notably older women with limited mobility – had restricted their energy intake for years and continued to do so. I scrutinized my calorie calculations and signed up for a course on behaviour change skills. When it still did not add up, that is, when I simply did not get the outcomes the textbooks and guidelines said were inevitable (read: weight loss), I decided to read the primary data cited in the current guidelines (then 'SIGN 1996) to see where I was going wrong. This was an eye-opener. I worked through the document retrieving research papers cited in support of the health benefits of weight management and lifestyle change as effective treatment for weight loss. It was with a rising sense of confusion and alarm that I realized the citations quoted neither reflected what I was finding on the ground, nor suited my purpose. Then, as now, I noticed a distinct absence of reporting on harmful effects. As summarized in the National Institute for Clinical Evidence (NICE) obesity guidelines for adults as 'no evidence statements can be made as reporting of harms and adverse events was rare and ad hoc' (NICE 2006: 560, 579). Equally perturbing, where harmful effects are known, the ensuing advice does not fully reflect them, with short caveats being buried within documents dedicated to encouraging the very pursuit it cautions against;

> Weight cycling is a common condition as only a minority of people who lose weight through weight management interventions are able to maintain their weight loss. … Weight cycling is a risk factor for all-cause mortality and cardiovascular mortality (SIGN 2010: 18).

I was now on a path that a growing minority of overseas activists and/or nutrition professionals had been carving out for some time: that of finding ways to reframe dietetics' relationship to bodies, food and eating so that our interventions firstly do no harm and are potentially worthwhile. This meant shifting my focus away from weight loss to advocate the value of healthy behaviours *per se*, and I began researching in an attempt to make sense of the impact of social factors in conditions I had previously imagined to be largely due to health behaviours, such as diabetes, hypertension and heart disease. It was an isolated place to work from. I was then a junior member of the department and colleagues seemed more troubled by my questioning policy than my suggestion that we were routinely recommending a futile practice. Before long, I came across writing on 'health at every size'. The HAES paradigm presumes that adiposity is not a reliable predictor of health, although it may serve as proxy indicator for certain health risks, such as a history of weight fluctuation or prescription slimming drug use. The significance of HAES did not sink in immediately but I recognized it as an approach that was not prepared to settle for the conventions that supported the status quo.

We pick up the story three years later when I was the dietitian in a multi-disciplinary cardiac rehabilitation team, working under the leadership of a nurse manager, who encouraged an ethos of on-going learning and critical enquiry. The team's specialist nurses and exercise physiologists welcomed discussion on weight science and I ran a number of pilot programmes over several years that were designed to incorporate and communicate several key tenets of a HAES philosophy. The HAES programme started as a six-week course, *Well Now*, and was open to English speaking adults who wanted to feel better about eating and their bodies. It was free at the point of access, and quickly developed to offer a range of community engagement activities, of which 'walks and more' and 'move to music' classes were the most popular. Ad hoc days out, drop-in art, relaxation, a support group and a choir were also available. Small group cooking sessions were also very popular, but unfortunately had to be withdrawn due to funding restrictions. *Well Now* was evaluated by asking participants to complete pre- and post-course self-reports and by inviting people for interview or to take part in focus groups. It was encouraging that questionnaire (n=70) data showed statistically significant short-term improvements, and no adverse effects, in a range of health behaviours. But it was the changes in participants' 'being-in-the world' that most keenly illustrated the power of acceptance and compassion in improving people's relationships with themselves and Others, and it is to this that I now turn.

Beyond Behaviour Change

In this section I (LA) explore four main tenets of a HAES philosophy in practice through the voices of participants who attended three different *Well Now* courses. These are: the HAES philosophy is based on respect and acceptance; the practice challenges the role of external expertise in knowledge creation about body weight and about health; the philosophy promotes compassionate self-care and fourthly, the philosophy seeks to advance social justice. One of the most striking themes to emerge is how strongly people valued being treated with respect and the pivotal role they ascribed to this acceptance and regard in catalyzing self-care behaviours.

Unbeknown to me until we spoke post-course, Jayne had a history of anorexia and had previously attended an Eating Disorders Clinic for treatment. She was receiving on-going support from a dietitian for bulimia when she first attended *Well Now*. When asked to talk about her experiences of the course Jayne said:

> I was really impressed because [the changes] did start straight away. I went home the first week and I started to think, thinking about what I was doing … funnily enough it was the chocolate biscuit thing, and I thought, you know, it's not the chocolate biscuit that's the problem, it's the way I think about the chocolate biscuit. So straight away I started to change things … And now that's

> how I look at food. If I want it I can have it, but I don't have to. And if I've eaten, and I'm not hungry, then I can do other things. And that's what I do and I don't purge now. I think well why am I purging? I've had something to eat, that's fine. I'm entitled to something to eat, so let's think about doing something else. It has really changed the way I think about food and about myself as well.
>
> Interviewer: So can you say a bit about that?
>
> Well, I used to think, you know, with clothes really, I used to think just put anything on it doesn't matter because I look a mess anyway. You know I'm just a fat mess so what does it matter? Now I think well no I'm not a fat mess, I'm me, so put something on, and I have actually, you can see I have changed the way I dress and stuff. And now I'm going to the shops I think well okay that's not right for me, but that don't mean it's because I'm fat and ugly, that just means it don't do me any favours. So that's how I'm looking at things.

Here, Jayne describes a process where attending the course meant she changed her views on her own self-worth and how this greater self-acceptance translated into her finding peace with food and her body. Respect and acceptance was the first theme volunteered by Rob in his reply too. He began his interview by commenting on how pleased he was that he 'was spoken to like an adult and not spoken at like someone speaking to a child', which had been his previous experience with practitioners. He returned to this later commenting that, while practitioners had 'usually been fairly kind in the way they've put it [advice]', he had predictably been told what to do and not do. He spoke of receiving similar, plus new, information with *Well Now* and how this had been 'so much more, well, I can only say compassionate and understanding as to coming to my side, allowing me to be as I am and to think as I have thought, rather than telling me how I should be thinking or behaving'. As with Jayne, this acceptance in turn helped Rob change self-care behaviours where previous attempts based on didactic interactions and cognitive restraint had failed.

Emma's reply to my opening question about the course encapsulated this dynamic. She said: 'I found it helped give me confidence about me as a person and how I approached food.' Speaking as a chronic dieter, she insisted later in the interview that 'the first thing you've got to [do]' to help people overcome eating problems is to make them feel good about themselves, 'because if you don't say to the person you matter then I don't think you're going to get anywhere'. Emma expanded on this belief, several times referring to the need for interventions to consider 'the bigger picture' in order to help someone effectively: 'look at the person, the individual and make them feel good about themselves. Help them to get the support they need, help them to make friends'.

Helen explained that she had signed up for the course 'because eating has always been an issue for me, I felt I should take any and every opportunity to do something that might help'. Taken with the rest of her testimony, this statement indicates the role disciplinary medicine has in Helen's self-policing food decisions. Here, the 'moral eater' is someone who accepts medical authority and interpolates this authority into personal efforts to attain thinness. In the context of current policy, this practice gets co-constructed as her civic duty. Elsewhere in her story, the spectre of surveillance, another instrument of governmentality, looms large. She continued:

> And for the first time ever I've found someone who is on my side, who is not getting on to me, not telling me what to do, not criticising me for the way I eat or what weight I am or anything but completely accepting me and saying, you're okay but you've just got a bit of a problem with dealing with this and that and I'm here to help you. Nobody has ever done that for me before and that was such a relief as almost made me cry really.

Her poignant testimony shows us the shortcomings of client-practitioner relationships conducted within a traditional 'control discourse' (Gringas and Brady 2010), and we read above how interactions with practitioners adopting different paradigm approaches impacted on how and why she ate – not least through teaching Helen to read her body as medically deviant and morally deficient.

At the end of each session participants are invited to make a note of something they want to think about, or act on, during the week. Helen said her first goal was to be completely open with her husband who she felt was 'on my case like everyone else' but explained how he 'didn't get much of a chance to understand [her difficulties with food] because I wasn't telling him because in the same way that I was eating in secret, I was keeping my thoughts and feelings secret'. Helen said that she told him everything and received his total support, remarking what a difference this acceptance made to her wellbeing. Her health project thus started to shift from a private enterprise designed to fix the infinitely malleable body towards one that allows for more dimensions of wellbeing, including relational agency.

As with the majority of course participants, Helen, Rob, Emma and Jayne attended to try and gain control of their eating as a route to weight loss. It is significant that, when asked about their experiences of *Well Now*, these participants all began their reply by talking about the transformational power of respect. Weight control – the impetus for their struggles – gets a mention but is no longer the organizing principle of their days. What comes across was how the historically persistent attendants of food- and body-guilt, shame and failure had begun to assume less significance in their stories and the positive ramifications this had for their day-to-day lives.

Two of the strategies taught in a HAES approach are mindful (or intuitive) eating and active embodiment. Both are grounded on the premise of advocating body respect for people of all shapes, sizes, fitness levels and health status. Both strategies also incorporate the philosophical premise of acceptance, including size acceptance, and compassion. Mindful eating and active embodiment encourage people to use their own lives and embodied knowledge, accessed as hunger, appetite, emotions, mood, tiredness levels and so on, to guide them in self-care. This turns the idea of external expertise on its head. Rather than ignoring people's real-life experiences of dieting and telling them to try again, as often happens with mainstream approaches, it means honouring this experiential knowledge, perhaps providing scientific back-up, and offering a safe alternative for those who are interested. In fact, for Rob 'the old way' of trying to adapt behaviours to meet guidelines and advice was counter-productive as it, 'promotes sort of ideas of failure … makes you feel stupid, worthless, useless and destroys your sort of self-belief, whereas the HAES approach has done quite the opposite'. Jayne commented that she had found it really positive that, unlike at the Eating Disorders Clinic, there were no scales to be seen, and how she subsequently got rid of her scales at home. The act of weighing herself or being weighed was detrimental to her developing a healthy relationship with food and fuelled anxiety, disturbed eating patterns and self-rejection. Jayne had begun paying attention to the links between when and what she ate and her energy levels and would have liked to learn more on nutrition. She also highlighted that she now paid attention to the visual attractiveness of what she was eating, as an act of self-care, which meant she was eating more salad.

Helen remembered intentional calorie restriction as a child and had had a lifetime of dieting. She said '… for a long time, I didn't know I couldn't do it, I just kept trying, and I thought next time, next time it'll be better'; illustrating how dominant discourse on dieting invalidated her own embodied realities. She recalled a dietetic appointment in which she had been told that 'if she must eat sausages' then she should grill rather than fry them. Recalling the encounter Jayne mentioned her circumstances at the time and her responsibilities for feeding the family saying:

> I just hadn't the oomph to do it, and she banged on about these sausages and in the end she said, well, if you're not going to cooperate there's no point in coming, and I just felt that she was part of Big Brother, everybody was on my case … and if that's how you feel, stuff it.

Interventions such as these are rooted in a mechanistic understanding of human metabolism and overlook the relational and social aspects of living in an eating body. Moreover, this practice paradigm serves to legitimize dietitians' authority as food and nutrition 'experts', thereby enabling intervention into the lives of others. Concomitantly, as we have previously argued (Gringas and Brady 2010), control discourse withholds power from the other by undermining clients' knowledge of and relationships with their bodies. Sitting snugly within a neo-liberal discourse that constructs health as the by-product of people's volitional choices, control discourse eclipses the science and narratives that speak to the impact of inequality and oppression in determining population health and illness. Not surprisingly, far from finding support in these encounters, Helen was left with a feeling of professionals 'just getting on to her'. This approach alienated her from her healthcare providers, leaving her feeling very isolated. The clear injunctions to follow externally imposed rules made with little regard to her current situation were disempowering and detrimental to self-care. Interventions such as this, emerging from neo-liberal agendas and that are not sensitive to social and structural factors, will ultimately increase health inequalities as they tend to meet the needs of those best served by the existing status quo (Rogers 2010).

Paradoxically, Helen then sought external assessment of her eating even though she expected approbation. Despite her best efforts to change, she still thought her eating 'wasn't good enough', and was 'very happy' when the HAES facilitator 'said it was and she should know'. Helen first experienced knowledge creation as a community level when she attended a Fat Studies/HAES seminar that was held locally. She was really impressed 'that it's not just airy, fairy wishful thinking. Related to her feeling that the HAES stance 'just makes sense', Helen described the fat women's dance group, the *Roly Polys*, as 'having rolls and rolls of fat' and 'leaping around like sticking tapes' (that is, apparently defying gravity) to illustrate the shortcomings of the idea that fat people are always unfit. Her story highlights the possibilities for challenging orthodoxies around power and knowledge, not least the fact that she was welcomed as a lay participant to an academic conference (Evans 2010), and that lay participation was noted in feedback as adding value to the discussion by the (mainly professional) audience.

Helen's experience speaks to how a paradigm that is not centred on weight loss and seeks to promote growth-fostering relationships (between practitioners and participants and between participants and food/eating/bodies) challenges the conventional dynamics of power, and the narratives available, between health care provider and clients (Brady et al. in press; MacLellan et al. 2011). In mainstream practice – where the professional is seen as objective ('the expert'); clients' needs are seen as dependencies, not strengths (clients not exhibiting dietitian-prescribed change may be labelled as non-compliant, resistant, and/or unwilling); and vulnerability among professionals is discouraged – professional authority arises from maintaining a power-over stance; the emphasis of dietetic practice is on fixing problems, not fostering relationships, and cultural and material differences among clients and professionals are unlikely to be acknowledged. Helen's story reveals the impact of this Foucaldian 'confessional' ritual (1978, in Murray 2008) in typical mainstream weight management that situates the practitioners' (clinical) knowledge as superior. Whereas her interaction with the HAES dietitian offers new possibilities as the dietitian derails the enactment of the confessional stance by questioning Helen's belief that her eating practices amount to failure (Murray 2008: 74).

Emma described the consequences when she took support within a traditional paradigm and attended a class as part of a community mental health recovery programme: 'I was at the gym doing exercise. I absolutely hate the gym … I just broke down in tears and said I can't shift the weight'. She depicted a long history of making efforts to change her weight and seeking support within a dominant paradigm approach that had consistently backfired. It was at this point that Emma's story took a new turn. She attended the *Well Now* course and 'just connected straight away'. Emma's new experience of listening to her body changed how she ate. She reintroduced fruit and vegetables to her meals and had 'learnt how to do what suits me best, like eat mild fruits, cook it a little bit … You said find out what suits me, so I had to look at me, you know'. Similarly she was more aware of the food she was buying and eating, describing herself as previously 'blind shopping'. Another difference was that before Emma 'wasn't prepared for eventualities. It was like I was hungry and I had to eat so, and I ate anything', and she now kept a store cupboard and shopped with meals that she would enjoy in mind. She had also cooked in bulk and frozen homemade meals which increased her options when money was tight. She felt she had been able to make these changes because of the new knowledge coupled with a 'can do' attitude she had gained from the course: 'whereas before I'd buy rubbishy things or I wouldn't eat at all until I could afford nice things'. Poverty continued to restrict Emma's choices but she was eating food she enjoyed more often, and enjoyed her new relationship with food; themes confirmed later when she said the course had been about 'not thinking about food too much and eating for enjoyment rather than for the sake of it and sitting there and just piling it all there, you know'. Her explanation for these profound changes returns us to the core premise of respect and acceptance: 'I think that's because of the way I think about myself that I matter more now. Whereas before I didn't think I mattered so it kind of didn't matter what I was putting into my body. So I'd eat rubbish, which wasn't good for my diabetes'.

Emma talks of previously feeling out of control because she didn't have any guidelines to help her, reminding us that the guidelines on which she had relied added to her problems with food. The 'choices' presented in mainstream policy, as epitomized by *Healthy Lives, Healthy People* are illusory, the individualistic mandate of cognitive control is predicated on a Cartesian mind-body split, which drowns out others ways of being and effectively constrains people's choices in how they relate to food (see also Brooks et al., this volume). Rob likewise recounted how he stopped eating 'thoughtlessly' when he rejected control discourse and began listening to his body. He found that shifting from cognitive restraint to mindful eating helped him engage meaningfully with food choices and spoke of feeling 'more accountable' to himself. He broadened this comment later saying that the lack of prescription in a HAES approach offered a 'more fundamental sort of change' that felt 'more genuine and sustainable'. Rob's experience that troubled eating emerges from diet mentality drivers of deprivation and judgment and improves with a HAES approach to self-care is reflected in the literature (Bacon 2005). Rob later spoke of a shift towards seeing weight and food as part of something much wider – his 'approach to life and living, really'.

The significance of participants' lived realities as a wider determinant of health is further illustrated by Emma who reported that the opportunity to think about herself, and her health, meant 'you're going to feel good about yourself, physically, mentally. It just gives you more energy for life, help you enjoy life. So it gives you more kind of oomph really … Gets you out of bed'. Emma described a process where learning new information, feeling supported and being treated with respect synergistically enhanced her motivation and coping around self-care. Likewise, Helen had also started 'seeing fitness now as a whole picture thing than as just to do with food' and had also changed her ideas about thinness and fitness explaining that 'there was never any option other than to believe [the stereotype]' beforehand. Challenging weight stereotypes had also helped Emma move towards compassionate self-care. She said her weight was still an issue for her but she was better able to protect herself from hurtful comments:

'I think now that's other people's viewpoint ... now I just think they're the people with the problem' reminding herself that she was doing her best, the rest 'was people's judgment'. In her opinion, compassion was essential in helping people change: 'A hard approach doesn't work for people: we're human beings. So I think loving kindness is the only way'. Acceptance and compassion extended beyond Jayne's eating and sense of self, as she went on to associate these shifts with a change in attitude to other people, explaining how she had begun to recognize and challenge her prejudiced thoughts about fat people eating in public: 'It's like I was saying to myself you can only eat if you're slim. If you're not slim then you shouldn't be eating'. Rob also spoke of wider shifts in thinking: 'It was an education in lots of ways. In terms of explaining, oh, nutritional stuff and sort of cultural and societal attitudes, beliefs ... it's changed the way I hear people'.

Participants' voices powerfully illustrate the interconnections between self-care, compassion and acceptance, and reveal the impact of practice that seeks purposely to respectfully affirm people's embodied experiences, promote self-care, and challenge the stronghold position of adiposity as a reliable predictor of health. This impact is felt beyond eating 'choices' and moves towards more equitable social relations and processes. These are a hallmark of a HAES ethos. HAES thinking takes it for granted that, like any other physical feature (i.e. eye colour, height), human bodies are naturally diverse sizes and rejects the reductionism that buttresses BMI. Adopting a HAES approach is a political statement against control discourse and the positivistic worldview in which these are rooted. In contrast to control discourse 'the focus [of HAES] is on the day-to-day activities that help individuals of any size to flourish' (Burgard 2009: 43). Rather than coercing people to follow rigid diet and exercise regimen, for (usually short-term) weight loss, to fit their bodies into very narrow weight range targets, our alternative approach seeks to support people in questioning the ways they relate to self and Other. A common goal for participants of programmes such as *Well Now* is growth toward more meaningful connections with their bodies. That said, delegates at the 2011 conference of the Association of Size Diversity and Health stated their commitment to rooting out any overtones of healthism in their work that would imply (fat) people had a moral obligation to pursue sound diet and exercise patterns, including HAES-friendly ones. In order to be meaningful, the work of helping people learn to value (their) bodies has to go hand-in-hand with an agenda to advance social justice that includes, and goes beyond, size diversity. As reflected in feminist work on public health ethics, HAES practice recognizes that 'economic and material disadvantage are important dimensions of inequity in the genesis of ill-health; however, the less tangible aspects of inequity are equally important. These include lack of power, oppression, diminished opportunity, and discrimination …' (Rogers 2010: 351).

Conclusion

Both Critical Dietetics and a HAES philosophy actively encourage involvement from people beyond the boundaries traditionally set by professional/academic affiliations. Both movements also call for a re-evaluation of personal and contextual knowledge. Consider, for example, the fact that most people who signed up for the *Well Now* course have a history of chronic dieting. So they already know what happens for them when they cut calories. And yet they are prepared to override, or rationalize away, this embodied knowledge when faced with the contradictory expert knowledge that says weight loss diets work. A HAES approach encourages people to use their personal eating narratives as valuable knowledge and to take this embodied experience seriously. It invites them to build a new relationship with food through reconsidering nutrition science, obesity discourses, a role for their own body knowledge and size equality. We provided data that illustrated how the HAES conversation is often the first time a health professional has discussed eating in terms of enjoyment and estrangement from food, and for self-care to be presented as a nurturing act rather than one experienced as punitive and blame-based.

Preparing the grounds for this shift means engaging people in fresh thinking about the complex interrelationships – and indeed, separations between – health, body size and eating. What helps make this criticality potentially life changing is that it is offered from within a philosophical stance of compassion and acceptance, or the compassionate-mind. The narratives shared above eloquently demonstrate how compassion and acceptance by/for the self/Other offer a transformational practice as people live new narratives of self-worth. The transformational element of this practice is thrown more acutely into relief when *Well Now* participants contrast their HAES experiences with those of mainstream medicine and control discourse. Drawing on Murray (2008), who argues that perception is productive, we suggest that these testimonies demonstrate a key role for the compassionate-mind in the process of translating critical thinking on weight into real-life change. So this constellation of a compassionate-mind approach and probing apparent givens about weight, food and wellbeing, together enables a resignification of bodies and practices as people find new ways of relating to food and self, in short, new ways of 'being in the world'. At the same time as these new ways of relating recognize and are necessarily touched by incorporation of the social code fat = bad, the compassionate-mind approach offers a position outside of the system of judgment that oppressive thinking is predicated on.

Anti-obesity policy is operative in people's lives before we meet as client-practitioner. As professionals we cannot disimplicate ourselves from these pre-existing narratives but, as we have shown, we can work with people to transform stories: we can question the authority of the clinical gaze, challenge science's claim to neutrality and the value of objectivity, and problematize binary, 'all-or-nothing' thinking that is the backbone of diet mentality thinking. In discussing people's experiences of a HAES approach, and of dieting, we have highlighted the need to take account of both eating and the complex relationships and embodied realties that are entangled in it, as shown by our data. In theorizing HAES we build on work from the HAES community and fat activists that develops a coherent framework for enhancing effective understanding of the problems that surface around anti-obesity discourse, problems of privilege, body-hatred, eating distress, health disparities and hierarchy. These issues have their roots in philosophies of individualism and domination – not in calories in and out.

References

Aphramor, L. and Gingras, J.R. 2008. Sustaining imbalance: Evidence of neglect in the pursuit of nutritional health, in *Critical Bodies: Representations, Practices and Identities of Weight and Body Management*, edited by S. Riley, M. Burns, H. Frith, S. Wiggins and P. Markula. London: Palgrave, 155–74.

Aphramor, L. and Gingras, J. 2009. That remains to be said: Disappeared feminist discourses on fat in dietetic theory and practice, in *The Fat Studies Reader*, edited by E. Rothblum and S. Solovay. New York: New York University Press, 97–105.

Aphramor, L., Asada, Y., Atkins, J., Berenbaum, S., Brady, J., Clarke, S. and Yarker-Edger, K. 2009. Critical dietetics: A declaration. *Practice* 48, 2 [Online]. Available at: http://www.criticaldietetics.org/Critical%20Dietetics%20Declaration.pdf [accessed: 25 October 2011].

Avenell, A., Brown, T.J., McGee, M.A., Campbell, M.K., Grant, A.M., Broom, J., Jung, R.T. and Smith, W.C. 2004. What are the long-term benefits of weight reducing diets in adults? A systematic review of randomized controlled trials. *Journal of Human Nutrition and Dietetics* 17(4), 317–35.

Bambra, C., Fox, D. and Scott-Samuel, A. 2005. Towards a politics of health. *Health Promotion International* 20(2) [Online]. Available at doi:10.1093/heapro/dah608 [accessed: 25 October 2011].

Bacon, L. 2005. Size acceptance and intuitive eating improve health for obese, female chronic dieters. *Journal of the American Dietetic Association* 105(6), 929–36.

Bacon, L. and Aphramor, L. 2011. Weight science: Evaluating the evidence for a paradigm shift. *Nutrition Journal* 10(9), 1–13.

Blair, S.N. and Church, T.S. 2004. The fitness, obesity and health equation. Is physical activity the common denominator? *Journal of the American Medical Association* 292, 1232–4.

British Heart Foundation. 2009. *So You Want to Lose Weight for Good*. London: BHF Publications.

Burgard, D. 2009. What is health at every size?, in *The Fat Studies Reader*, edited by E. Rothblum and S. Solovay. New York: New York University Press, 41–53.

Butler-Jones, D. 2011. *The Chief Public Health Officer's Report on the State of Public Health in Canada*. Ottawa: Health Canada. [Online]. Available at: http://www.phac-aspc.gc.ca/cphorsphc-respcacsp/2011/pdf/cpho-resp-2011-eng.pdf [accessed: 27 October 2011].

Campos, P. 2004. *The Obesity Myth: Why America's Obsession with Weight is Hazardous to Your Health*. New York: Gotham.

Campos, P., Saguy, A., Ernsberger, P., Oliver, E. and Gaesser, G. 2006. The epidemiology of overweight and obesity: Public health crisis or moral panic? *International Journal of Epidemiology* 35(1), 55–60.

Chan, J., Rimm, E., Colditz, G., Stampfer, M. and Willett, W. 1994. Obesity, fat distribution, and weight gain as risk factors for clinical diabetes in men. *Diabetes Care* 17(9), 961–69. Available at: doi: 10.2337/diacare.17.9.961 [accessed: 25 October 2011].

Department of Health. 2010. *Healthy Lives, Healthy People: A Call to Action on Obesity in England*. London: HMSO.

Evans, J., Evans, R., Evans, C. and Evans, J.E. 2002. Fat free schooling: The discursive production of ill-health. *International Studies in Sociology of Education* 12, 191–212.

Evans, B. 2010. *Fat Studies and Health at Every Size ESRC Funded Seminar Series. Seminar 2: HAES in the Clinic*. 13th–14th May, 2010, Warwick University.

Flegal, K.M., Graubard, B.I., Williamson, D.F. and Gail, M.H. 2005. Excess deaths associated with underweight, overweight, and obesity. *Journal of the American Medical Association* 293, 1861–7.

Gingras, J.R. 2006. Throwing their weight around: Canadians take on health at every size. *Health at Every Size* 19(4), 195–206.

Gingras, J.R. 2009. *Longing for Recognition: The Joys, Contradictions, and Complexities of Practicing Dietetics*. York, UK: Raw Nerve Books.

Gingras, J.R. and Aphramor, L. 2010. Empowerment, compliance, and the ethical subject in dietetic work, in *Configuring Health Consumers: Health Work and the Imperative of Personal Responsibility*, edited by R. Harris et al. Basingstoke: Palgrave Macmillan, 82–93.

Gingras, J.R. and Brady, J.L. 2010. Relational consequences of dietitians' feeding bodily difference. *Radical Psychology* 8(1) [Online]. Available at: http://www.radicalpsychology.org/vol8-1/gingras.html [accessed: 25 October 2011].

Grace, C. 2008. *The Dietetic Weight Management Intervention for Adults in the One to One Setting. Is it Time for a Radical Rethink? BDA Professional Consensus Statement*. Birmingham, UK: BDA.

Ikeda, J., Hayes, D., Satter, E., Parham, E.S., Kratina, K., Woolsey, M., Lowey, M. and Tribole, E. 1999. A commentary on the new obesity guidelines from NIH. *Journal of the American. Dietetic Association* 99, 918–20.

Lyons, P. and Miller, W. 1999. Effective health promotion and clinical care for large people. *Medicine and Science in Sports and Exercise* 31, 1141–6.

Maclellan, D., Lordly, D. and Gingras, J. 2011. Professional socialization: An integrative review of the literature and implications for dietetic education. *Canadian Journal of Dietetic Practice and Research* 72(1), 37–42.

McKibbin, L. 2009. *The Food for Thought Pyramid*. [Online]. Available at: http://www.food-for-thought-pyramid.com [accessed 20 January 2012].

Miller, J.B. and Stiver, I.P. 1997. *The Healing Connection*. Boston, Massachusetts: Beacon Press.

Montani, J.P., Viecelli, A.K., Prevot, A. and Dulloo, A.G. 2006. Weight cycling during growth and beyond as a risk factor for later cardiovascular diseases: The 'repeated overshoot' theory. *International Journal of Obesity (Lond)* 30(Suppl 4), S58–66.

Murray, S. 2008. *The 'Fat' Female Body*. Basingstoke, UK: Palgrave Macmillan.

National Institute of Clinical Evidence (NICE). 2006. *CG43 Obesity: Full Guideline, Section 5b – Management of Obesity in Clinical Settings (Adults): Evidence Statements and Reviews*. [Online]. Available at: http://guidance.nice.org.uk/CG43/Guidance/Section/5b/pdf/English [accessed: 22 November 2011].

National Association to Advance Fat Acceptance. [Online]. Available at: www.naafa.org [accessed: 25 October 2011].

O'Dea, J.A. 2004. Prevention of child obesity: First, do no harm. *Health Education and Research: Theory and Practice* 20(2), 259–65.

O'Reilly, C. 2011. *Interrogating the Dominant Obesity Discourse*. 1st International Critical Dietetics Conference, 19th–20th August. Toronto, Canada: Ryerson University.

Pischon, T., Boeing, H. and Hoffmann, K. 2008. General and abdominal adiposity and risk of death in Europe. *New England Journal of Medicine* 359, 2105–20.

Puhl, R. and Brownell, K.D. 2001. Bias, discrimination and obesity. *Obesity Research* 9, 788–805.

Razak, F., Anand, S.S., Shannon, H., Vuksan, V., Davis, B., Jacobs, R., Teo, K.K., McQueen, M. and Yusuf, S. 2007. Defining obesity cut points in a multiethnic population. *Circulation* 115(16), 2111–18.

Rogers, W.A. 2010. Feminism and public health ethics. *Journal of Medical Ethics* 32, 351–4.

Rothblum, E. and Solovay, S. (eds) 2009. *The Fat Studies Reader*. New York: New York University Press.

Scottish Intercollegiate Guidelines Network. 1996. *Management of Obesity. A National Clinical Guideline*. Edinburgh: SIGN.

Scottish Intercollegiate Guidelines Network. 2010. *Management of Obesity. A National Clinical Guideline*. Edinburgh: SIGN. [Online]. Available at: http://www.sign.ac.uk/pdf/sign115.pdf [accessed: 28 October 2011].

Solovay, S. 2000. *Tipping the Scales of Justice. Fighting Weight-based Discrimination*. Amherst, New York USA: Prometheus Books.

Thomas, B. and Bishop, J. (eds) 2007. *Manual of Dietetic Practice. Fourth Edition.* Oxford, UK: Blackwell Publishing.

Vitaliano, P.P., Scanlan, J.M., Zhang, J., Savage, M.V., Hirsch, I.B. and Siegler, I.C. 2002. A path model of chronic stress, the metabolic syndrome, and coronary heart disease. *Psychosomatic Medicine* 64, 418–35.

Chapter 5
Eating and Drinking Kefraya: The Karam in the Vineyards

Elizabeth Saleh

Introduction: Springtime in the *Karam*

As the wind blows across her *janani* (garden), the sun starts to set and Hajjeh Nabhane is thankful that spring has arrived in the village of Kefraya, Lebanon. Her walking stick is perched next to her and she shouts up to her grandson who is standing on the veranda above to pick one of the sharon fruits growing from the tree that rises out of the ground near where she sits. Hajjeh Nabhane explains to me that the juicy orange fruit is one of Samah's favourites: 'Abdo! Pick one of these fruits for your sister. She will be home soon'. Abdo, whose name is short for Abdel Helim, does as he is told and goes over to the tree. He touches several of the fruits, squeezing each one gently while turning them ever so slightly to see the shades of its colours against the sun's fading rays. He picks one and takes the fruit into the kitchen. He enters the house and passes the framed black and white photo hanging on the wall. This is of his grandfather, also named Abdel Helim. There, from the entrance the long since passed away Abdel Helim stands watch, while his *misbaha* (prayer beads) hang from the right hand corner of the frame. Next to the photo is a framed cutting from the *Al Nahar* newspaper. The article is about the Chateau Kefraya winery located less than a mile down the road, where Abdo's grandfather once worked, as his father does now. A photo of the Chateau takes up most of the page, surrounded by fields of wheat and vineyards. There is a woman bending over the wheat, her arm is raised and she clutches a sickle. She is covered in layers of clothing to protect her from the sun and it is difficult to see who she is. This is the late Abdel Helim's daughter, Samia, who lives across the road with her husband. Much has changed since she reaped the wheat, which is now replaced with vines and, while Abdo is too young to remember these changes, he is mindful of the shifts across the Kefraya landscape; narratives passed down through the generations, but spoken through a language of sensory experience, that the *kouroum* of Kefraya is part of the cosmology of his village and so too it belongs to the genealogy of his family.

This chapter explores the intimate everyday engagement residents in Kefraya have with certain types of *kouroum* and their produce, and looks to elucidate the ways such engagements are entwined in affective relations, kinship and memory. I say 'types' because the word *kouroum* in Lebanese Arabic and its singular, *karam*, refers either to an olive grove, a fig tree orchard or a vineyard. An olive grove is known as *karam al zeitoon* and its plural is *kouroum al zeitoon.* A fig tree orchard is *karam al teen* and the plural, *kouroum al teen. Kouroum al anab* signifies vineyards. Of these types, fig tree orchards and olive groves are in the minority, compared to the panoply of vineyards. Wild edible plants such as *hindby* (wild dandelions) grow seasonally in small clusters among the vines that traverse the lands owned by most residents in Kefraya village (Edgecombe 1970). Despite the prevalence of the vine and the economic significance it holds as a site of commodified wine production and livelihoods for Kefraya residents, the *kouroum* is also, I argue, imagined and nurtured as a site where grapes, olives and figs are all cultivated, harvested and, in turn, eaten among intimate acquaintances and kin. Moreover, these two spheres of reproduction and production are not distinct domains but, rather like the grapes, figs and olives of the *kouroum*, are entangled and woven together through the embodied experiences and social relations of Kefraya's residents and workers. Giving attention to these entanglements, both within and outside the body, can, I suggest, illuminate the ways through which self, others and places are socially reproduced through eating, as well as demonstrating how eating is a simultaneously sensuous and symbolic experience.

I argue that the reproductive notion of the *kouroum*, outside of wine production, evokes a sense of place in Kefraya similar to what Seremetakis describes as 'alternative perceptual epistemologies' (Seremetakis 1996: 22). This refers to the ways in which history and memory are narrated within everyday life through attachments to sensory experiences in order sustain a continuous relationship with the past. As Seremetakis suggests, memory can be stored in 'specific everyday item' that 'form[s] the historicity of a culture' (ibid.: 3) and I draw on this by demonstrating that the articulation of certain senses, such as the familiarity of touch to ensure the smoothness of the vines that will be used in cooking, brings to the fore the *kouroum* as a site for the reproduction of familial relations, while simultaneously situating the *kouroum* as a place for wine production to the background. I thus show how narrative is at once a site of 'political and cultural tensions' (ibid.: 19). The sensory practices of the people of Kefraya are also coterminously enmeshed into the etymological practice of the word *karam* itself. Here *karam* goes beyond the physicality of a place where grapes, olives, figs and *hindby* grow. It enters a realm of the abstract where life genealogies and social relations (and particularly kinship) are in turn imagined, remembered and nurtured through such sensory practices. Yet why is so much social significance given to this understanding of the *kouroum* instead of that which is centred on wine production? This question is particularly pertinent in Kefraya as the *kouroum* was initiated with the intention of producing wine for the market. Consequently, I ask in what instances precedence is given to such reproductive meanings of the *kouroum* and explore how affective relations are reproduced through such conceptualizations.

In order to think about these questions further, this chapter examines how such a reproductive concept of *kouroum* converges, and at times diverges, with *kouroum* as a site for the production of wine. In order to do this, the following sections continue to follow a theme of seasons and generations. Structuring around the seasons is less related to ecological processes as it is to 'prospective memory' and the planning process that takes place 'in the present to remember food events in the future' (Sutton 2001: 19). This was how knowledge of what the *kouroum* evokes and invokes is intertwined with other senses that are, in part, founded upon an understanding of the climatic conditions that are to come and those that have past. In turn the notion of generations recounts both the passing of time and the changing of the landscape. The intergenerational transmission of cultural memory is therefore, I argue, intrinsically embedded in the act of eating and the embodied engagement of Kefrayan actors with the *kourum*. Yet this process is not impervious to change and disruption; villagers are migrating and new knowledges – and grapes – are being introduced. This has led, quite possibly, to a displacement of certain sensory experiences and their transformation into a historical narrative for the younger generation. Thus, I not only show how eating establishes and maintains relations across time and space, but also look to move beyond symbolic accounts of exchange and commensality and suggest ways in which the senses and narrative are mutually constitutive.

Returning to Spring: The Bountiful *Kouroum*

Back in the house of Hajjeh Nabhane, Abdo enters the kitchen and Han'a (Abdo's mother) directs him with her eyes to put the sharon fruit on the kitchen counter. She is sitting with other women from the village encircling vines piled upon on a plastic sheet in front of them. The women are sipping *kahwa* (Arabic coffee) and are talking amongst themselves when one of them loudly tuts:

> Did you see her coming back from the *kouroum* with that huge sack of *hindby* on her shoulders? I cannot believe that she did not tell us where all the *hindby* was growing! She does this every time! She goes to pick *hindby* alone and does not tell any of us!

Some of the women nod in agreement. Han'a says nothing but offers her guests more coffee. While some accept a refill, others decline politely and continue their work, shortening the stems of the leaves with their fingers and then running their hands across each leaf to check for irregularities. They barely look down at what their hands are doing. There is no point really. Checking the colour of the vine leaves cannot be done indoors. Instead it is done in the *kouroum* whilst picking. Like Abdo with the sharon fruit, the vine leaves are twisted delicately from side-to-side in order to check their nuances of green under the spring sun. If the leaves are dark green, they are no good for stuffed vine leaves; likewise, if they are too light in colour. The women make separate piles and Han'a stands up to reach for the empty plastic water bottles into which the vine leaves are stuffed and then stored until they will be used as the edible outer wrappings of stuffed vine leaves.

Before returning to the vine leaves, Han'a washes her hands and inspects the sharon fruit. Her legs brush past another pile of plants lying on the floor. This is the *hindby*, a dandelion plant which grows wild this time of year inside the *kouroum*. She had been out earlier with the other women in her kitchen and the clutter of kitchen knives, with blades still covered in soil from where they had pierced the ground and uprooted the plants, lie in a kaleidoscope of colours amongst plastic gloves caked with mud and the scarves that protect the women from the sun.

As their work slowly comes to an end, Han'a steps out of the kitchen onto the veranda. She looks west, out towards the plains of the Kefraya region of the Bekaa Valley in Lebanon. Not too far in the distance, the tower of the Chateau of Kefraya looms out from its surrounding protective poplar trees. The *sahal* (plains) surround the Chateau Kefraya winery and are covered in vineyards, some of which are situated within the company's estate. The other plots of vineyards belong to the residents of Kefraya who sell their grapes to Lebanese wineries. Han'a's husband, Nabhane Nabhane will be on his way back home from the *kouroum* of the Chateau. He is responsible for the winery's viticulture and supervises the workers in the vineyards of its estates. So, like his wife, Nabhane has also been out in the *kouroum* that day but, rather than picking vine leaves and *hindby* for personal consumption, he has been overseeing the maintenance of the vine for commodified wine production.

The *kouroum* yields food during every season of the year. Figs come out before the grapes. Some can be eaten fresh but the juicier ones are dried for winter. Depending on the seasons, the grape vines offer more than its grapes. Now in spring, their sour but sweet stems are ready to be chewed and sucked upon, and the leaves will be wrapped delicately around rice, meat and mint to make stuffed vine leaves. In late summer, the grapes are eaten or dried in the sun to produce raisins. Grape juice is transformed into vinegar. The olives are ready after the grape harvest and some are sent to the *makbas* (press) in the neighbouring village, while others are sealed in jars with olive oil from the previous year's harvest. As *hindby* illustrates, the *kouroum* is also the place where many wild edible plants grow according to season and these are usually found just under the vines, fig and olive trees. These too are plucked from the earth, cooked and then placed on kitchen counters and dining tables for the family to eat. Some are also parboiled and frozen. As the seasons change, slight shifts in what the *kouroum* offers can thereby be perceived.

Yet there are certain aspects that remain constant. As indicated above, the singular for *kouroum* is *karam* which, in Arabic, spelt the same way, can also mean bountiful and generous (Cowan 1974). The *kouroum* of Kefraya, a place where cultivated and wild plants grow, speaks continuously throughout the seasons of such qualities. *Karam* in the sense of bountiful is implied through the variety of foodstuffs that are reaped from the *kouroum*. It evokes kindness and generosity because these foods are never sold, but are rather shared and exchanged through acts of reciprocity. Not sharing the produce, as the woman who left the *kouroum* without telling the others the secret location where *hindby* grew in abundance, ruptures these links while also highlighting to newcomers, like myself, the subtle interrelatedness of these narratives of *karam*. Understanding these connections of *karam* can, as Carol Delaney's (1991) ethnography also shows, draw attention to the ways in which bodies and places are socially produced through a referential system that brings together – within an imagined cosmology – bounded notions of the body, gender, family, home, village and nation. Delaney suggests that through symbolic usages of terms such as 'seed' and 'field', ideas of procreation and nurture, sustenance and nourishment are evoked within the daily activities of those living in a Turkish village. While this may have certain parallels to practices within the *kouroum* that are metaphorically attached to notions of gender, it may perhaps be more conducive for the case of Kefraya to consider notions of *karam* as allegorical narrative instead of simply symbolic. This is especially the case given that narratives of *kouroum* in Kefraya are distinctive in the sense that they are fairly recent and entwined with historical narratives of the modernization of wine production in Lebanon.

These potential intimacies, estrangements and ambivalences are all the more evident given that the Kefraya region is continuously subject to the entrepreneurial strategies deployed by wineries across Lebanon. The founder and current chairman of the Chateau Kefraya winery began his wine-producing enterprise in the late 1940s on lands inherited from his father. Michel de Bustros comes from a Beiruti aristocratic family and told me that, before he started planting vines, Kefraya was just another 'property' for his family (see also de Bustros 1983). Yet this conception shifted and in his autobiography, *D'Haute d'une Breteche* (2002a), de Bustros writes not only about his experiences of transforming Kefraya into a region of extensive viticulture, but also about his personal visions and aspirations. In his efforts to establish such an enterprise on his family property, located far out in the rural hinterlands, he intended to support the vision of the newly independent Lebanese state. A brand new nation with freshly planted and rooted vines; Lebanon would have its very own wine region in the heart of the agricultural Bekaa Valley. De Bustros explains that he began to sell grapes to wineries across Lebanon before eventually opening his own winery in 1979. Kefraya has since emerged as the wine grape-producing hub of Lebanon. Including Chateau Kefraya, there was at the time that I conducted my fieldwork (between late 2006 and early 2008), one other established winery in Kefraya and another under construction. Cave *Kouroum*, the second winery in Kefraya was founded in the 1990s and is run by the Rahal family, who come from the village of Kefraya. The new winery, established after my departure, is managed by members of the Saade family; another elite entrepreneurial family who owns estate, distributing and shipping companies. Each year, over 900 tonnes of grapes are harvested from the *kouroum* of the Kefraya region and sold to wineries situated both within the municipality and across Lebanon.

Despite the region's reputation as a wine growing hub of Lebanon, many in the Kefraya village do not drink wine. This is also the case with *araq*,[1] which is produced by wineries and locally in other neighbouring villages, where it is consumed frequently. This could be attributed to the religious practices and beliefs of the local residents in Kefraya, many of whom are Muslim. Yet some local residents do drink other alcoholic beverages, such as beer and whisky, and thus there appears to be an estrangement between the commercial product of the vineyards and those who own, work and live amongst the vines. But in spite of this apparent disconnection, it cannot be said that the *kouroum* is not in the minds of the people of Kefraya village.

Summer in the Vineyard of the Olivish

Every morning before Hajjeh Nabhane's son, Nabhane goes to work at Chateau Kefraya, he takes his breakfast; a small cup of *kahwa* (coffee) and '*el zeitoon*' (olives) that come from the *karam* down the road, which he eats with some bread. He then departs for the laboratory of the winery to meet with the French oenologist and plan the day. Occasionally, if Michel de Bustros has arrived from Beirut, Nabhane will drive to his office at the chateau to discuss the *kouroum*. This is Nabhane's daily routine without fail. The only difference is that now in summer time when the grape harvest begins, he rises much earlier so to arrive at the winery's *kouroum* before dawn. On his way home during this harvest time, Nabhane usually picks some of the grapes from his family's *kouroum* and, once home, his wife Han'a will set some aside for the large fruit tray on the veranda. Based on their ripeness, the other bunches of grapes are laid out under the sun where they dry into raisins. The *Cinsault* grapes are full of juice and sugar and Nabhane informs me are the best of the different kinds to make *zbeeb* (raisins) and grape juice. Most families in Kefraya have vineyards mainly of the *Cinsault* variety, known locally as *Zeitooni*.[2] When late summer arrives and the harvest is around the corner, the plump *Cinsault* grape berries, with their deep hues of purples are olivish in character and hence the name *Zeitooni*. The *Cinsault* grape varieties I am told have been in the *kouroum* since *karam* began in Kefraya. The story goes that before Michel de Bustros, or *Khawaja*,[3] as he is known in Kefraya, planted the vines, there was nothing really of significance in the village. Local residents only grew legumes such as lentils and chickpeas. Following the first plantation on the *Khawaja's* family lands, his interests expanded to the villagers'

1 *Araq* is an alcoholic beverage produced by the distillation of wine into a spirit flavoured with aniseed.

2 In Arabic grammer *Zeitooni* is a form of an-nisbah; a suffix is added to a noun in order to form an adjective. The suffix iyy is masculine and iyya(t) is feminine.

3 Residents in Kefraya inform that this term means 'esteemed sir' and its usages can be traced to the Ottoman period.

lands. By the 1950s he began to offer to buy, on their behalf, vines of the *Cinsault* variety imported from France. Payment was not required until the grapes were at the optimum levels to be harvested. This usually took a minimum of three years. The profit the viti-farmers made by selling the grapes to other wineries in Lebanon, such as Chateau Musar and Chateau Ksara, was then used to repay de Bustros. The region was thus transformed into the ordered landscape that it is today because of the *Khawaja's* enterprising visions (see also Karam 2005).

Kefraya is now well known, across the region, for its beautiful *kouroum*. At first glance, this landscape might appear static; but the *kouroum* is continuously changing. Wineries are demanding that *Zeitooni* grapes are pulled out to make way for the noble kinds, such as *Cabernet-Sauvignon* and *Chardonnay.* Despite the wineries coming together in order to agree upon paying a lower price for their grapes, most in the Kefraya village continue to grow their *Zeitooni.* During this harvest, tension arose once more, because the French oenologist at the Chateau Kefraya winery had let it be known in the village that, after this harvest, there would a dramatic reduction in the amount of *Cinsault* bought from the village. This way, he explained to me, vineyard owners in Kefraya would have the incentive to pull out these table wine producing grapes and plant the more noble varieties. This was not the first time that such initiatives had been taken by the wineries that source their grapes from Kefraya. The Chateau Ksara winery located in central Bekaa had done the same in the mid-nineties. They had set up contracts with landowners in the other parts of the Bekaa and planted grapes such as *Cabernet-Sauvignon*. Once these vines had started yielding their grapes, the Chateau Ksara winery stopped buying the *Cinsault* grapes from Kefraya. The village was left with over 700 tonnes of grapes and nobody to buy them. This was when the Rahal family of Kefraya decided to open their winery, Cave Kouroum. Since then, some of the villagers started to plant the grape varietals on demand. However there are still *kouroum* where the *Cinsault* varietals have been left.

The landscape of Kefraya is therefore not the scenic, static picture of Lebanese wine production that it initially appears. Competing visions of the *kouroum* have transformed it at times into a discordant object that belongs simultaneously to different temporal dimensions, visions and actors. The potential disconnections in the way *kouroum* is used and experienced by those living and working in Kefraya are brought into sharp relief when wineries seek to harmonize the wine making and viticulture productive spheres in order to align the styles of their wines with global standards of production. It is not only the same space – that is the *kouroum* – but also the same plant that is, at times, multivocal. With new pressures now coming from Chateau Kefraya for a change in the varieties grown by Kefraya residents, it seemed that *karam* was once more in transition. What would happen if the wineries no longer wanted the *Zeitooni* grapes? As Nabhane explained to me, the wineries did not just want new vines to be planted. It was expected that the new vines would be planted in a different way, along wires rather than in the current goblet style. With this would come an unfamiliar way of pruning the vines, tending to them, and nurturing them. The grapes would also be smaller in size and there would fewer of them. These were not *Zeitooni* and thus no longer *karam*.

Encounters with the *Zeitooni* grapes are part of a perceptual consciousness among Kefraya residents where on the one hand, the apparent mundane daily activities such as picking the grapes or drying certain kinds under the summer sun (re)create and sustain a seasonal connection to the generosity of *karam*. Thus evoking a seasonal association with what Sutton (2001: 19) describes as 'prospective memory'. On the other, *karam al Zeitooni* is also a place that allows for the commemoration of a historicity of *karam*. The breakdown of *kouroum* through the removal of the *Zeitooni* grapes could result in a discontinuity of *karam* and thus also in a 'sensory experience of history' (Seremetakis 1994: 3). This was particularly so for the Nabhane family.

As many in the village informed me, the knowledge Nabhane has of the vineyards of Kefraya derives mainly from his experiences of working with his father when young, a couple of decades or so ago. It is largely due to this knowledge that Nabhane is well respected in both the winery and the village. Perhaps it was also because Nabhane's father had a special relationship with the *Khawaja*. It was Abdel Helim who had planted the first of these vines in the region. And while Abdel Helim is rarely mentioned in the autobiography of the *Khawaja*, there is still a level of care to maintain the bonds with the Nabhane family through exchange and reciprocity. For one, Nabhane is now the head of 'agricultural affairs' for the Chateau Kefraya estates. This position has now become an established part of the everyday life of the Nabhane family, but there are also other moments where both the *Khawaja* and the Nabhane family reinforce their social relations and become intimately entwined, with food and commensality playing a particularly significant role.

On a couple of occasions during the hot summer months, the *Khawaja* de Bustros sent his driver up to the Nabhane household to deliver some fish that had been freshly caught and cooked. During these occasions, Hajjeh Nabhane, who rarely joins everyone for lunch, would also sit down to share the meal with the rest of the family. The fish is her favourite dish, and it would take precedence on the table, being served solely with the rice that was delivered with it, without any other accompaniments. Exchange relations also flow in the other direction, and the Nabhane household also made sure to send food stuffs to the *Khawaja*. Significantly all this came from their own *kouroum*. Olives and dried figs were sent to the *Khawaja*, despite the fact that he had his own *kouroum* of olives and figs. The same was for the food made from the vine. Delicately stuffed vines leaves made from the smallest of leaves were wrapped by Hajjeh Nabhane's youngest daughter, Dunya who lives in the same household. She has especially nimble hands and fingers for rolling the little leaves and once a pile had been made, her brother would take them to the chateau during work the next day. The best stuffed vines leaves I am told by both Dunya and Nabhane are made from the *Zeitooni* vine leaves.

The temporality of these exchanges is seasonal and thus both cyclical and continuous. In this regard they evoke a sense of 'food generosity'. In this way it is much like those food exchanges observed by Sutton in Kalymnian that articulated ideals of personhood through the sustainment of a shared historical consciousness (2001: 16). At the same time, *karam* (generosity) wove a narrative that linked the prestige food brought in from outside by the *Khawaja* with prestige food returned from inside Kefraya. *Al karam al zeitooni* was therefore also a 'lieu de memoir' where the Nabhane family and Michel de Bustros were both able to lay claim to ideas of personhood and status. This process resonates with Palmer's (2002) work on food practices and food identity among the Bedouin and *fallahin* in Jordan in which she identifies the ways in which they are embedded historically within hierarchical social relations. Drawing on Palmer, and more widely Mauss's (1990) theory of the gift, the commensality of sitting at the table with only the food sent over by the *Khawaja* can be seen to not only symbolize his social status, but is also indicative of his political and economic influence, with his gifts informing the way 'his' food is eaten in his employee's home. The Khawaja's influence, then, is not contained to the productive sphere of the winery, but rather extends into the domestic, reproductive domain.

Yet these acts of remembrance that lay claim to status and honour are mainly metaphorical; the relationship between the *Khawaja* and the villagers of Kefraya has gradually changed as the winery has gone from a predominantly family-orientated company to a share-holding organization, which was made open to other investors by the end of the 1980s. With the arrival of new prominent businessmen into the company, Michel de Bustros is no longer solely in charge of the winery business. New management structures have been introduced and specialists, such as French oenologists, rather than the *Khawaja*, are now in charge of selecting the types of grapes for the wines. Consequently the *Khawaja's* relationship with the *kouroum* of Kefraya is continuously changing. Nevertheless he still holds his title and this, the villagers say, is because of Hajjeh Nabhane's husband, the late Abdel Helim and the important relationship he had with the *Khawaja*. 'Food generosity' therefore collapses space and time as it called forth previous actors such as the late Abdel Helim who had an important relationship with Michel de Bustros. The prestige food and the ways in which commensality was practiced in the home was a process that also narrated and commemorated the memory, life and honour of Abdel Helim Nabhane, and reinforced inter-household relations across the generations.

Al Karam al zeitooni, or the vineyard of the olives, is part of an ontological account that, although attached to the historical narrative of wine in Kefraya, diverges from more contemporary viticulture. It speaks of a genealogy of kinship and relatedness (cf. Carsten 1997, 2000) and the spaces in which it is nurtured within Kefraya. Here the intimacies of the *kouroum*, where grapes are connected to olives through etymological terminology, bring forth other senses of production and consumption related to the embodied acts of eating. The breaking down of these connections, as Nabhane suggests, could also lead to other notions of *karam* coming apart. The replacement of the *Cinsault* vine could, as Nabhane explained above, break up different notions of *karam* or generosity, and with it could come a fragmentation of the transmission of knowledge and cultural memory; perhaps not too dissimilar to the way in which the *Zeitooni* grapes serve as a reminder of the olive harvest to come in autumn.

Autumn in the *Karam al Zeitoon*

Dunya is sipping *kahwa* from her little cup. She inhales her cigarette quite deeply and lets out a puff of smoke before taking another sip. It will be a long day for Dunya, as she will be working in the family's *kouroum*. Although the family hire extra labour to work in their *kouroum*, Dunya still prefers to attend personally to the vines, olive and fig trees. Throughout the year, she wakes up before dawn and dresses in her attire of thick jogging trousers, long-sleeved shirts and jumpers, rubber boots, the all-important *kafiyeh* and a long straw hat. If it is summer, then she will stand or sit in the veranda and look out across the Kefraya plains onto the anti-Lebanon ranges in the distance. As autumn is now underway, she will take her coffee and cigarette by the *soobieh* (hot stove) that is situated in the middle of the family kitchen. She is usually the first one to light this for the others, who will soon awaken. As she looks out, lights begin to flicker on these mountain ranges. Perhaps other early risers are also brewing coffee. Some of these lights belong to the United Nations soldiers who are perched strategically on top of Jebel Al Sheikh (Mountain of the Sheikh). From where Dunya is standing, she is also able to see the contested Golan Heights, an area also known for its extensive viticulture. These somewhat distant undulating lands have been an ongoing part of her perspective, memories and experiences. They vacillate through history, yet still remain part of her view from here, where she remains.

The grapes have just been harvested but autumn is still a busy time in the *kouroum*. The olives are now ready but, unlike the harvest of the grapes when mainly seasonal migrant workers work in the fields, now it is the time for Dunya, Han'a, and Han'a's two grown up daughters Samah and Salam, to come out into the *kouroum*. They bring clothes and plastic sheets, which are placed under the trees, and plastic beach combs to brush out the olives. Just like the spring and summer seasons, there are many others from the village out in the *kouroum* this time of year. The cool breeze is welcomed and the trees are rustling less from the wind and more from the people shaking their branches. The olives are sent off in buckets, ferried in either cars or tractors, to the olive press in the neighbouring village.

These olive trees began to appear in Kefraya a couple of decades earlier than the *Zeitooni* grapes and, like the grapes, the olive saplings were imported from Europe. But this time the plants came from Italy instead of France, and some residents in Kefraya still have the wooden crates with the date of the olive saplings' import – 1934 – stamped on the side. The reasons as to why Italian olives were chosen or favoured over the local Lebanese varieties are unclear and lost in the mists of time. Nevertheless the narratives of the olives describe the hands that planted these olive saplings as those same hands that planted the *Zeitooni* grapes – Kefraya's forefathers. Abdel Helim, Dunya's father had been one of those men to have planted the *karam al zeitoon* or the olive grove many years ago. Despite this patriarchal legacy, much of the work that takes place in the *kouroum al zeitoon* today is mainly done by the women of Kefraya. Apart from driving the olives to the *makbas* (press), the women are usually the ones out in the *kouroum* rustling the trees. They are also the ones at home plucking the olive pods from the stems, before allowing the olives to soak in the cold water for a period of time, usually up to a month. For unlike the grapes, that are mostly sold to the wineries, most (if not all) of the olives reaped from the *kouroum* are kept for consumption within the home. The villagers of Kefraya eat olives mainly in the form of olive oil. *Al zeit zeitoon* (olive oil) is poured into jars full of cured olives or strained *labneh* (yoghurt), the latter of which is rolled into balls. These jars are stored in the kitchen throughout the year and spoonfuls of the stored foods are scooped out into little bowls, which are commonly placed on the kitchen table or the large round tray carried out to the living room or balcony.

The harvest of the olive in this respect is quite different to that of the grapes. It is the men of Kefraya out in their *kouroum* supervising the workers cutting the grapes, which are then sent off to the wineries. There, only some of the grapes that are brought in by the men are kept for raisins or grape juice. In autumn, when women return to the *kouroum* and recommence the gathering process, the produce is once more kept and preserved entirely for eating in the home. I was told that the men used to be more involved in the olive harvest, and elder residents in Kefraya can recall their fathers working in the olive groves. While the women did help at times in the *kouroum*, they were mainly in charge of curing the olive and creating the *mounieh* (stored food produce). The elder residents lament the times gone by and the disconnection of the younger men from the *kouroum*. When I put this to Nabhane, he tells me that there is no future in the *kouroum* for young men. His nephew Omar, for example, who had expressed a desire to work in the *kouroum*, is trained as an agricultural engineer at university. Despite having gained knowledge of the techniques required to maintain the new vines, none of the wineries is hiring and Omar is unable to find employment. There is also very little income to be made from looking after either his family's *kouroum* or others in Kefraya. On being asked his thoughts on his son Abdo's future, Nabhane tells me that Abdo 'must become a man of the pen and not of the land.' He will most likely not take over the role of his father at the Chateau Kefraya winery. For, as the *Khawaja* aged and the authority he once wielded is weakened by new management structures, there is an increasing estrangement between the winery and the village. There is also, Nabhane reminds me, no son to take the place of the *Khawaja*: the Chateau Kefraya Company is, after all, now a share-holding company.

With socio-economic and also material transformations taking place in the *kouroum* of Kefraya, the narratives of *karam* entangled within it are increasingly reproduced through practices carried out by women. These practices do not only result in the availability of the foods throughout the seasons. The jars of olives or *labneh* preserved in olive oil are also sites for the storage of *karam* – just like the vine leaves stuffed into plastic bottles. Here, as the olive oil is soaked up by the bread dipped into the little bowls on the table, traces of the *karam al zeitoon* seep across the seasons and enter into daily life. Similar too are the *Zeitooni* vines leaves pulled from plastic bottles, bringing the spring of the past into the summer or autumn of the present. These jars and bottles of food thus invoke Sutton's (2001) 'symbolic vestiges' which, when opened, activate memory and reproduce an historical consciousness. That *karam* is predominantly associated with food-centred activities that begin in the *kouroum* but are finished within the home is suggestive of the ways in which continuity of narratives is practiced. This is especially salient considering the context in which the changes that have taken place (and continue to do so) in the *kouroum* of Kefraya are often beyond the control of those living in the village.

These changes do not only concern the strategies deployed by the wineries. Other, more dramatic events, such as the occupation of the region by Israeli and Syrian armies during the 1980s have also left a distinct impression within the minds of local residents (see also de Bustros 2002b). The village, I am told, has changed dramatically. Many people left, migrating to urban areas such as Beirut and others have emigrated abroad. Once again the *kouroum* is a constant in these narratives. Some recount how army tanks trampled across vines and trees and shot out shells. Others speak in great detail of the knives used to pierce the earth and dig out holes to bury parts of soldiers' bodies, killed by the opposing army. Such accounts are replete with the *kouroum* and of its importance, emphasizing how, despite all of the traumatic events, it still remains constant, albeit changing. These accounts carry more weight given that once more it is the plants picked from the *kouroum* that are sent outside Kefraya to relatives in Beirut or abroad as reminders of home. Again, it is the women of Kefraya who are active in weaving and (re)binding these ties.

Autumn is important in this regard as it is also the time for goodbyes. Summer is when those who have left Kefraya to live in the capital or abroad return for family visits and to oversee the grape harvest in their *kouroum*. Hajjeh Nabhane's eldest daughter had migrated to Argentina many years ago and was not able to visit her family in Kefraya that year. She worked and had her own family now, with two teenage daughters. Nevertheless, she still liked her *hindby*. Each year in spring, when Dunya went out to the *kouroum* to pick the *hindby*, she would make sure that some of the plants were parboiled and frozen immediately. These would be set aside, awaiting the next person who might be making a trip to Argentina. Now that someone in the village was about to return to Argentina, Dunya removed a bag of the *hindby* and also two bottles of vine leaves, before preparing for a visit to the neighbouring house. I am left sitting with Hajjeh Nabhane and tell her that I think sending the *hindby* to her daughter is a really nice idea. It must make her daughter happy to receive these plants, especially as they are picked by family members so far away in Lebanon. Hajjeh Nabhane nods in agreement. Of course it was. She tells me that her daughter 'can eat the hands that picked it'. So it was critical that Dunya and other family members had picked the plants. These moments of picking the plants and eating them in Kefraya – and perhaps across space and time with Hajjeh Nabhane's daughter in Argentina – go beyond symbolic efficacies that link family members through time. The hands that picked the plants and the mouths that ate them embodied and transmitted knowledge that allowed for the possibility of, and enacted, the reproduction of family ties.

The ability of these types of foods to generate what Seremetakis (1996 and also Sutton 2001) refers to as 'counter-memories' is done through an awakening of the senses. By preparing and cooking these plants, aromas are released and their textures change once more; the senses are thus revitalized and reawakened. The linking of the hands that pulled the *hindby* and vine leaves from the earth, and prepared and cooked them, with the mouths that eat and taste them in a location beyond Kefraya, intimately connects *kouroum* and *karam* with people and place. These multiple encounters between bodies collapse time and space as senses become stimulated and kinship relations enforced. Such gendering of the work around food in order to cultivate memories of lineage is also perhaps suggestive of their roles as 'agents of memory' (Holtzman 2006: 370). The amount of time and effort Kefrayan women take to ensure that such daily food activities are not forgotten help to provide a connection between the past, present and future.

Winter and the Fig Tree

Hajjeh Nabhane sits on the concrete step that extends out on to the *janani* (garden) of her family's home. The steps behind her lead up to the veranda of the Nabhane household, where the rest of the family sits indoors awaiting the end of winter. Spring is not too far away and Hajjeh Nabhane prefers to sit and watch the early morning skies and feel the warmth of the promising new sun. Now an 80-year-old woman, and a Hajjeh through a spiritual journey to Mecca in dream and meditation, she watches the distant regions beyond the garden, the village of Kefraya, and the *sahal* of the West Bekaa with eyes aged through sun and snow. I am sitting next to her and she recounts the days gone by, where different borders had once stood. We share a kettle of *mate* tea between us, cleansing the *bombija* that her daughter sent her from Argentina, with a slice of lemon, before passing it to the other. Hajjeh Nabhane's face is now weathered and her hands are wrinkled and leathered; traces of her outdoor labours and engagement with the land. She has witnessed many shifts and changes across these landscapes. Born on the eve of the end of the Ottoman Empire some thirty kilometres down the road from where Sykes and Picot had put pen to paper and signed their secretive agreement, Hajjeh Nabhane grew up during the French mandate. She recalls that the boundaries of the Kefraya village were not so rigid and clearly defined in those days. The main road of the Bekaa Valley had hardly been there and her father used to take his sheep to the markets in Beirut along the tracks of the mountain pass that no longer exist. When French surveyors began to visit, property papers had to be signed and registered, and the roads became clearly demarcated. The *kouroum* of Kefraya arrived not long after. Hajjeh Nabhane was barely a woman, yet she remembers the fig tree she planted not long before the vines covered the landscape. She speaks less to me and more to the wind when she asks out loud what will become of Kefraya when she passes.

Leaning against her stick, Hajjeh Nabhane pokes her finger into the rocky gravel by her feet. She scratches at the rocks before smelling the dust that has settled on her finger. It has been a cold winter with plenty of snow. This is good she confirms because the snow kept the *shersh* (roots) underground safe from the frost and ice. Now the earth awaits the warmth of spring and there will be an abundance of food to pick from the *kouroum* in the coming months. She tells me of the wild edible plants that grow next to the vines, such as *korad* (leeks), *hindby* and *sli'ia* (wild spinach). The exceptionally sweet figs will also soon blossom from the now-ageing fig tree.

Some weeks later when the vine leaves have started to emerge, her granddaughters, Samah and Salam, take me out to the *kouroum* to teach me how to pick the leaves. Samah maintains a watchful eye over me as I glide my hands over the vine leaves, trying to decide which are the most suitable: 'Take the top three or four. If they are too small, then count two down and then take three or four'.

We are not too far away from the Nabhane house where Hajjeh Nabhane and Han'a gaze out from the veranda in our general direction. Samah and Salam wave at them and we stop for a rest under the fig tree, its gnarly and twisted branches extending out above and across the vines below. This was the fig tree that Hajjeh Nabhane had planted and her two granddaughters look up in search of some fruits to take back to her at home. Their grandmother can no longer make the trip to the *karam*. They choose the softest of the figs and break each one off from the branch. Sweet translucent syrup gushes out from the tree and the two girls beckon me to taste the *attar* (syrup). This they say cannot be taken back to Hajjeh. But, Samah continues, she would want us to eat it for her, here in the *karam*. I ask the girls about the *karam*. Was it not confusing that a *karam* referred to a vineyard, a fig tree orchard or an olive grove? Samah pauses and turns to me to explain:

> A vineyard cannot be on its own, along with the vines must be fig trees and olive trees, such as my grandmother planted when she was a girl. It makes no sense to have one without the other. They are the same but different.

Conclusion

Activities, from picking the produce of the land to sharing and eating the food, offer sustenance to the body while also keeping mindful a historicity of the self that is attached to the *kouroum* itself. All the while, the *kouroum* remains a constant, a site for the storage of sensory memory awakened and reproduced through moments that lead to, and during acts of, commensality. The knowledge of how to pick these plants and other food from the *kouroum*, such as grapes, figs, and olives is a knowledge that is transmitted from one generation to the next. For the younger generation, who cannot recall when the *kouroum* began and the trees and vines were planted, the stories of their grandparents and parents become an historical narrative of self and its attachment to place. These stories are rarely vocalized but are, rather, expressed through the embodied acts of picking and eating. Commensality takes place in the *kouroum* and around dinner tables to generate a subjective commentary (Sutton 2001) of the self and others. It prescribes a sense of history and offers a shared narrative. These perceptions of the *kouroum* are therefore the deeply-embedded sediments of memory ascribed upon and within the bodies of those living in Kefraya (cf. Connerton 1989). The knowledge of waiting (to pick), knowing what and where to pick, as well as knowing how to eat it draws upon layers of embodied experiences; an engendered form of knowledge as it is increasingly transmitted and reproduced though time by grandmothers, mothers and daughters.

Simultaneously, the *kouroum* belonging to residents of Kefraya continue to have a significant role in wine production and the region remains the wine grape hub for the Lebanese wine industry. Yet as new grape varietals displace the older *Zeitooni* and the younger generations of men continue to seek work outside of the *kouroum*, there is the potential rupturing of *karam. Karam* here refers to how the *kouroum* is an interconnected site for the commodified production of wine and, at the same time, a place for the making and consumption of food. *Karam* therefore evokes and awakens sentimental connections across time and space. These are sentimentalities that might otherwise remain dormant, forgotten and could potentially become ruptured. Once the young men become disconnected from the *kouroum*, and unable to make a livelihood from the land, then the future of *karam* in Kefraya is put into question. In this regard, eating from the *kouroum* and sharing its food with others in order to bring forth a sense of *karam* (generosity) becomes increasingly essential in an attempt to keep younger generations of Kefraya mindful of *karam* and thus also of the ongoing link between kin, place and land. The sensorial characteristics of food revealed through eating thus narrate historicities of bodies and their relatedness to place. With such forms of narration, the *kouroum* is re(produced) as a space where one not only enacts these histories and shares cultural memories on an everyday basis, but also picks the very sustenance from which to survive.

Acknowledgements

Fieldwork was funded by an Emslie Horniman Fieldwork Grant from the RAI; the Centre for British Research in the Levant, the BRISMES Institute; and the Central Research Fund, University of London; Goldsmiths, University of London. This chapter is dedicated to the memory of Hajjeh Nabhane who passed away in the winter of 2011.

References

Carsten, J. 1997. *The Heat of the Hearth: The Process of Kinship in a Malay Fishing Community*. Oxford: Clarendon Press.

Carsten, J. (ed.) 2000. *Cultures of Relatedness: New Approaches to the Study of Kinship*. Cambridge: Cambridge University Press.

Connerton, P. 1989. *How Societies Remember*. Cambridge: Cambridge University Press.

de Bustros, N. 1983. *Je me Souviens*. Beirut: Dar-An-Nahar.

de Bustros, M. 2002a. *Du Haut D'une Bretèche*. Beirut: Dar-An-Nahar.

de Bustros, M. 2002b. *Tourmente d'une guerre dite civile*. Beirut: Dar-An-Nahar.

Cowan, M.J. and Wehr, H. 1974. *A Dictionary of Modern Written Arabic*. Beirut: Libraire du Liban.

Edgecombe, W.S. 1970. *Weeds of Lebanon*. Beirut: Heidelberg Press.

Delaney, C. 1991. *The Seed and the Soil; Gender and Cosmology in Turkish Village Society*. Berkeley: University of California Press.

Holtzman, J.D. 2006. Food and memory. *Annual Review of Anthropology* 35, 361–78.

Karam, M. 2005. *Wines of Lebanon*. Beirut: Saqi Books.

Mauss, M. 1990. *The Gift: The Form and Reason for Exchange in Archaic Societies*. New York: W.W. Norton.

Palmer, C. 2002. Milk and cereals: Identifying food and food identity among Fallahin and Bedouin in Jordan. *Levant* 34, 173–95.

Seremetakis, N. 1996. *The Senses Still: Perception and Memory as Material Culture in Modernity*. Chicago: Chicago University Press.

Sutton, D. 2001. *Remembrance of Repasts: An Anthropology of Food and Memory*. London: Berg.

Chapter 6
Negotiating Foreign Bodies: Migration, Trust and the Risky Business of Eating in Highland Ecuador

Emma-Jayne Abbots

Introduction

On returning to the city of Cuenca, situated in the Southern Ecuadorean Andes, after a two year absence, I was immediately struck by two notable changes. Firstly, the local restaurant scene had exploded, with the opening of a range of higher-end, self-styled fusion and cosmopolitan restaurants, a number of which promoted 'Gringo Nights'. The second change related to the gringos who comprise the target audience for such nights, and for the first time I observed a conspicuous presence of, predominantly, retired North Americans living in the city. This recent inward migration has not gone unnoticed by the incumbent Cuencano population and many are ambivalent about the their presence, highlighting that they keep to their own and do not integrate into broader society, while also acknowledging that they bring wealth, potential job opportunities, and open new markets and spaces of consumption. Integration, or lack thereof, in relation to food spaces is the subject of this chapter, and I explore the bodily encounters that privileged migrants, as I refer to this social group rather than using the politically and morally charged terms expatriate (cf. Fechter 2007) or gringo, both create and deny in their everyday eating practices.

In particular, through the lens of risk, trust and control, I highlight the disjuncture between migrant dreams of consuming a diet of simple, clean and local food and the reality of their practical eating experiences. A desire to minimize risk by reasserting control over their bodies, and ultimately their physical wellbeing, by escaping the industrial food chains of their home countries is one of the primary motivations privileged migrants cite for moving to Ecuador, and the valorization of short food chains, peasant producers and retailers, and the wide availability of healthy, fresh and pesticide-free foods at local markets and restaurants in Cuenca is a recurring motif. Yet, privileged migrants obtain the majority of their food in the clean and safe environment of the supermarket and modern retail outlets, and eat primarily at home and in cosmopolitan restaurants, a number of which are managed by members of their own community. My purpose in this chapter is thereby to explore this contradiction between discourse and practice, and I argue that migrant estrangements from Cuencano foods and food spaces are less founded on practical issues, and more related to migrants' broader fears over foreign bodies, in the shape of both local people and pathogens, and their ambivalence over dirty and risk-laden spaces, including markets. Thus, I highlight how privileged migrants are hindered in the realization of their lifestyle aspirations by their need to assert their own cultural and bodily boundaries, and their concerns over intimate engagements with foreign bodies. Moreover, privileged migrants in Cuenca are a distinct and self-identifying social group, due to their own foreign status, but they are by no means homogenous, and economic, political and geographical divisions that existed pre-migration are being usurped by new, more locally salient and contextual boundaries. This chapter consequently illuminates the role that eating plays in creating these emerging boundaries, not only between migrants and Cuencanos, but also among migrants; thereby elucidating how food and its spaces create new sites of 'distinction' (Bourdieu 1984).[1]

'Live Healthier … Live Longer …'

By moving to Cuenca. So advised a recent article in *International Living* (IL), a magazine targeted towards those, primarily North American and retired, individuals looking to enhance their lifestyle by migrating to another country.[2] IL promotes Cuenca as the ideal destination for those who are searching for a 'quality of life that just plain doesn't exist anymore in the States' (IL website 'Why Ecuador'), naming the city 'The World's Best Retirement Haven 2009', and putting Ecuador at the top of its list of preferred retirement countries in 2010 (ibid.). As perhaps can be anticipated, IL's representation of everyday Cuencano life is romanticized, synchronic and essentializing, with the local population being portrayed as 'fine craftsmen and attentive caretakers of their land' who 'adhere to the natural healing traditions their ancestors taught' (unattrib. IL website 'Why Ecuador').

---

1 In addition to food, new self-acknowledged distinctions are emerging between migrants who are investing in land and property and those who cautiously prefer to rent, and those who regard themselves as 'pioneers' in contrast to the more recent arrivals whose entries have been smoothed – and in part motivated – by the existing 'expat' community and the promotion of Cuenca by *International Living* (IL) Magazine.

2 IL also helps facilitate privileged migration through a range of advisory activities, including tours that show potential migrants real estate, arranges meet and greet sessions with 'successful' migrants and 'friendly locals', and highlights migrant spaces, for example, a gringo-owned coffee shop. For many of my participants, an IL tour was their first experience of the city and many cited the magazine as being a significant factor in informing their decision to move to Cuenca. Consequently, the magazine has an unusually marked influence and presence among the privileged migrant community in Cuenca and for many it is the definitive text and source of authority, especially if they do not speak Spanish and have little knowledge and experience of Ecuador.

Migrants are told they 'can sit on a front patio each morning in the clear mountain air, sip coffee, and bask in the equatorial sun', 'discover the natural rhythms of the world', and 'feel younger, years younger, every day' (op. cit.). Cuenca is thereby portrayed as the site in which the aspiration of living longer and living healthier can be realized, and it is represented as one of the 'few small, protected pockets of the world' where life can continue as 'Mother Nature intended and natural remedies have not been supplanted in popularity and accessibility by synthetic medicines' (ibid.). Consequently, living in Cuenca is 'like living in the 1950s in the USA' (ibid.).

Just as the backgrounds of privileged migrants are varied, the individual journeys that have led to their decision to leave their homes and move to Cuenca are multiple, complex and often deeply intimate, with a number recovering from chronic illnesses and/or becomingly increasingly aware of their ageing bodies. This often interplays with feelings of disempowerment and vulnerability vis-à-vis multinational corporations, Big Pharma and the state that are further compounded and crystallized by personal tragedies and health concerns.

Steve is an Italian-American from Minnesota, and he and his wife had been permanently living in Cuenca for approximately six months when we first met in a popular gringo café. The recent passing of Obama's healthcare bill by Congress was our first topic of conversation, and Steve was feeling very emotional at the 'bittersweet' news. Two years previously, he had been diagnosed with an illness that had 'financially wiped [him] out', yet he downplayed his economic motivations for leaving the US, explaining that his primary reason was to live a 'more simple way of life'. Asked to elucidate what he meant by this, Steve invoked a *New York Times* article that stated 'a hamburger is more well travelled than most Americans' and he said 'I just want to get away from this complexity of the food-chain in the US, you know, with all its pathogens, salmonella and mad cow disease'. He continued 'for me, well, simple means eating in a more simplified and less complex manner; you know, cutting out some of the links in the food chain, and eating clean food that hasn't been produced on these great big industrial farms'.

His desire for simple food chains thus arose from a concern over bodily health, and Steve explained that he found it impossible to follow this philosophy back in the USA, primarily because of the intensity and complexity of industrialized food production in North America, and the number of links in, and lack of transparency of, the food chain. He had located 'alternatives' back in Minnesota in the form of farmers markets and organically certified produce, but had found the comparatively high retail prices of these foods prohibitive. In comparison, it was 'easy to be a locavore in Cuenca: the food just hasn't travelled so much so, well, it's cleaner'.

Kiki has not been through the same life-threatening health scares as Steve, but her reasons for migrating are similar, albeit more overtly politicized. A strict vegetarian who has a strong interest in yoga, alternative remedies and natural healing practices, she consistently referred to the disproportionate political and economic power of Big Pharma, and the ways in which alternative health practices, and their practitioners, are consistently undermined and marginalized in the United States. A significant proportion of my conversations with Kiki, and her close circle of friends, were thereby given to the relative merits of alternative diets and eating practices, and the manner in which multinational pharmaceutical and food production companies, with the support of the state, attempted to limit these. Like Steve, Kiki is also searching for 'a simple life', and she explained that she was trying to forge a new way of living in which she would be able to control her food and water sources, and know the 'clean' provenance of the food she ingested, concluding 'in cases of political, economic and/or natural disasters, people in the cities go crazy, but in the country I'll be able to live quietly, away from this mess and the political and corporate powers that try to control our lives'.

Steve and Kiki are not alone in their ambitions to lead a more simple life. For, although the privileged migrants with whom I worked had different life-stories and personal catalysts for making the life-changing decision to migrate to Cuenca, all cited their aspirations of eating a diet free from pathogens and pesticides, and were looking to source food locally grown by smallholders rather than that which was industrially produced. The consistent valorization of organic foods is a recurring theme, and the range and forms in which this finds expression is testament to the value that pesticide free food has in everyday migrant lives. Organic, local and small-scale production are not, of course, synonymous and, far from being diametrically opposed to industrial agriculture, organic farming, in some contexts, has been appropriated by major players in the food industry and can replicate the agricultural relations of production it originally set out to challenge (cf. Belasco 1989; Buck et al. 1997; Guthman 2004; Marsden and Arce 1995). Yet, as Coles (this volume) also notes, consumers often collapse these categories into each other, and the privileged migrant community in Cuenca are no different: organic = local = smallholder = simple.

I first came across the unquestioned value attached to organic food in an IL article in which a privileged migrant extolled the virtues of living in Cuenca by stressing that the city 'is a place where you can reconnect with the Earth' and 'eat organic fruits and vegetables … [sic] because that's what sold in the markets' (unattrib. IL Website 'Why Ecuador'). This perspective was further reflected in my interviews with migrants, as Ian explained:

> I like the fact that fruits and vegetables here do not contain pesticides and insecticides because of the elevation. Also many restaurants serve grass-fed beef. Because they are raised freely and not fed with chemicals I would eat [meat] here, whereas I wouldn't eat it back in the US.

Likewise, on meeting Zelda for the first time and being introduced as somebody 'who knows about food in Cuenca', I was immediately asked about grass-fed beef and the agricultural use of pesticides. Zelda understood that most, if not all, fruit and vegetable production was organic, while her partner Quentin was excited about the 'abundance of pesticide-free, organic fruit and vegetables that you can find in markets and being sold by those peasant women in the street'. Yet it was not just in interviews that I found organic food being championed; on ordering a *mojito* cocktail during a 'Gringo Night' drinks evening, Jimmy's biggest concern was to ask 'is the mint organic?', and a favourite theme of discussion during social dinners was the difficulties of procuring certified and trustworthy organic produce. Consequently, during the course of my research I found myself being drawn into a social network in which tidbits of information about possible sources, and the trustfulness, of organic foods were exchanged. These networks of information exchange become critical for privileged migrants in their attempts to eat simple produce as, contrary to IL's assertions and migrants initial images of Cuenca, organic food, as migrants understand it and need it to be presented, is not as readily available as they wish.

The realization of the scarcity of organic food is often cited by privileged migrants as one of their greatest disappointments and one of their most pressing practical concerns. Consequently their definition of a simple life is being constantly renegotiated, as they look to reconcile their expectations and desires with the reality of everyday eating experiences. This process of renegotiation requires, in the language of Beck (1992) and Giddens (1991), the management of risk that, in turn, involves the assessment of knowledge – and their sources – and the allocation of trust. As Caplan states, 'eating has become a risky business' (2000: 187). For privileged migrants, who have left their homes and established networks and are entering new, often unknown, food spaces, the quotidian decisions of what to eat, where to eat it, and from whom to procure it are, I suggest, particularly risk-laden, and the question of who and what to trust are at the forefront of their everyday encounters with food.

Many migrants are unprepared for this negotiation of new risks, in part because they are unaware of the realities of living in Cuenca before arrival and have created an image of the city – and its foods – on idyllic representations in the media, but also because their initial move to Cuenca from the USA can be construed as attempt to better their social risk position (Beck 1992: 23). Both Kiki and Steve's commentaries suggest that privileged migrants comprehend their subject position in the United States as relatively weak in relation to the state and capitalist interests, and that these asymmetries are felt through their bodies. This is particularly acute, as Steve's experience indicates, during times of ill health. Thus the personal and political become entwined: Steve wants to avoid ingesting 'pathogens, salmonella and mad cow disease', but finds this difficult back in the US due to the levels of industrial food production, a lack of transparency and the economic cost. Similarly, Kiki feels subjected to the ambitions of Big Pharma and industry that, in conjunction with the state, restrict her alternative wellbeing and clean food choices. Both, therefore, like Ian, Zelda and Quentin, amongst others, are attempting to gain more control over what goes into their bodies in the form of food. Thus, their migration to Cuenca can be understood as an expression of agency in which they are attempting to alter the terms of their engagement with food, and its providers, by relocating to a place that, as they understand it, has an abundance of clean, simple produce.

In this respect, privileged migrants echo the reflexive consumption politics underpinning natural food (cf. Belasco 1989), and the 'Not In My Body' movements (cf. DuPuis 2000), as they seek to reassert control over, and minimize the risk to, their own bodies, thereby improving their physical wellbeing. Concerns over others, in the form of long-term ecological sustainability and a desire to establish more equitable relations of production, as commonly located in fair trade, organic and community supported agriculture movements, may be less pronounced in migrant discourse, which appears predominantly concerned with the self, but it is no less political. As Goodman's (1999) actor-network approach to food studies suggests, eating is embedded in an everyday politics in which the claims and counter-claims of actors are evaluated and contested and, as DuPuis (2000) notes in reference to organic milk, consumers do not have to be social activists to shape the relations of food production. DuPuis charts how eating involves consumers negotiating what they will or will not put into their bodies (ibid.: 289) and by refusing to consume non-organic milk, due to the concerns over rBGH (recombinant bovine growth hormone), shows how they contest notions of risk; a practice that subsequently draws them into the formation of the organic milk industry. These politics of refusal can also, I suggest, be identified in migrants' relocation to Cuenca. Some, for example Kiki, may formulate their move in more explicitly political terms than others, but all, in looking to minimize their risk position, participate in networks in which claims about foods are made, evaluated and enacted; a process that, in turn, constructs Cuencano food spaces and actors as either 'safe' or 'risky', and potentially leads to the formation of new markets.

Not Alongside My Body: The Antithesis of Commensality

At first glance, the city of Cuenca and its rural environs appear to be the perfect site in which to pursue aspirations for simple, clean food. Set at an elevation of 8,200 feet, the city's climate is relatively temperate, the infrastructure is good, and an absence of biting insects helps ensure the region does not suffer from the outbreaks of seasonal disease which plague its lowland neighbours. The area further boasts a thriving dairy sector and the surrounding countryside is fertile and productive. Low cost fresh produce, including corn, potatoes, cabbages, peas, beans, blackberries and apples, when in season, are widely available in local markets, as are a range of tropical fruits imported from the coastal regions and Amazonia lowlands. This is, however, only one half of the story.

Due in part to its historical position at the heart of the export-orientated Panama hat industry, the greater Cuencano region has been outwardly orientated since the 1930s, and the region has been the site of persistent outward migration and remittance-fuelled 'development' since the 1960s (Jokisch and Pribilsky 2002; Kyle 2000). Moreover, IMF/World Bank-led restructuring policies which encouraged the rolling-back of the state, together with national economic crises in the 1980s and late 1990s, were keenly felt in the rural economy, and the resulting proletarianization of the peasantry, consolidation of landholdings, and opening up of markets to international capital have inexorably altered patterns of food production and consumption in the region. Today small agricultural plots are making way for the semi-urban sprawl of new housing developments, new malls containing global fast-food chains and supermarkets are opening, and the state is looking to reverse the trend of diminishing agricultural yields by incentivizing the use of chemical fertilizers (el Tiempo 2010).

As I have argued elsewhere, Cuencanos are generally rather relaxed about these changes and are skilled in negotiating a path between reasserting their 'traditional' and 'local' food practices and embracing 'modernity' and the 'exogenous' in the form of a Kentucky Fried Chicken (KFC) bargain bucket (Abbots 2012). Global fast food chains, and their local counterparts, are popular with a range of Cuencano social groups and across all generations, yet the one social group who remain conspicuous by their absence are privileged migrants, who roundly condemn this perceived 'McDonaldization' (cf. Ritzer 2010). The presence and local popularity of KFC and Burger King (BK) are a particular source of consistent dismay for migrants, as Xenia explained, 'the local people, well, they're just not as far along as us in realising the dangers of these foods' and Carol exclaimed 'we're creating monsters in these countries; people don't realize what they're eating, what harm they're doing to themselves with all those fats and poisons'. Consequently, although they commonly frequent the malls in which the chains are located, privileged migrants never, to my knowledge, purchase or consume food from these restaurants.[3]

Although it is tempting to assume, based on scholarship on the role food plays in reaffirming migrant identities and sense of belonging (cf. Abbots 2011; Harbottle 2000; Kershen 2002; Ray 2004), that North Americans will gravitate to the food of their home country, the self-distancing of privileged migrants from global fast-food restaurants is not unsurprising when their views on complex, industrial foods are considered. The relations of food production within these chains may differ significantly from the image circulated in popular discourse (Striffler 2005) but, for privileged migrants – as expressed by Steve's 'well-travelled' hamburger and Carol's assertion that eating the foods of KFC and BK is tantamount to self-harm – they are the antithesis of simple, clean food and symbolically epitomize the industrial chains, and its associated pathogens, from which they are trying to escape. The dangers to which Xenia refers are not, however, limited to the foods and the bodies that ingest them, but are conceived as being multi-dimensional: they erode the sites of simple food production, displace the local, and supplant tradition[4] Privileged migrants' rejection of the global fast-food chains thus appears consistent with their food politics and their

3 The global fast food chains are one of the few areas in which privileged migrants discourse and their practices were consistent.

4 See Abbots (2012 and forthcoming a) for a broader discussion of the perceived 'dangers' and the attractions of fast-food chains in the greater Cuenca region.

aspirations to manage risk. Yet their self-removal from the shared eating spaces of these restaurants has broader social consequences, by creating bodily distance from the incumbent Cuencano population.

The role that the exchange of food and commensality plays in establishing social relatedness has been well documented (cf. Carsten 1997; Saleh this volume; Strathern 1988) and, although these discussions have commonly been concerned with the domestic sphere and kinship relations, and have centred on the active sharing of the same food, rather than the incidental sharing of the same food space, their suggestion that food constitutes coterminous bodies can be extended to a restaurant setting. For these works all demonstrate that eating forms persons and enacts social relations. Broadening this argument, I suggest that eating bodies which share the same food space thus encounter each other and, in doing so, become related. These relations have different levels of engagement and may be fleeting; eaters within the restaurant space may never acknowledge their relatedness, or seek to develop it within or outside of the restaurant space. Yet the act of consuming and ingesting the same food in the same space establishes an incidental encounter between bodies, and by removing their bodies from this potentiality, privileged migrants are denying the possibility of relations with other consumers, much in the same way as refusing a gift of sharing food is to deny a social relation (cf. Mauss 1990). Thus, migrants' politics of refusal does not just pertain to the products of fast food chains, but extends to the consumers of these products.

By consciously removing their bodies from the same sites of eating as many Cuencanos, privileged migrants are consequently the architects of their own social estrangement. This does not occur in all food spaces, and elsewhere I have discussed the emergence of cosmopolitan eating spaces which are shared by middle- and upper-class Cuencanos and migrants (Abbots, forthcoming a), and later in this chapter I discuss the clean coterminous bodies that move through the sterile environment of Supermaxi. But first I turn my attention to the ways that migrants further establish social distance through their rejection of foods from markets and local restaurants, and explore some possible motivations informing this process. For, the rejection of global chains and industrial food does not automatically equate to privileged migrants seeking out 'alternative' food spaces situated at the other end of the spectrum – the restaurants and markets seemingly selling local and simple food. Migrants may commonly extol the virtues of Cuencano markets and local produce, but a survey of their shopping habits demonstrates that they do not acquire their food, or eat, in these spaces. Rather, the majority of their everyday food purchases are made in Supermaxi, the national supermarket chain (whose outlets are located in the same malls as KFC and BK): a practice that appears to be somewhat contrary to their food aspirations and politics.

Who and Where to Trust?: Evaluating Risk through the Body

Caplan (2000) has argued, in the context of BSE scares in Wales and England, that consumers look to localize their eating practices. In other words, when faced with risks, in the form of toxic food that has the (unknown) potential to damage bodily health, social actors attempt to minimize the danger by purchasing and eating food of which they are confident that they know the source, either in the form of producer or retailer. She thereby notes that it is not just knowing where food comes from, but also knowing whom one buys it from that is of import (ibid.: 192), and thus she draws attention to the centrality that established social relations play in evaluations of safe and dangerous foods, as well as indicating the critical role trust plays in these evaluations. The issue, however, that privileged migrants in Cuenca face is that, unlike the members of Welsh farming communities discussed by Caplan, migrants do not have established social relations with local producers and retailers nor the wherewithal to develop them. The majority of my research participants, particularly those whose relocation to Cuenca has been orchestrated by an external agency, have very little knowledge and experience of the city, Ecuador, and even Latin America more generally. Consequently they are commonly lacking the language skills, social networks, and cultural knowledge required on which relations of trust can be built.

This lack of cultural knowledge further extends to understanding and recognizing the food for sale at markets, a point made by Abi who explained:

> I wish they [International Living] offered tours around the markets: I don't know what all these strange fruit and vegetables are; I've got no idea how to cook them or eat them, what I should pay for them – or even if they're safe! We need someone to show us – we don't know this food. I get all my food from Supermaxi; that way I know what it is and how much I'm paying for it. I recognize the food there.

Relations of trust with Cuencano food producers and retailers are therefore difficult, although not impossible, for privileged migrants to establish, and the issues they face are further compounded by a lack of transparency in market negotiations. Cuenca's food markets appear distinctly confused; non-organic produce is, at times represented as organic (and vice-versa) and industrially produced foods are often marketed and sold as if they were the products of local and small-scale agriculture. This confusion stems, in part, from the long resale chain that culminates in the produce of multinationals being sold by street and market vendors,[5] as well as

5 Ironically, this produce may also have originated from small farms and been purchased by multinationals through contract farming arrangements. See Hellin and Higman (2003), Lyon (2010) and Ormond (this volume) for further discussions of contract farming.

a less rigorous system of regulation than privileged migrants are accustomed.[6] It is also exacerbated by vendors' own marketing techniques: until recently, the term 'organic' was rarely, if ever, in the lexicon of small retailers but now Luis the barman responds to Jimmy's concern over the mint in his *mojito* with smiling assurances and confirmation that 'yes, of course it's organic; I know how to look after my customers', and it is not uncommon to be stopped in the street by a fruit seller, be offered strawberries of extraordinary size, and be told 'they're organic, from here; the best quality with no chemicals'. Thus, while Quentin regards the women standing on the street corners dressed in the 'traditional' clothes of a Cuencana peasant selling fresh fruit from wheelbarrows and baskets as symbolic of an uncomplicated food chain,[7] a closer look commonly reveals the branding of Del Monte and Dole lying beneath the façade. To add to the confusion, fruit and vegetables produced on smallholdings without the use of pesticides, that may well be organic in essence, are not constructed and represented as such in the formalized and certified manner that privileged migrants recognize and are also, at times, sold alongside imported, non-organic produce.

Deciding in whom to invest their trust is thereby a potential minefield for privileged migrants, with many feeling that sellers were taking advantage of their cultural and social incompetence. As Steve explained when discussing street sellers:

> I feel, in my heart, that it's [their produce] probably not organic; I mean strawberries shouldn't be that size! But these local producers, they need to produce as much as possible, they need to make money, so you can't blame them for telling you what you want to hear.

Yet Steve was also one of the few participants in my research who occasionally ate fruits purchased from street vendors, telling me 'I want to believe it's organic, so I tell myself it is. My head won't quite let me at times, but I keep trying to believe 'em!' This Cartesian battle between head and body is not uncommon among privileged migrants, with Yvonne also explaining that she 'wanted to buy into this whole nature's basket and it's all natural thing, but it's a struggle when you know it's not', and it is, in many ways indicative of the disjuncture between migrants' aspirations and the reality in which they find themselves. Yet, as Steve

---

6 Systems of organic certification do exist in Ecuador, but the majority of producers who are engaged with certification are generally orientated towards the export market, for example Dole Organic. Hence, the availability of organic produce that is certified as such is not only extremely limited in Cuenca, but is also commonly the produce of agro-industry. See, for example, Belasco (1989), Buck et al. (1997), Guthman (2004), Marsden and Arce (1995) and Lyon (2010) for further discussions on the politics of organic production.

7 I have discussed the image of the *Chola Cuencana* and the ways she embodies country – city relations in more detail elsewhere (Abbots forthcoming b). See also Weismantel (2001 and 2003).

indicates, a number of migrants, albeit a minority, persist because, 'we've made this life-choice now'; a point neatly surmized by Yvonne who concluded; 'I'm here, I've got to eat – and the alternative [back in the USA] is so much worse!'

Steve and Yvonne's eating practices are based on an evaluation of relative risks, but their comments also suggest that this process is informed by their current social position, both real and imagined. Both felt that, after investing so heavily (economically and emotionally) in their dreams of 'a simple life', they had little option but to follow the path they had 'chosen', even when it challenged their instincts and intellect. This battle with the self further indicates, however, that migrants, at times, are making conscious, reflexive decisions to establish relations of trust with Cuencano food retailers, even when it potentially carries a risk of ingesting chemicals. This is further demonstrated by Bryan, one of the pioneer privileged migrants who relocated to Cuenca over four years ago. While still being a keen supermarket shopper, he, along with his wife, Carol, were keen to stress the frequency of his market visits, with Carol proudly telling me that 'the women know Bryan in there now'. Bryan himself continued:

> I go to the same two or three stalls, and have been for some time. They even let me fill the bag now, you know you get four avocados for a dollar, but they fill the bag; well I like to select which ones I want – and that might mean me giving them a dollar for three – but they're the ones I want. I choose each individual strawberry too.

Over time Bryan has developed relations of trust with two to three market vendors not only by repetition and familiarity, but also by changing the terms of engagement. He asserts control in the market; selecting the fruits he wants at the price he is willing to pay. The couple repeatedly told me how comfortable Bryan was in the market and emphasized his close personal connection to the vendors, using this as the means by which to demonstrate their integration into Cuencano society and openness to new cultural experiences which, in turn, from their subject position, distinguishes them from other 'more recent – the International Living' migrants. To borrow from Caldwell (2004), Bryan has thereby 'domesticated' the foreign space of the Cuenca market, transforming the unfamiliar and risky into the known and trustworthy. Yet he has not just done so by symbolically appropriating Cuencano food spaces and using them as a place for intimate celebrations, nor by bringing Cuencano foods into the private space of his home. Rather, he has physically domesticated Cuencano markets, and their produce, through his embodied engagement with them; drawing the food of the market into his body and making it his own.

Establishing relations of trust with local food spaces and retailers may, in some contexts, be a case of mind over matter, but privileged migrants also discover that they cannot negate their bodies altogether in their evaluations of risk. For the body plays a significant role as a test-bed for the relative safety of foods: Bryan continued shopping in the market as he had suffered 'no ill effects' from the fruit purchased from his trusted vendors, whereas Josie's story is somewhat different. She avoided all local foods, telling me that this avoidance was due to the manner in which her body had gone into 'a tailspin' since arriving in Cuenca. Josie and I first met during dinner at a higher-end Italian restaurant, and I was initially surprised when her order only comprised bottled water. She proceeded to remove from her bag, and eat from, a plastic container filled with a seaweed and cider vinegar mix, and explained she was following a restrictive 'body ecology' diet:

> When I arrived, I ate loads of fruit – I mean I concentrated on the fruit you could peel, or I washed it – but my body just went into a tailspin; I guess it's just that there are so many germs here and my body isn't used to it. So now, the only way I can get by is by being on this diet. I need to sort out my intestines, y'know, the balance of them – they've been put completely out of whack here.

With the exception of quinoa, which is produced in the Andes but rarely consumed by Cuencanos, Josie is unable to source the food her diet requires her to eat, and consequently her husband made frequent trips back to the United States in order to fill empty suitcases with sea vegetables and supplements. This practice of importing food, especially supplements, is not uncommon and forms part of migrants' exchange networks, but I first wish to address Josie's concerns over germs.

The 'Dirty' Market: Avoiding Pathogens and Pollution

A dominant theme in the food discourse of privileged migrants is the manner in which they perceive Cuencano food markets – and often, by extension, the people contained within the market – as dirty and dangerous sites;[8] a perspective typified by Toni, who stated that 'we have been to a couple of the open-air markets but they are crowded, confusing, and really dirty. They are interesting but we do not buy anything'. She continued to explain that she had suffered severe food poisoning, which she attributed to the one occasion she had eaten food purchased from a local vendor, and, as a result, would 'never eat anything sold in those places again – it's just not safe'. Like Josie and Bryan, Toni judges the risks of certain foods through the effects on her body, but her evaluations are premised not only on *what* food she is eating, but also *where* it has been purchased. She consequently drew attention to the dirt of the market and, through this framework, renders the unseen

8 Markets are often seen as dangerous sites that present a risk to personal safety (cf. Carrier 1997). This extends beyond food to the more immediate risks of robbery and violence, and there is some evidence to suggest that privileged migrants imagine Cuencano markets in this way. However, while this may influence the eating practices of some, as Valerie suggests, others visit the markets 'out of interest' while still continuing to refuse to eat any of the food sold there.

intangible threat of germs visible and material. Food from markets is thus deemed unsafe because it is from the market; a place of dirt and chaos. However, not all food receives the same treatment, and some are understood to be less susceptible to the dirt of the market, or in other words potentially less risky, than others. This is seen in the eating practices of Bryan and Steve. Both will, through varying acts of trust, purchase fruitsthat can be washed or peeled from markets, but when asked if they would purchase meat from the same sellers, they responded with a resounding 'no'. As Carol told me; 'we're fussy about meat'. Trust, it seems, can only extend so far.

Privileged migrants are not, however, avoiding Cuencano markets solely because they understand them to be dirty and the food potentially harmful to their bodily health. Rather, their estrangement from these spaces, I argue, indicates a deeper cultural aversion to the local population and a desire to create bodily, and consequently social, boundaries. This is indicated in the broader concerns that migrants have about the transmission of disease and germs, which results in their extensive use of hand washes and sprays, bleaches and cleaning products. As Kelly-Mary stated; 'I always use one of those anti-bacterial hand washes after being in any public space because, well, you could pick up anything there'. Kiki agreed, explaining that she always carried a spray with her and used it while on public transport, after handling money and always before eating outside of the home. The implication is that local bodies, as well as local food and markets, are dirty and, hence, risky. The fear of foreign bodies thereby extends beyond pathogens to include the bodies of Cuencanos, with migrants erecting barriers – in the form of anti-bacterials – to protect their own bodies from 'contamination'. As Douglas (2002) has eloquently argued, these fears of bodily pollution symbolize broader concerns regarding the social body, and the reassertion of bodily boundaries can be interpreted as a desire to affirm social boundaries and maintain purity. Thus, while the factors informing migrant estrangement from Cuencano bodies in markets follows a different, pathogen-laden path from their distancing from those same bodies in fast-food chains, their core aspiration – that of controlling the risks to their own bodily health – not only leads to the reproduction of bodily boundaries, but also, by reaffirming social boundaries, can be understood as producing their own social categorization.

Not all bodily reactions to Cuencano foods are as visceral and immediate an experience as Toni's and not all foreign bodies can be rendered material through dirt. As Steve's quandary demonstrates, pesticides present privileged migrants with a more oblique risk than pathogens: they remain hard to discern and any potential ill effects on the body are felt over a longer time frame. Yet, they are not totally imperceptible. In making the statement 'strawberries shouldn't be that size', Steve draws on his experience of eating organic strawberries from his most trusted source in the USA – farmers' markets – and draws on this experience and knowledge to evaluate Cuencano 'organic' strawberries. A large part of this process is evidently visual, with the fruits being judged on size, but the embodied act of eating also continues to play a role. Steve told me he was disappointed with the sweetness of the strawberries, finding them 'tasteless' and 'watery', and to make these quality judgements he used his body, measuring the taste of the strawberries against known organic produce, and ultimately finding it lacking. Thus taste, in addition to the look of the fruit and his relation with the seller, becomes a mechanism for judging whether it is, in his view, organic or not, and consequently the practice of ingesting and tasting food is a mechanism through which risks, in conjunction with social relations, are evaluated, however invisible they may be.

Strawberries are not unique in this context, and food doesn't have to taste 'good' to be deemed organic by migrants. For the converse occurs with the local grass-fed beef, a common symbol, as indicated earlier, of a simple food. Yet migrants commonly find it 'tough' and 'unpleasant', a perspective illustrated by Bob who complained, 'it's so chewy, it can take you half an hour to get through a piece; it's because they don't feed the cows corn, and you need corn to put some fat on that meat and make it nice and tender'. In migrant discourse on beef, being chewy and tasteless is equated to being grass-fed which, in turn, is associated with simple food chains and the absence of pathogens and industrialization. Thus, while it may not be to the personal tastes of the privileged migrant population, being unpleasant and chewy bestows beef with a particular 'simple' quality that indicates lower risk. As such, trust and an absence of risk become associated with particular material qualities and sensorial experiences.

As suggested in my discussion of markets, the risk that grass-fed beef poses increases significantly when purchased from a Cuencano market, and thus far I have to attempted to tease out the multiple ways through which privileged migrants negotiate risk through their eating practices, tracing the interplays between foods, the sites in which they are purchased and consumed, social relations, and embodied reactions. It has become evident, I hope, that a tension exists between migrants' aspirations and their eating practices, as they discover that eating 'simple' food involves the necessary engagement with, and often the introduction of, foreign bodies (in all their forms). The resultant risks to their own (also foreign) bodies, both real and imagined, has to be managed – migrants do, after all, have to eat – and I close my discussion by turning to the foods and food spaces they consume, rather than those they avoid, as they seek to regain control over the substances they ingest.

Clean, Sterile and Safe: The Lure of Supermaxi

Privileged migrants are thus faced with a dilemma. They left the United States in search of simple food that would reduce their exposure to risk, but this relocation produces new, often unforeseen, risks that also potentially endanger the body. Consequently, they turn to familiar safe territory and the food spaces that appear to present the relatively lesser risk: the supermarket. Whether it is Bryan buying selected foods from trusted women in the market, Steve trying to trust the words and produce of a local vendor, or Josie following a restricted diet, the one food space that all my migrant research participants had in common was Supermaxi, the national Ecuadorean supermarket chain. Commonly situated in malls located on the inner fringes of the city, the stores would not be out of place in the USA, and they boast well-stocked shelves, wide aisles and checkouts, and a bright and spacious environment. All the self-serve foods available, with the exception of some fruits and breads, are prepackaged, all are clearly labelled, and their prices prominently displayed. It is therefore tempting to view Supermaxi's attraction for migrants purely in terms of convenience which, as Reardon and Berdegué (2002) have identified, is one of the primary reasons, together with time pressures, urbanization and increased wealth informing the rise of supermarket culture. Yet, while not disregarding this self-evident convenience, as well as the transparency of pricing that is, for example, so important to Abi, I suggest there are other factors influencing Supermaxi's popularity among privileged migrants.

Toni buys nearly all her food in Supermaxi although her other 'favourite place is a very modern and clean Italian deli' and in our conversations she contrasted these two food spaces to the market, creating a striking dichotomy between the sterile safe surroundings of the former to the dirty risky environment of the latter. The association between Supermaxi and sterility was further made by Mitch, who explained the attraction of the supermarket as having 'something to do with the type of architecture: malls and supermarkets, well, they have that wonderfully sterile environment, don't they?' He continued to elucidate his patronage by challenging the 'convenience' thesis: 'it's crazy really, I have all the time in the world now, I can shop every day and have no pressures on my time, and yet I still go to Supermaxi – what can I say? I'm Western!' The supermarket thereby appears to invoke a certain localized familiarity in migrants, and this, following Caplan (2000), helps facilitate relations of trust. This is reinforced by the perceived cleanliness and modernity of the space, although I suggest the sterility to which Mitch refers is not solely premised on Supermaxi's scrubbed tiles, neat displays and tightly wrapped foods, but is also related to the lack of foreign bodies, and the chaos they are perceived to create. Cuencano bodies are not totally absent from supermarkets – they are evidently present in the form of customers and staff – yet they are more ordered and clean in this context; smartly uniformed check-out assistants replace the peasant, lower-class market vendor, and customers glide between the aisles behind their shopping trolleys and wait in line instead of jostling through narrow spaces and loudly bargaining. Hence the bodies that are present are more familiar, or in other words, less foreign, to privileged migrants. Familiarity, and the associated diminished risk, in this context does not just refer to the produce and space, but also the bodies who handle the produce and move through this space.

Relations of trust with Supermaxi are further enabled by migrants' desire for organic produce, as the chain is one of the few sources of certified organic produce, as Carrie highlights. Carrie is a big fan of the supermarket and has 'struggled' with the lack of certified food available in Cuenca, telling me that she didn't 'trust the street-sellers' and didn't know 'where else to go to get the food I want; I want to know it's organic and that's impossible here unless I go to Supermaxi'. Carrie is therefore looking for guarantees and for her this takes the form of a recognizable brand that certifies the food's organic status through recourse to regulatory agencies and the state. Her position is therefore somewhat paradoxical: in order for her to trust her simple food she needs it, and its producer, to be subjected to a complex system of governance and regulatory procedures. In evaluating her risk position, Carrie places more trust in formalized certification and Supermaxi than the word of a Cuencano market-seller. Yet only a very limited range of produce is certified organic and visible regulations do not extend to all the foods on the supermarket shelves. Hence the fear of pathogens, and the need to control it, persists and Carrie, like other migrants, looks to minimize the risks associated with fresh fruit and vegetables through sterilization.

The best methods for sterilizing food are commonly shared between migrants, and many used the 'special wash sold in Supermaxi' and a peroxide solution for this purpose. They thus look to depollute produce by similar methods as those used to ensure their bodies are not 'contaminated' by germs in public spaces, and revert to chemicals as the means of cleansing. However, not all foods lend themselves to being cleaned in this manner, meat being especially problematic, and a number of migrants, like Josie above, rely heavily on multi-vitamins and diet supplements, often imported from the United States, to counter-balance the potentially harmful effects of ingesting risky foods and pathogens.[9] In addition, a small number advocate vitamin C injections, of which Penny is a fan. She explained she 'have never felt so good' after following a course of injections that were designed to counteract the pollutants in her body: she would not, however, eat a local Vitamin C-rich mango. When viewed through the lens of risk this preference is understandable; the mango can potentially increase the pollution levels inside Penny's body with pesticides and pathogens whereas the pure injections defend the body from such harm.

Penny's practices are reminiscent of Strathern's (1988) account of gift-giving in Melanesia, in which women feed yams to male migrants in order to strengthen and replenish their bodies, which have been weakened by the consumption of the exogenous foods of rice, tinned fish and corned beef. In this work, Strathern shows that symbolically resonant foods can counteract the potentially dangerous foreign foods consumed by migrants, helping minimize the risk that they will assimilate into their host country. Her argument thus points to the social and cultural risks that consuming foreign foods can pose and, in indicating the ways these dangers can be neutralized by ingesting home substances, she elucidates the interplay between the body and the social. In this chapter I have attempted to trace this relation between the individual and social body by exploring how the practice

9 The high rates of import duty that they have to pay on multi-vitamins and diet supplements are a consistently common complaint among the privileged migrant population. A number hire 'baggers' to carry vitamins across the border on their behalf.

of eating is fraught with dangers for privileged migrants. Some of these dangers are visible, while others are not. But by examining the everyday encounters between foods and migrants', and others', bodies I have aimed to provide an insight into migrants' broader concerns over intimacy with the foreign, as well as the ways in which their establishment of bodily boundaries indicates a process of social and cultural distancing and estrangement. It is perhaps not surprising that many migrants turn inwards, towards the non-foreign (from their subject position) bodies of other migrants, and the trends of restricted networks of food exchange, communally held land and food co-operatives (for self-consumption) are beginning to emerge. In addition, a number of migrants are opening up their own commercial food spaces in which to satisfy their ever-increasing demand for safe food from a trusted source.

Yet, to portray privileged migrants as inherently insular would be a misrepresentation. Many have made a significant, emotional as well as financial, investment to realize their aspirations of a simple life and, as Bryan and Steve indicate, a number make considerable efforts, at times contesting their bodily instincts, to live that dream and engage with Cuencanos and their food spaces, even when it involves risky eating. This engagement entails a process of domestication of Cuencano food sites and the consequences of this are yet to be seen. There are hints, in the exclusive 'Gringo Night' dinners and the emergence of cosmopolitan restaurants, that migrant intimacy with Cuencano food spaces may be a case of appropriation rather than integration, but these new sites, while excluding some, commonly lower-class, Cuencano bodies also provide a space that migrants and, higher-class, Cuencanos intimately share.[10]

Conclusion: Who, What and Where to Trust?

Or, in other words, who, what and where presents the lesser risk?

As I indicated at the start of this chapter, privileged migrants, while appearing from the outside as a distinct, homogenous social group, are, to some extent, heterogeneous in their food practices. All may shop consistently in Supermaxi and avoid the multinational fast food chains, but in between these two poles of engagement and refusal, there are multiple relations of trust. Some, like Steve, consciously decide that, as they have invested so much into their new lives, they will make concerted efforts to trust Cuencano retailers and their produce, despite what their bodies, instincts and intellect tell them. Others, for example Bryan, domesticate unfamiliar spaces and foods, making them familiar and trustworthy by establishing social relations with vendors and ingesting their wares. Whereas some, as exemplified by Josie, turn inwards towards restrictive diets, imported foods, supplements and migrant exchange networks. There are therefore varying degrees of engagement with Cuencano food spaces, and multiple relations are

10 See Abbots (forthcoming a) for further discussion.

simultaneously created and denied through migrant eating practices and food preferences. Yet these relations are not mutually exclusive, and privileged migrants are in a constant process of renegotiating their notion of a simple life with the realities of their everyday, embodied lived experience. Eating is a risky business, and assessments of the relative risks are informed by conceptions of people, places and foods: the decisions of who, what and where to trust are not only founded on experience, knowledge, and personal politics, but are also shaped by everyday social relations. Evaluations of trust are further mediated through the body, either in its visceral reactions to foreign food or through the less dramatic experience of taste, texture and smell. Deciding what, where and how to eat is thus a complex practice in which the body and the social interplay. Food decisions, moreover, create intimacies and encounters between coterminous eating bodies, while also negating social relations, or their possibility, with others. In some contexts, as in the case of migrants refusing to consume in KFC and BK, these politics of refusal can be explicit, although the negation of a social relation with other consumers can be incidental. In other scenarios, for example distancing one's body from a market because of fears over dirt and pathogens, or using a hand-spray, the politics of refusal and self-estrangement may not be as clearly articulated, but, I suggest, are still very much present. As Goodman (1999) highlights, eating is a political act and, through their evaluations of safe and unsafe foods, vendors, and spaces, privileged migrants are not only creating new social boundaries and relations, but in doing so are also participating in, and helping shape, the changing foodscape of Cuenca.

References

Abbots, E.-J. 2011. 'It doesn't taste as good from the pet shop'; Guinea pig consumption and the performance of class and kinship in Highland Ecuador and New York City. *Food, Culture and Society* 14(2), 205–24.

Abbots, E.-J. 2012. The celebratory and the everyday: Guinea pigs, hamburgers and the performance of food heritage in Highland Ecuador, in *Celebrations: The Proceedings of the Oxford Symposium on Food and Cookery 2011*, edited by M. McWilliams. London: Prospekt Books.

Abbots, E.-J. forthcoming a. The fast and the fusion: Class, creolization and the remaking of *comida típica* in Highland Ecuador, in *Food Consumption in Global Perspective: Essays in the Anthropology of Food in Honour of Jack Goody*, edited by J. Klein and A. Murcott. Basingstoke: Palgrave Macmillan.

Abbots, E.-J. forthcoming b. Embodying country-city relations: The figure of the *Chola Cuencana* in Highland Ecuador, in *Food and Foodways in the Country and the City* (working title), edited by N. Domingos, J. Sobral and H.G. West.

Beck, U. 1992 [1986]. *Risk Society: Towards a New Modernity.* London: Sage Publications.

Belasco, W.J. 1989. *Appetite for Change: How the Counterculture Took on the Food Industry, 1966–1988*. New York: Pantheon Books.

Bourdieu, P. 1984. *Distinction: A Social Critique of the Judgement of Taste*. New York and London: Routledge.

Buck, D., Getz, C. and Guthman, J. 1997. From farm to table: The organic vegetable commodity chain of Northern California. *Sociologia Ruralis* 37(1), 3–20.

Caldwell, M. 2004. Domesticating the french fry: McDonalds and consumerism in Moscow. *Journal of Consumer Culture* 4(1), 5–26.

Caplan, P. 2000. Eating British beef with confidence: A consideration of consumers' responses to BSE in Britain, in *Risk Revisted*, edited by P. Caplan. London and Sterling, Virginia: Pluto Press, 184–203.

Carrier, J.G. 1997. *Meanings of the Market: The Free Market in Western Culture*. Oxford: Berg.

Carsten, J. 1997. *The Heat of the Hearth: The Process of Kinship in a Malay Fishing Community*. Oxford: Clarendon Press.

Douglas, M. 2002 [1966]. *Purity and Danger: An Analysis of Concepts of Pollution and Taboo*. London: Routledge and K. Paul.

DuPuis, E.M. 2002. Not in my body: rBGH and the rise of organic milk. *Agriculture and Human Values* 17(3), 285–95.

El Tiempo (Cuenca). 2010. *Entrega de la urea se hará en Riobamba*. 29/01/10, page A8.

Fechter, A.-M. 2007. *Transnational Lives: Expatriates in Indonesia*. Aldershot, UK and Burlington, USA: Ashgate.

Giddens, A. 1991. *Modernity and Self-identity: Self and Society in the Late Modern Age*. Cambridge: Polity Press.

Goodman, D. 1999. Agro-food studies in the 'age of ecology': Nature, corporeality, bio-politics. *Sociologia Ruralis* 39(1), 17–38.

Guthman, J. 2004. *Agrarian Dreams: The Paradox of Organic Farming in California*. California: The University of California Press.

Harbottle, L. 2000. *Food for Health, Food for Wealth: The Performance of Ethnic and Gender Identities by Iranian Settlers in Britain*. New York and Oxford: Berghahn Books.

Hellin, J. and Higman, S. 2003. *Feeding the Market: South American Farmers, Trade and Globalization*. London: ITDG Publishing.

International Living '*Why Ecuador*'. [Online]. Available at: http://www.internationalliving.com/Countries/Ecuador/Why-Ecuador [accessed: 18 March 2010].

Jokisch, B. and Pribilsky, J. 2002. The panic to leave: Economic crisis and the 'new emigration' from Ecuador. *International Migration* 40(4), 75–101.

Kershan, A.J. 2002. *Food and the Migrant Experience*. Aldershot: Ashgate.

Kyle, D. 2000. *Transnational Peasants: Migrations, Networks and Ethnicity in Andean Ecuador*. Baltimore: The John Hopkins University Press.

Lyon, S. 2010. *Coffee and Community: Maya Farmers and Fair-Trade Markets*. Boulder, Colorado: University Press of Colorado.

Marsden, T. and Arce, A. 1995. Constructing quality: Emerging food networks in the rural transition. *Environment and Planning A* 27, 1261–79.

Mauss, M. 1990 [1950]. *The Gift: The Form and Reason for Exchange in Archaic Societies*, translated by W.D. Halls. London: Routledge.

Ray, K. 2004. *The Migrant's Table: Meals and Memories in Bengali-American Households*. Philadelphia: University of Pennsylvania Press.

Reardon, T. and Berdegué, J.A. 2002. The rapid rise of supermarkets in Latin America: Challenges and opportunities for development. *Development Policy Review* 20(4), 371–88.

Ritzer, G. 2010. *McDonaldization: The Reader 4th Edition*. LA: Pine Forge Press.

Strathern, M. 1988. *The Gender of the Gift: Problems with Women and Problems with Society in Melanesia*. Berkeley: University of California Press.

Striffler, S. 2005. *Chicken: The Dangerous Transformation of America's Favorite Food*. New Haven, Conneticut: Yale University Press.

Weismantel, M.J. 2001. *Cholas and Pishtacos: Stories of Race and Sex in the Andes*. Chicago: University of Chicago Press.

Weismantel, M.J. 2003. Mothers of the *Patria*: la Chola Cuencana and la Mama Negra, in *Millennial Ecuador: Critical Essays on Cultural Transformations and Social Dynamics*, edited by N.E. Whitten. Iowa City: University of Iowa Press, 325–54.

Interlude

Reflections on Fraught Food

Jon Holtzman

> Marry, and you will regret it. Do not marry, and you will also regret it. Marry or do not marry, you will regret it either way. Whether you marry or you do not marry, you will regret it either way. Laugh at the stupidities of the world, and you will regret it; weep over them, and you will also regret it. Laugh at the stupidities of the world or weep over them, you will regret it either way. Whether you laugh at the stupidities of the world or you weep over them, you will regret it either way … Hang yourself, and you will regret it. Do not hang yourself, and you will also regret it. Hang yourself or do not hang yourself, and you will regret it either way. Whether you hang yourself or do not hang yourself, you will regret it either way.
>
> Soren Kierkegaard (1843: 38)

Imagined interjections of the great Danish pessimist on this section's papers: Abbots: 'Migrate to Ecuador to seek out simple, organic food and you will regret it. Do not seek out simple, organic food in Ecuador and you will also regret it. Whether you seek or do not seek simple organic food in Ecuador you will regret it either way'. Aphramor et al.: 'Attempt to lose weight by achieving a calorie deficit through eating less and being more active and you will regret it. Do not attempt to lose weight by achieving a calorie deficit through being more active and eating less and you will also regret it. Whether you do or do not attempt to lose weight by achieving a calorie deficit you will regret it either way'. Saleh: 'Orient your vineyards around commodified wine production and you will regret it. Do not orient your vineyards around commodified wine production and you will regret it. Whether you do or do not orient your vineyards around commodified wine production you will regret it either way'.

Food is fraught. If food gives life, brings joy, and brings humans together in numerous forms of commensality, it may do exactly the opposite in equal measure. Eating can give birth to and express key facets about the self and key facets of how we in varying ways relate to others, rendering food – as this group of papers brings to the fore – an uncannily intimate personal and social object. Yet with intimacy comes both ambivalence and danger. As Wilk (2010) shows, the 'family meal' is not only a time-honoured rite connecting us to those with whom we are closest, but can be the arena for the most trying, horrific memories. Similarly, Batsell (2002) demonstrates that the most enduring, childhood 'flashbulb' memories involve episodes of 'forced feeding', when those who care for us attempted to compel us to eat meat, or spaghetti, or the now hackneyed spinach, likely in the belief that eating those foods would be best for us. In my own research among Kenyan Samburu

(Holtzman 2009) I have explored how a high quality and socially meaningful diet centered on pastoral products has largely given way to nutritionally poor and symbolically empty purchased agricultural products, this dietary change harkening cultural and social transformations that are greeted to a great extent with disdain at the same time that they are welcomed. The new foods – demeaned as 'gray foods' 'government foods', foods of the enemy or simply 'dirt in which is put this and that' – are seen as having brought on a whole host of problems ranging from physical and behavioural problems in children, liver problems in old men, promiscuity in women, patterns of disrespect among murran (bachelor warriors), and a general breakdown of love among community members. Yet access to these new foods has also ended the toll of death that famines brought by drought or livestock disease portended in the past, and has – in the words of one woman who had moments before condemned the dietary changes – made the everyday pursuit of food, made *life*, 'so easy'.

Developments such as those experienced by Samburu are, then, neither strictly good nor bad. They are both and neither, their deeply felt ambivalence emanating not because the issues at stake mean so little but because they mean so much. What then, of a young Samburu man, whose very identity as a moral person remains defined by prescriptions to eat food only outside of the gaze of women, preferably on livestock slaughtered in the bush, but now in an era of pastoralist poverty frequently eats maize meal porridge cooked with his age-mates in his own mother's house (sometimes after requesting she leave or be in a separate room that women have taken to constructing for that purpose)? Eating in the bush in an era with a dearth of livestock strikes many as an unnecessary, and sometimes prohibitive inconvenience; eating in the house renders one in the eyes of many a quintessentially pathetic figure. Thus, if a Samburu murran does eat food in the house he will regret it. If a Samburu murran does not eat food in the house he will also regret it. Whether he does or does not eat food in the house he will regret it either way. Such are the fraught predicaments of food.

Now, it is not my intention to reduce either this rich topic or these three fine papers to a Kierkegaardian bon mot. If one side of the coin of ambivalence is regret (or worse), the other side may be joy, satisfaction, happiness or pleasure. I emphasize the darker side here mainly because this aspect, made to varying degrees explicit in these papers, has been relatively absent on the splendid table that has disproportionately constituted the scholarly study of food. Relatively few studies emphasize negative aspects of food (apart from its absence or overabundance), much less bad food (cf. Stoller 1989). Possibly an anthropologist today should not complain if food remains a bit of a Pollyannaish bastion within a discipline that has shifted in recent decades from a celebration of human diversity to obsession with themes such as the Foucauldian power eminent in all things or globalized exploitation in its varied forms – the sinister aspects of human existence. If food can make a twenty-first-century anthropologist smile, and make our readers smile, perhaps we should savour that. We should. Yet food is one of the most central and complex objects of human life, full of joys, but also deep contradictions and ambivalences, a complexity that remains far from fully explored. Sadly, the celebratory, epicurean tone that too often surrounds the scholarship on food can, thus, obscure or foreclose the rich analytical possibilities that food offers.

The three papers in this section are anything but simple, and easy to categorize. Each is laden with its own set of ambivalences and contradictions, bringing to life both the joys and regrets that their subjects experience through food. The most clearly fraught of these topics is raised by Aphramor et al.'s rather pointed intervention into the debate on obesity, and strident advocacy for a Healthy at Every Size (HAES) approach to dietetics in the UK. As a minor critique, the extent to which Aphramor et al. take such a categorical stance on the correctness of their approach and the scientific and moral flaws of the 'body as bomb calorimeter' partially undermines the intrinsic tensions in her subject matter. It might seem upon reading this paper that fraughtness is, in Sartre's terms, 'Other People' – dieticians who impose a flawed and harmful model on an unsuspecting public – rather than something that is located in each and every one of our intimate personal and social relations constructed through food. Even were it the case that the science refuting the conventional emphasis on obesity/weight loss is as definitive as Aphramor et al. insist, the conversion of all dieticians tomorrow would not put an end to the guilty regret that most of us have probably experienced in intending to eat two cookies and realizing that we somehow ate the whole bag, nor to believing one looks better and feels healthier after having lost weight, nor put an end to society (and to be honest, ourselves) judging the attractiveness of others partially based in the extent to which their shape and size conforms to cultural ideals. In this sense, dieticians become a kind of straw man that distract us from what is most nuanced and powerful in Aphramor et al.'s subjects' experience with food.

Indeed, it is the voices of Aphramor et al.'s subjects that offer the most compelling intervention into the intimate ambivalences surrounding food, particularly in relation to weight and physical/psychological well-being. One of their subjects expresses relief at deciding to give up the 'dieting lark' she had been pursuing for 30 years, only to be struck with the prospect of guilt upon shortly thereafter developing diabetes, guilt that it would be her fault, or be perceived to be her fault, if in time the disease made her toes fall off and led to blindness. The poignancy here of the psychological toll on Aphramor et al.'s subjects is that the prospect of this 'guilt trip' might feature at least as prominently in their minds as the prospect of losing eyesight and appendages. The fraught intimacies of food are brought to the fore in their subjects' discussions of the HAES approach's emphasis on 'feeling good about one's self'. Whether this is significant from a clinical health/weight loss perspective may be an entirely different issue, but it does speak with great effect to the darker power of food, at least in contemporary Euroamerican societies. If HAES can help one feel *good* about oneself it is only because food can already make one feel *bad* about one's self, as in the words of another of Aphramor et al.'s subjects who thought herself an 'ugly, fat mess' unworthy of nice clothes. Another woman intends to become completely open to her husband, from whom she had been hiding her eating because she felt that he was, like everyone else, 'on her case' about food. In this sense food is actually a barrier to intimacy, furtively hiding eating – as if it were an extramarital affair – from the person she is closest to.

A peculiar intimacy brought out in the accounts of Aphramor et al.'s subjects is the degree to which food becomes quite literally fetishized, in the classical sense of an inanimate object that becomes animated almost to the point of personhood. We repeatedly hear in the discourse of HAES that patients develop a new 'relationship to food', as if a salad becomes the lover who they are true to because of ties that reach to the heart, that they would not consider cheating on with a plate of fried sausages. This is contrasted by Aphramor et al. with what they term the 'control discourse' of traditional dieticians, characterizing this as a Foucauldian ritual of confession. Yet I am struck by ways that the HAES approach may be rather more totalizing on the psyche than the traditional approach, in the same sense as Zizek (1999) argues that the 'Post-modern father' is more pernicious than the traditional one. The traditional father was (like the controlling dietician) simply authoritarian, telling the child that he must visit his grandmother whether he wants to or not. The 'Post-modern' father tells the child that he need only visit the grandmother if he loves her, if he wants to make her happy, manipulating not only the child's actions but conniving to colonize the mind, to insist that he love what the father expects him to do or otherwise bear the guilt. What then, of a HAES subject who does not in truth love their salad or their fruits, or who is more inclined to ravage their food than caress it and consume it mindfully. Face the fact that it was no accident that Aphramor et al.'s subjects were previously eating fatty meats – they ate them because they liked them – and one may find the element of control intrinsic in the new 'relationship to food' far more totalizing. As a final note, irrespective of physical health it would be useful to see evidence that HAES subjects are, indeed, psychologically healthier at whatever size they are. We may find it unlikely that their patients have truly become completely new people, forgetting the inner experiences and rejecting societal attitudes that have not only caused them hurt from perhaps their earliest childhoods but which also continue to surround them at every turn in society at large.

Saleh's lyrical movement through both seasons and changing relationships in the *kouroum* (orchards/vineyards) of Kefraya, Lebanon constructs an entirely different set of intimate relationships to food. Her account successfully captures a sensuousness that is one of the most compelling aspects of food scholarship – one can almost smell the freshly brewed Arabica as women chat around the afternoon table, feel vine leaves twist delicately between our fingers, and taste the tingling of the juicy sharon fruit on our tongues – to an extent that the tensions experienced by her subjects can almost become lost behind a wave of sensual rapture. Yet, in the end, they are not. Saleh artfully shows the *kouroum* to be many things: a cornucopia of fruits and vegetables, food that is as socially meaningful and it is succulent; a vector of layers of individual and social memories; and a place where commensality and social relations are lived, felt and reproduced. Yet these are also laden with ambivalences and contradictions. The past is never, of course, what it used to be, and its meanings are always remade, contested and recontested. Where there is commensality there is also its absence, where social relations are forged there are jealousies, slights, spites and vulnerabilities.

If these more fraught aspects are not at the forefront, there are rich examples and sub-texts that mesh fluidly with the intimate tensions found in the companion papers. Women snipe about a relative who had failed to share where she had found a bounty of *hindby* – an edible plant that grows wild in the *kouroum* – the spoils of nature seemingly outweighing the social relations that were soured and ruptured by her apparent stinginess. The seemingly timeless, picturesque and socially vibrant landscape of the *kouroum* is a product of competing visions for local or export-oriented grape production, pressures to uproot longstanding varieties in favour of more 'noble' ones favoured by French oenologists, and tensions and anxieties about the possibility that either this variety or that one will fall out of favour and no longer be marketable. And if there can be a deeply felt intimacy created between producers and consumers of the harvest – one woman extols that her daughter can 'taste the hands' that picked the plant – one wonders whether these emotions are as deeply felt on both sides of the interaction, as younger people find no work opportunities in the valley's vineyards, find new opportunities in cities, or find their home areas to be less an idyllic landscape than one disrupted and scarred by years of war and insecurity.

Of the three pieces Abbots presents us with perhaps the most striking, surprising set of ironies and contradictions. Let us start simply with the ethnographic particulars: older North Americans are flocking in conspicuous numbers to Cuenca in the southern Ecuadorean Andes, using it as a sort of New Age culinary retirement community. This simple fact borders on exploiting anthropology's historical role as peddlers of the bizarre, and yet it is decidedly not, as Jennifer Cole described surprising aspects of her own fieldwork, '... a world that Meyer Fortes (1945) and Evans Pritchard (1956) might well have invented for an essay exam for Oxbridge undergraduates ...(2001: 3).' Indeed, the concussive effect of the world Abbots paints, of this fact that reads like fiction, is that our sense of surprise is predominantly rooted in the fact that the actors here are utterly mundane. The subjects who are creating this world previously beyond our ethnographic imagination are not a barely touched people in some highland New Guinea valley but might rather be my own frumpy neighbours living until recently just a few houses down the way. My response to the news – if it got to me from another neighbour before it got to me from Abbots – could only be: 'They did what?!?!?' That they are motivated to reforge their lives in this utterly unexpected way by anxieties and hopes about food adds another unlikely twist: the foods that should be homey and familiar, that they have been eating their entire lives, are (at least in large part) the things that anxiously send them on pilgrimages to the exotic, hoping to find to find in Cuenca a place somewhere on the dreamy edges between an organic Garden of Eden and a bargain basement Whole Foods. Though surprising, this fact could not better exemplify the ways in which food's power can be cast in the deepest, most intimate corners of the psyches of affluent twenty-first-century Euroamericans. That the organic mango is a lie, a false hope that sent them on a quest to a world where eating is a riskier business than the world they left, provides us with the fraught ending that we would have had to have expected if we could have, in the first instance, imagined this scenario at all.

Abbots's essay is full of circumstances that strike at the deep, if troubled intimate meanings and experiences of food. A woman's stomach rejects the local fruits she had come to Ecuador to cleanse herself with, eating instead out of suitcases packed with dried sea vegetables and supplements, brought by her husband on regular flights to the U.S. undertaken to refresh her larder. Local markets, largely epitomizing the simple food these expats came to find, come to be viewed as sites of danger and ambiguity, many turning instead to the big, industrial supermarket, Supermaxi, or to 'Gringo Nights' at cosmopolitan fusion restaurants. On a positive side, it is through intimacy that this sense of danger can be ameliorated. By developing a sense of connection to the sellers of market produce, with those who sell expats this foreign food, human relationships can lead to these otherwise dangerous items becoming domesticated and largely seen as safe, even if this sense of safety can only be extended just so far. Moreover even those local foods that may be eaten without anxiety can bring with them other contradictions. Her informants' discussions of local beef are, for instance, reminiscent of Chekhov's famous short story *Gooseberries*. In Chekov's account a bureaucrat invests years and endless resources into the development of a country estate, symbolic of his rise in position as landed bourgeoisie, and which is epitomized in the prospective delight of gooseberries grown from his own bushes. When the plan finally comes to fruition the gooseberries are rather unpleasantly sour, but he greedily consumes them nonetheless, their actual taste – as sour as the life he has led to reach that moment – meaning nothing in comparison to their symbol as what he feels they show he has achieved. The parallel between Chekov's gooseberries and the Ecuadorian beef of Abbots's subjects is not a perfect one. The Cuenca expats seem far more benevolent, and their culinary anxieties and imaginaries rooted more in human frailty than in moral flaws. Yet what they share is the irony that something that actually tastes bad can, because of what it represents, actually taste good. If the Ecuadorian beef is tough and tasteless it is because it is grass fed, natural, its deficiencies part and parcel of precisely what expats came to Ecuador to seek. One may well wonder in reading Abbots's account if in Ecuador her subjects found any food that was good to eat, though the foods that were 'good to think' may have fared marginally better.

As a set, these essays offer us much to think about – and indeed things that scholars of food would do well to think about more, both in regard to the scope of intimacies that can be created through food and the sense in which these intimacies can be sometimes ambivalent and sometimes just plain bad. Just as our growing interest in what Stoller (1989) has termed 'the taste of ethnographic things' has too often pushed us to what I elsewhere (Holtzman 2006) labelled 'the ethnography of tasty things' – focusing on foods that will intrigue the palates of readers in a way too closely aligned with lay interest in food – so to can the emotions we describe and ascribe to food replicate rather pleasant, convenient and familiar motifs. Just as in everyday speech we speak, and journalistic food writers write, of 'comfort foods' food scholars evoke the ways in which foods may soothe us, whether in nostalgia for Mom's meatloaf or in the aroma of a pungent cheese returning us to a homeland we left long ago. Scholars are dead right in emphasizing the raw sensual power of food to conjure these affects (Seremetakis 1996; Sutton 2001), for these tell us some of the most important properties of food's capacity to affect us as humans, yet we must not abandon the necessity of also looking beyond now familiar paths. As these papers brilliantly illustrate, food can construct and conjure a fantastic range of intimacies, yet closeness is not always good. If we may think of the things we eat sometimes as comfort foods, perhaps some of what we have seen in these essays are dis-comfort foods: a meatloaf may not stand only for the mother who lovingly cooked it, but perhaps for the mother who forced you or guilted you to eat it (if only out of love); your 'relationship' with a spring salad may not be that of a fresh, new love but an old, conflicted one still able to touch the deepest parts of you but not without simultaneously invoking the pain of all the things that should have never come to pass (a break up, an indiscretion, the trousers that don't fit after a year of daily visits to a new French bakery); a cornucopia of new Ecuadorian foods may not be a gift of new friends to enrich your life, but rather cause a claustrophobia-inducing, crowd standing way too close to you, your heart racing, sweating, knowing you cannot escape feeling.

The Samburu herders I study in Kenya typically prefer their food unmixed and unadorned, emphasizing the pastoral triad of milk, meat and blood. A rare food in which they deliberately mix various ingredients – meat, blood, honey, fresh milk and sour – is a soup consumed at ceremonies to give blessings to a new marital bond. Its name, *lmutuchu*, means essentially 'as long as life' and the reason that so many thing are mixed together in the soup is because as they see life, as they see the future life of those two people who have been intimately bonded together, it is a mixture of many things, some as sweet as honey others with the pungency of sour milk. This is, indeed, the mixture of the experience of the intimacies that food constructs that this set of papers serves to illustrate, an arena to create and evoke joys in things that are so close to us but also a site for tensions, anxieties and regrets.

References

Batsell, W.R., Brown, A., Ansfield, M. and Paschall, G. 2002. 'You will eat all of that!': A retrospective analysis of forced consumption episodes. *Appetite* 38, 211–19.

Chekhov, A. 1898. Gooseberries, in *The House with the Mezzanine and Other Stories, by Anton Tchekoff*, translated by S.S. Koteliansky and G. Cannan. New York: Charles Scribner's Sons.

Cole, J. 2001. *Forget Colonialism?* Berkeley: University of California Press.

Holtzman, J. 2006. Food and Memory. *Annual Review of Anthropology* 35, 361–78.
Holtzman, J. 2009. *Uncertain Tastes: Memory, Ambivalence and the Politics of Eating in Samburu, Northern Kenya*. Berkeley: University of California Press.
Kierkegaard, S. 1843. *Either/Or*. Reprinted 1988. Princeton: Princeton University Press.
Seremetakis, C.N. 1996. *The Senses Still*. Chicago: University of Chicago Press.
Stoller, P. 1989. *The Taste of Ethnographic Things*. Philadelphia: University of Pennsylvania Press.
Sutton, D. 2001. *Remembrance of Repasts*. Oxford: Berg.
Wilk, R. 2010. Power at the table: Food fights and happy meals. *Cultural Studies: Critical Methodologies* 10, 428–36.
Zizek, S. 1999. You may! *London Review of Books* 21(6), 3–6.

PART III
Contradictions and Coexistences: What We Should and Should Not Eat

The chapters in this section have a united interest in the ways in which eating, and the 'choices' we make about what is 'good', 'healthy' or 'ethical' to eat, are continually contested and reframed. As such, they explore how competing paradigms of 'how' and 'what' we 'should' eat are (re)produced, contested and (re)formulated by an array of social actors and agencies. The paradigms analysed are situated within diverse, although often intertwined, value systems – from public health debates and alternative food networks to environmental protection measures and nationalist discourses. All the chapters demonstrate that the paradigms in which eating bodies are entangled do not simply conflict, but are drawn into uneasy relationships by the embodied practices of eating. There is thus a critical engagement across these chapters with the myriad ways in which choice and necessity, moralities, and ethics, are negotiated and transacted on a day-to-day basis by eating, and each explores how the actors caught up in these transactions range from consumers to producers, scientists to activists, and food policy makers to multinational corporations. As such, this section further elucidates the scales of eating, not only in terms of local and global, but ranging from the micro-organism to supra-national agencies and markets. In doing so, they highlight the multiplicity of co-existent bodies and knowledges that are encountered and engendered by eating. One element of eating that particularly comes to the fore across these scales is taste, in both its literal and Bourdieusian (1984) sense, with social actors making trade-offs and assessments between what they *should* eat and what they *want* to eat. Our attention is thereby drawn to the ways in which consumer desire for foods is ideologically manipulated, regulated and constructed, while also being simultaneously embedded in the material, sensorial and experiential. Through its attention to this simultaneity this section, as Anne Murcott points outs in her interlude, raises challenging methodological questions about the ways in which it is possible to reflect on, and analyse, the practices of eating.

References

Bourdieu, P. 1984. *Distinction: A Cultural Critique of the Judgement of Taste.* New York and London: Routledge.

Chapter 7
Chewing on Choice

Sally Brooks, Duika Burges Watson, Alizon Draper,
Michael Goodman, Heidi Kvalvaag and Wendy Wills

Introduction

The concept of 'individual choice' has become central to contemporary understandings of the relationship between food, health and wellbeing. Drawing on four research projects in which the authors have recently been engaged (Brooks 2010; Goodman et al. 2012; Kvalvaag 2012), this chapter explores how and why the concept of choice has become so central to explanations for 'why we eat how we eat'. We explore the multifarious ways and means through which discourses of 'food choice' have been deployed and gained political and material 'real world' salience in a number of different contexts; locating 'choice' theoretically – as a concept borrowed from neoclassical economics to augment biomedical theories with a thinking subject – and politically, as an indeterminate and 'slippery' concept adaptable to shifting policy platforms. The multiple manifestations and consequences of this slipperiness are explored through the cases of: food and nutrition policymaking in the UK over the last 25 years; an international nutrition system generating policies and programmes for 'beneficiaries' in the developing world; and the strategies of an increasingly high-profile alternative food movement.

This chapter is organized as follows. We begin by tracing the emergence of choice as a core concept guiding food and health policy; finding its origins in the a priori separation of the thinking subject from the physical body in early medical science. In this case, the slipperiness of 'choice' is a result of its conceptualization as 'clean thought', disconnected from both embodied experience and societal context. Secondly, we examine the operationalization of the concept in UK food and nutrition policy between 1976 and 2010 through a critical discourse analysis of selected key policy documents published during this period. Several 'frames' of choice, are identified, all of which accommodate discourses of choice that are complex, overlapping and contradictory. Furthermore they have changed over time; reflecting the shifting balance of influence between different stakeholders in food and nutrition policy.

Thirdly, we explore the repackaging of choice for export to low and middle-income countries via international development programmes that seek to engineer choice in the direction of predefined development goals. In this case, nuances in interpretation observed in UK contexts are contrasted with globalized programmes that construct beneficiaries as passive objects of policy. Finally, we trace attempts of alternative food movements (AFNs) to reframe food choice as an ethico-political act from numerous angles and with varying outcomes. Considering the discourses of choice articulated by AFNs, this section explores the politics embedded in individual food choice and the various rationales for opening up and/or closing down food choice in AFNs. In this case, recent developments in what might be called the 'taste turn' in food studies converge with new developments in human biology, which are only touched upon here, to provide a starting point for the re-mapping of the conceptual and political terrain of food choice, health and eating through a consideration of the socialized and 'visceral' aspects and geographies of food (Goodman 2011). We conclude by considering, briefly, what a 'chewing over' of individual choice might mean for further research and scholarship.

Forming Choice: Theorizing Action Without a Body

In this section we explore the origins of 'choice' as a core concept informing contemporary policies on food and health. The concept of choice, we argue, is the product of the co-evolution of three distinct bodies of knowledge concerning 'the body', 'food' and 'eating' (Kvalvaag 2012). We begin by exploring how 'the body' came to be understood within modern medicine from the Enlightenment era onwards. Secondly, we trace the development of nutrition as the science of food and food-body interaction. Thirdly, we turn our attention to how the act of 'eating' has been conceptualized. This analysis highlights 'individual choice' as an explanatory concept for action able to co-exist with established sciences of body and food premised on the analytical severance of the thinking subject from the material body (Kvalvaag 2012).

The production of knowledge about the human body is the mandate of medical science (Porter 1996). Medical science, or biomedicine, emerged in late fifteenth-century Europe; at a time when modern science was developing in search of 'the' true knowledge, cleansed of all myth and superstition. Such knowledge could only be accessed through systematic empirical method (Porter 1996), which required the isolation of both scientist and object of study from all personal and contextual disturbances (Haraway 1997). This was consistent with the Cartesian separation of mind and body, which made it possible to study 'the body' as an object independent of 'the person'. A dualistic model became established which divided modern science into natural science (the study of nature and objects) and human science (the study of subjects and meaning) (Hawson 2004). Thus the study of the human body – understood as a physical object – was defined as a natural, not a human science.

This disciplinary demarcation co-evolved with technological advancements enabling the accumulation of more detailed knowledge of the body. The most important of these was the microscope (Amerman 2010), which made it possible to map, in detail, the body and its component parts – its anatomy. It also enabled close examination of its functions in terms of biochemical processes – its physiology. Thus the science of the body was defined in terms of two complementary disciplines – anatomy and physiology – with the laboratory as its central arena (Shier et al. 2008). From this knowledge of 'normal' anatomy and physiology, it became possible to identify and treat disturbance and abnormality (Shier et al. 2008). Thus 'medical treatment' was understood as acting upon an identified abnormality in order to restructure the anatomy (through surgery) and/or restore the physiology (through medicine).

Food, together with oxygen, is essential for bodily existence and development (Shier et al. 2008). Formal knowledge of food is generated by nutritional science, through the study of chemical and biochemical aspects of food (Andersen and Drevon 2007). As with medical science, the study of processes through which food and body interact once food has entered the body is bounded by the parameters of natural science. The science of nutrition emerged from biomedicine and is, to a large extent, constituted by the same configuration of theory, method, equipment and laboratory as the science of the body. Nutrition science can be summarized in terms of three types of research enquiry (Andersen and Drevon 2007). The first of these is basic research on the chemical content of food; in terms of proteins, carbohydrates, fat, vitamins etc. This is the foundation of nutritional science, on which other, more recent branches of nutritional research are based (Andersen and Drevon 2007).

The second type of enquiry in nutritional science examines interactions between food and the body after ingestion (Aas 2008). From these studies, scientific knowledge about how, why and where biological decomposition of food occurs in the digestive system – how different foods affect the body and what the body does with food that has been digested – is derived (Andersen and Drevon 2007). The third area of study is the mapping of diets. Translating diets, reported or observed, into chemical compounds analysable in terms of anatomic and physiological variables (BMI, blood sugar, cholesterol etc.) enables physiological correlations between food and the body (or in the case of epidemiology, food and populations) to be made. These studies generate knowledge about what kinds of food promote health and cause illness (Andersen and Drevon 2007).

Food, like oxygen, is located outside the body, where it does neither good nor harm. Unlike oxygen, however, which is found everywhere, food needs to be eaten and thus brought into the body through individual action. Food needs to be accessed, selected from among alternatives, prepared for consumption in suitable contexts and using appropriate tools, and consumed. Eating therefore requires both will and skill. With 'the body' and 'food' defined as physiological objects, it has therefore been necessary to identify theories of action to explain 'eating'; and for this medical and nutritional scientists have had to look to the social sciences. Two types of action theory can be identified; derived from neoclassical economics and behavioural psychology respectively (Montano 1995). While both disciplinary perspectives are located in the broad tradition of methodological individualism in the social sciences, the key difference between them is their understanding of meaning and will (Gunnerius 2003).

In neoclassical economic theory the unit of analysis is the autonomous, rational individual who chooses whatever brings maximum benefit – as long as s/he is provided with the right information (Cooper et al. 2010). Behavioural theory also focuses on the individual. This branch of psychology, however, developed in a positivistic tradition, mirrors natural science. For behaviourists there is, in principle, no difference between studying materiality (objects) and studying human behaviour (Gunnerius 2003). As with economic theory, the psychological mechanisms in behaviourism are to avoid discomfort and achieve reward (Gunnerius 2003). What distinguishes behaviourism is that action is theorized as an automatic (rather than a calculated) response to external stimuli. Individuals do not choose – they react (Teixeria 2011). Despite these differences, however, understandings of 'food intake' are in both cases premised on a dualistic model that has separated the natural world (of body and food as objects) from the human world of taste and preference, skill and action. As such, both theories serve to bridge the analytical gap between food and body without disturbing the established paradigm.

It is in light of the prior analytical treatment of the 'body', 'food' and 'eating' that we now come to the question of 'choice'. Firstly, we showed how established physiological understandings of body and food lack both a subject to act and context of interaction. In this context, theories of action have been imported, highly selectively, from those social science disciplines best placed to deliver an individual subject independent of body, food and societal context. Individual choice as a concept derived from neoclassical economic theory serves this purpose. Premised on the existence of the rational, sovereign, choosing subject – the neoclassical perspective severs 'the individual' from all embodied knowledge and experience (e.g. taste, texture) as well as societal influence and constraints (e.g. culture, social class). 'Choice' is thus valorized as 'clean thought' (Kvalvaag 2012).

This severing of the thinking subject from body, food and society implies that s/he is 'free' to choose. Herein lies the contradiction. With no connection to body and food, the subject is utterly dependent on external sources of information in order to know how to act. The established conceptual scheme described here does not provide the tools for exploring the embodied nature of food choice (Kvalvaag 2012). Access to the right (i.e. scientific, evidence based) knowledge about body and food thus becomes a prerequisite for particular kinds of choice, e.g. 'healthy' choices. Which raises the question – why, in modern societies in which such information is said to be freely available do people make 'unhealthy' choices? The inability of scientists and policymakers to answer this question has created an ambiguous role for 'choice' as a concept informing contemporary food and health policy, as the following section highlights. In practice, and despite ubiquitous references to 'choice' as a guiding concept, distinct shifts in policy and practice can be detected which occupy the space between two extremes set by, on the one hand, a sovereign subject constructed by economists as free to choose and, on the other hand, a physiological determinism endorsed by behavioural theories that deny a role for choice.

Operationalizing Choice: The Case of UK Food and Health Policy (1976–2010)

Individual 'choice' has been operationalized in one political, discursive arena; that of recent UK food policy. As commentators have observed, the dominant approach in food and nutrition policy in the UK (as in many other high income countries) over the last 20–30 years has been a focus on achieving better public health outcomes via behaviour change; specifically by changing food choices (Coveney 2003, Caraher and Coveney 2007). Given the growing evidence of the limited effectiveness of such approaches, particularly in the context of widening health inequalities, the continued reliance on an approach privileging choice as a pathway of change is puzzling. While some studies have explored choice in UK health policy and found it to be an indeterminate, but a nonetheless important organizing principle (Clarke 2005; Clarke et al. 2006; Greener 2009), there has been no critical examination of the concept of choice in UK food policy.

Food and nutrition emerged as a public health priority in the UK in the 1970s. Previously, food policy had been primarily concerned with agricultural production, reflecting a post-war preoccupation with food security. During the 1970s, however, alarm about rising oil and food prices converged with new concerns about diet-related non-communicable diseases, and, very gradually, food started to appear on the health policy agenda (Murcott 1994). While there were no significant developments during the 1980s,[1] from the early 1990s onwards there has been a succession of policy documents linking food and health. How has the concept of choice been put to work in these policies spanning 25 years? Using a critical discourse analysis approach (Fairclough 2001; Shaw 2010) we explored the uses and meanings of the term 'choice' in a series of policy documents.[2] From this analysis we identified five frames (Schön and Rein 1994), each of which represents a distinct articulation of the relationship between subject, body and food (see Table 7.1).

1 Whilst not a policy document, the National Advisory Committee on Nutrition Education (NACNE) discussion document published in 1983 caused quite a stir with its recommended intakes of fat, salt and fibre.

2 This paper analyses the different ways in which the following UK policy documents construct choice: *Prevention and Health* (1976); *The Health of the Nation* (1992); *The Scottish Diet Action Plan* (1996); *Food Standards Agency: 'A Force for Change'* (1998); *Choosing a Better Diet* (2005); *Food Matters* (2008); and *Healthy Lives, Healthy People* (HLHP) (2010).

Table 7.1 Frames of choice in UK Food Policy (1976–2010)

Choice as …	Comments
Personal responsibility	e.g. a civic duty to choose 'well', must choose
An instrument for change	e.g. a means to achieve policy goals
An editing tool	e.g. because of an over-abundance of things to choose from we need someone else to choose for us
A problem	e.g. the 'wrong' choice by particular groups e.g. young people; those who are obese
As freedom	e.g. choice as sovereign, as a right and policy goal

It should be noted that these frames are by no means commensurate. Choice is framed variously as an action that we do (e.g. because we must, or something we do improperly), as a pathway to achieve change (e.g. via individual choices or others choosing for us) and as an object (e.g. freedom of choice as a policy goal). Secondly, while all frames were identified across all the documents, the extent to which different frames were emphasized has varied between documents and over time. These dimensions of variation are manifestations of the indeterminacy of the concept of choice identified earlier.

It is interesting to note how these policy documents position the role of personal responsibility. Only the oldest document we considered, *Prevention and Health* (1976), frames making healthier choices as an issue of civic responsibility. For example, this document includes statements such as 'the weight of responsibility for his own state of health lies on the shoulders of the individual himself' (p. 38). Documents from the 1990s and 2000s illustrate a move away from choice as individual responsibility towards an acknowledgement that consumers might (or should) desire healthier choices and that they need help in order to do this. Regulatory bodies (like the Food Standards Agency), the private sector (as seen in *Healthy Lives, Healthy People*) and government itself (in *The Scottish Diet Action Plan*) are each highlighted as in some way responsible for helping consumers make better choices.

The corollary of this discourse of responsibility is that someone, usually the individual consumer, is perceived as a 'problem chooser' who has failed to self-govern and make the 'right' choices. *Prevention and Health* (1976) speaks bluntly of 'public apathy', 'self poisoners' and positions some individuals as 'reckless' in the light of their choices. By 2010 when *Healthy Lives, Healthy People*, the most recent of the documents analysed, was published the language had been tempered but certain groups of individuals, notably teenagers and young people, were still viewed as problem choosers because of their 'harmful lifestyles' (see also Aphramor et al., this volume).

Most of the documents looked at cited 'freedom of choice' as an important concept although, notably, this applies not only to individuals but also the food industry as a sector. When the Food Standards Agency (FSA) was conceived of in the early 1990s it was argued that freedom of choice should be constrained as little as possible. Freedom of choice for both consumers and the food industry was an a priori condition for the terms of reference of the new FSA (Food Standards Agency 1998). Later policies maintained this position and in 2008 *Food Matters* highlighted that individuals enjoyed greater freedom to choose food from a wider range of retailers. Paradoxically, of course, it is this freedom that is also considered a problem. 'Problem choosers' exercise too much freedom, whether individual consumers making 'unhealthy choices' or the food industry developing too many 'unhealthy' or unsustainable products from which to choose.

The Scottish Diet Action Plan (1996) was the first policy document to be explicit in allocating the role of 'choice editing' to the food industry. Retailers, in particular, are highlighted as being well placed to guide consumers towards healthier food items, through point of sale materials, for example. Even small independent retailers, it is suggested, can edit consumer choices. This choice-editing role is developed further in the cross-governmental approach advocated in *Food Matters* (Cabinet Office 2008). This document highlights the need for choice editing, not only to reduce the 'burden' on consumers in making healthier choices, but also to guide them towards broader food sustainability goals. Referring to 'evidence that consumers are looking to retailers to make some of the more difficult environmental and ethical trade-offs on their behalf' (ibid.: 60), it suggests supermarkets adopt environmental and ethical screening criteria in their product selection. Here the term 'choice editing' is used interchangeably with 'screening' (ibid.: 61). Moreover, this document goes further in acknowledging the limits to individual responsibility for choice, through its overarching frame of choice as an 'instrument for change' in the context of a cross-governmental initiative to 'facilitate a public debate about food that fosters cultural and behavioural change' (ibid.: 36).

The most recent of the documents analysed; *Healthy Lives, Healthy People* (2010) features a new strategy, that of 'nudging' consumers towards better choices. 'Nudging' is a relatively new concept that has been taken up by the Obama administration in the USA and attracted the attention of 'big society' advocates in the UK (Hunter 2011; Thaler and Sunstein 2008). It implies a greater role for the private sector that involves using the techniques at its disposal to identify flaws in individual decision-making and make use of those flaws to shape choices (Hausman and Welch 2010: 126). The example presented in Box 7.1 is illustrative. While new to UK policy discourse, this approach could be interpreted as an extension of the industry's 'choice editing' role. However, a shift from making healthy choices easier to making (albeit 'unhealthy') choices impossible by exploiting human flaws is no small step. While framed by an overarching – and enduring – discourse of 'choice', the concept of nudging appears to owe more to behavioural theories than neoclassical formulations of rational choice.

Box 7.1 A nudge in the right direction? 'Bigging up healthier favourites'

A recent article in *The Grocer* magazine, aimed at the food industry, highlights a new initiative to 'big up' the fruit and vegetable content of 'some of the nation's favourite dishes'.

> 'From spaghetti Bolognese to chicken korma, plans have been drawn up by leading supermarkets and suppliers to boost the fruit and veg content of their products at the expense of fatty, high energy density ingredients'.
> 'In some cases the radical plans will even see consumers encouraged to eat bigger portion sizes that satisfy their appetites while providing more low energy density food'. This strategy, it is argued, 'will have a broad appeal as many customers view low-calorie foods as a major turn-off'.
> This initiative 'is also aimed at satisfying the Department of Health, which is drawing up plans for an obesity White Paper and seeking commitments from the industry to slash calorie intake in the next phase of the Responsibility Deal'.

Source: 'New obesity plans "big up" healthier favourities' by Ian Quinn, *The Grocer*, 10 September 2011.

In summary, this analysis highlights that choice is, indeed, a dominant theme within UK food policy discourse, but it is neither a monolithic nor a stable concept. Rather it is indeterminate and slippery. Despite identifying five frames of choice, the discourses they reveal are complex, overlapping and contradictory. These contradictions betray an unresolved tension between two parallel 'explanations' for individual action. Neoclassical economics, purportedly the hegemonic social science discipline of our day, posits a rational, choosing subject. Our analysis of policy documents is revealing of the attempts by governmental actors to explain the gap between such simplistic constructions and the daily, lived, embodied decisions and actions of individuals living in modern society. In the process, a drift towards behaviourism is discernible. While retaining the language of choice, policies increasingly defer to actors best placed to shape the choices of those apparently unable to do so for themselves; even if these actors represent a food industry largely responsible for the range and quality of 'good' and 'bad' choices available.

Exporting Choice? Engineering Choice in Pursuit of 'Development'

We now shift our focus to the developing world, tracing the ways in which 'individual choice' has been reframed for export from 'the North' to 'the South'. This has been a relatively recent development within an international nutrition system[3] traditionally oriented towards 'needs' rather than 'choices'. International nutrition (the branch of nutrition science concerned with nutrition-related research, policy and practice in low and middle income countries) has historically been concerned with closing the gap between the (deficient) nutritional status of disadvantaged groups in developing countries and an 'ideal' nutritional standard of some kind.

While precise definitions and measures have been a subject of contestation and debate over the years (for example, see Sommer and Davidson 2002), the international nutrition community has maintained its commitment to what Pacey and Payne (1981) call the 'fixed genetic potential view' of nutrition, which is based on the premise that 'there is an optimal or preferred state of health, fixed for each individual, and determined by his or her genetic potential for growth, resistance to disease, longevity and so on' (Pacey and Payne 1981: 37). However, given the challenges inherent in measuring 'genetic potential', the default position has been to 'assume that the standards of body size and food intake observed in 'well-fed' and 'healthy' populations approximate to this optimum' (Pacey and Payne 1981: 37–8). In other words, the field of international nutrition is based on a model that accepts aggregated data on 'well fed' bodies in industrialized nations as the yardstick for assessing nutritional 'deficiencies' of individuals and populations in developing countries.

Given this starting point, it is not surprising that international nutrition policy discourse has tended to downplay (individual) choice in favour of more pressing (generic) needs. In practice, however, the interventions employed – from community-based education for behaviour change to national policies enforcing mandatory salt iodization – are clearly based on implicit assumptions about the relationship between individual choice and desired public health changes to nutrition and health. The example of micronutrient programming is illustrative. Since the 1990s, vertical micronutrient delivery initiatives, such as industrial food fortification and pharmaceutical supplementation, have been favoured by international development agencies and donors with their eye on the millennium development goals (MI 2001). In marked contrast with the UK policy context discussed earlier, a key characteristic – even selling point – of these vertical programmes is their explicit removal of individual choice as a potential obstacle

3 The 'international nutrition system', while far from cohesive, comprises actors from 'international and donor organizations, academia, civil society and, increasingly, the transnational, private sector' that collectively set the agenda for policy, programming and funding allocation aimed at reducing the global burden of malnutrition (Morris et al. 2008: 608–9).

to the achievement of 'impact at scale'. On the other hand, many NGOs advocate community-based behaviour change strategies (such as the promotion of market gardening) as a more sustainable alternative (Delisle 2003). In each case, however, 'beneficiaries' are constructed as 'problem choosers': The difference lies in whether the solution is to 'improve' people's choices or obviate choice altogether.

Recent developments in international biofortification research suggest that the question of how to influence individual choice is becoming a more central concern in international nutrition policy. Biofortification is a new and evolving interdisciplinary science bridging agriculture and public health (CIAT and IFPRI 2002) in which a network of international agricultural research centres (the CGIAR) has assumed a key role. Based on an assumption, clear evidence for which remains elusive, that resource-poor farmers in developing countries cannot access a balanced diet and therefore have no choice but to subsist on the staple crop that is most readily and/or cheaply available, global biofortification initiatives, such as HarvestPlus (the Biofortification Challenge Program of the CGIAR[4]), are developing technologies to increase the micronutrient content of a series of staple crops through biological methods (plant breeding and/or genetic engineering).

These global biofortification initiatives continue the well-established tradition of setting programme-wide goals with respect to an 'ideal nutritional standard' (Brooks 2010). HarvestPlus, for example, is organized around a matrix of 'breeding targets' that specify the required nutrient level by crop (e.g. rice, wheat, maize, sweet potato, cassava, bean) and micronutrient (e.g. iron, zinc, pro-vitamin A). These targets indicate the minimum level of nutrient required to achieve 'impact', regardless of context. In addition, the use of biological rather than chemical methods has enabled promoters to present biofortification as a one-time investment that capitalizes on the multiplier effect built into seed (re)production systems. As such, these initiatives take the logic of pre-existing, large-scale micronutrient delivery systems a step further by embedding nutrition 'in the seed' in a method conceived as inherently scalable across space and over time (Brooks 2010). The parallel advocates draw with water fluoridation is illustrative: 'The [required nutrients] will get into the food system much like we put fluoride in the water system. It will be invisible, but it will be there to increase intakes' (Bouis 2004: 8).

In the policy discourses surrounding these global biofortification initiatives, human bodies are invisible. Instead, benefits are presented as accruing directly to 'nutritionally disadvantaged' populations in non-specific locations (CIAT and IFPRI 2002: 5). In this context, the growing body of empirical research on micronutrients and choice carried out under the auspices of HarvestPlus is noteworthy. Of course the range of choices considered in these studies is already narrowed since, as mentioned earlier, target populations are assumed not to have the luxury of choosing among diverse dietary items, only between different varieties of a specified staple. Furthermore, the problem of 'choice' is

4 http://www.harvestplus.org/ [accessed: 15 December 2011].

conceptualized by a community of agricultural economists firmly rooted in the neoclassical tradition (Brooks 2011) as a question of how to induce poor (but nevertheless rational) consumers to 'switch' from non-biofortified to biofortified varieties (Stein et al. 2005). Various methods for assessing 'willingness to pay' for biofortified crops are currently being tested in Sub Saharan Africa in this vein (De Groote et al. 2011; Meenakshi et al. 2010). A consistent feature of this work is a view of 'user' engagement as necessary for securing acceptance for pre-defined products (Brooks 2010; cf. Ashby 2009).

The development of sophisticated methods for engineering choice for biofortified crops belies a reliance on simple causal pathways linking (the right) consumer choices with desirable public health and socio-economic outcomes (for example, see Stein et al. 2005). Such a formulation denies the bio-cultural diversity that still exists in many developing country agri-food systems (Johns and Sthapit 2004), in which 'individuals' are both consumers and producers, and local markets display an array of seed and crop varieties adapted to diverse agro-ecologies, seasonal conditions, farming systems, tastes and cultural occasions (for examples see Asia Rice Foundation 2004; Castillo 2006). Research partners from the international nutrition community have yet to draw attention to this point, perhaps because the approach does not represent a significant shift of paradigm given the widespread acceptance of large-scale micronutrient delivery programmes that claim large-scale impact (Brooks 2010), despite the dearth of evidence in support of these claims (for example, see Latham 2010).

The current configuration of global biofortification research reflects its membership of a new generation of centralized programmes featuring public-private partnerships whose shared aim is to extend the reach of an increasingly privatized formal seed sector at the expense of informal institutions adapted to local economies, cultures and agro-ecologies (cf. Brooks et al. 2009). Meanwhile, evidence exists that some of the 'traits of interest' pursued by the plant geneticists employed by these programmes have often been there all along, in the form of traditional varieties maintained by farming communities, often specifically for their nutritional benefits (for example, see Frei and Becker 2004). That findings such as these do not register in official biofortification policy discourse is indicative of a tendency to conflate variety (as represented by an expanded range of certified seeds available through commercial channels) with genuine diversity that 'reflects the many dimensions of difference inherent in the heterogeneity [that] exists in particular places' (Brooks and Loevinsohn 2011: 3, see also Stirling 2007).

In summary, this analysis shows that the concept of choice has been exported from the industrialized North, though the globalized programmes of an expanding international nutrition community, to diverse countries and communities in the South. Here we find that choice is a shifting and indeterminate concept able to accommodate yet more contradictions. The example of biofortification reveals a centralized approach to engineering choice between pre-selected options, while constructing 'choosers' as members of homogenous populations who have no choice. As in the UK example, the provision of information and market signals are the chosen mechanisms through which, it is believed, people will be induced to make the 'right' choices. As such, these programmes contribute to the broader trends in which the choice as diversity emergent from human-environment interactions over time in particular places is being gradually displaced by mechanisms designed to extend choice as variety to individualized consumers in an abstract 'marketplace'.

Reframing Choice through Tasti-ness? Articulations of Choice in/by AFNs

In addition to theorizations of choice and its embeddedness in national and international food policy, choice is also 'put to work' by the movements and in the politics of Alternative Food Networks (AFNs) (Goodman et al. 2012). Indeed, choice is always present in the discourses of AFNs and here we present three of these discourses to illustrate wider points about how choice, eating and politics are embedded in these networks. Although the ways in which these movements frame 'choice' differ from those found in public policy they are, nevertheless just as slippery, ambiguous and contradictory. Furthermore, they are also shifting, particularly with the transformation of 'alternative' foods from being fringe items to becoming familiar supermarket fare.

The politics of choice, in one way or another, greatly inform and indeed motivate AFN movement actors and, as far as can be understood, consumers. Yet, there is also a great diversity of interpretations of what choice is, should be, how it should be articulated and to what ends by these movements and their academic commentators. At one extreme, there are parts of the movement that champion individual choice as the seemingly only, but also 'right' and 'best' way to articulate AFNs. This first discourse is encapsulated in the words of Harriet Lamb, the executive director of the Fairtrade Foundation, in a biopic of the travels of the 'queen of fairtrade':[5]

> She energetically mimes out British supermarket shoppers, whizzing round in a hurry, loading up their trolleys at breakneck speed. 'Imagine this is a shop', she says. 'And I'm going shopping. Shopping, shopping [she wails like a baby] and I'm quickly taking tea, coffee, sugar from the shelves. Quick, quick, quick. Then I'm looking for cheap tea, cheap coffee. If I'm only buying cheap coffee then the price for you is also low.' Suddenly she raises a hand, and her voice, and addresses in absentia the great British shopper. 'STOP!' she exclaims. 'STOP! Don't buy cheap coffee! If you buy cheap coffee then it is bad for the workers. Look for Fairtrade. Ah, 'Fairtrade. From Rwanda'

5 This quotation is taken from a biopic entitled 'On the road with the queen of fairtrade' which was first published in *The Independent* on 28 February 2009. It is available at http://www.lalettredelacheteur.com/LDAENG/archives/539 [accessed: 24 May 2012].

Choice here takes on a moral-ethical, political and economic function in that it is the 'right' thing to do as an individual consumer, signalling to the supermarket that consumers do not wish to buy 'cheap' coffee and buying into fair trade provides economic development to the farmers at the other end. For these tropes, 'real' and 'fair' food comes at a cost that needs to be borne by consumers out of their desire and obligation to pay the 'real costs' of these often higher-priced foods. Food labelling is crucial here, as a means to provide consumers with the information they need to make the 'right' choice. In a very real way, this deployment of choice in AFNs seems to combine many of the policy approaches described above – especially that in the consumer-facing 'nudge' model of change, often used in addition and parallel to the industry-led nudge model described above – whereby more information provided to consumers (on labels here) sees them making the right choice as responsibilized and rational thinking consumers.

At another extreme are those AFN movement actors who work to take choice *out* of the equation, articulating that food should be healthy, safe, accessible, and 'fair' for everyone. Much of this rhetoric is about 'transformations' towards a socially and environmentally just food system, most often through regulatory and governance structures that work to change the provisioning of food from the outset. This second discourse is exemplified in one of the 'key messages' that resulted from *Food Justice: The Report on the Food and Fairness Inquiry* that was executed and published by the Food Ethics Council (2010). In summarizing the committee's findings, the report states that; 'business as usual is not an option' in creating a more just and ethical food system, instead, 'we must fundamentally change the way we live' (ibid.: 80). In this, '… solving social justice problems in the food sector generally pointed towards wider social and economic policy, for example, unemployment, benefit levels, competition and finance' (81). Here, '"ethical consumption" is just one of the ways in which people can potentially act upon their values in relation to food and farming' (83); rather, there is … scope for promoting social justice through food policy' (81) and, seemingly most importantly, 'to enable people to change their behaviour, we need to address the inequalities that underpin their behaviour' (83).

This suggests that, at a deeper level, we will not be able to choose our way to healthier, safer and fairer foods. Indeed, many activists and movement actors in this camp are suspicious and rather critical of the power of choice as a form of politics. 'Choice editing' is nevertheless entertained here as in food policy (see Lang 2010), but more as an element of this second discourse in AFNs in that it is about removing opportunities for choosing 'bad' foods and/or other commodities based on social, environmental and other criteria.

A number of scholars occupy the space between these two extremes and critically explore the complexities and contradictions inherent in 'choice' as a form of politics in AFNs. Julie Guthman (2007; see also 2008a, 2008b) highlights an 'anaemic' politics of alternative food choice which merely replicates the inequalities of consumption already embedded in consumer capitalism and bolsters already powerful mechanisms of neoliberalism. Raj Patel (2007) concurs with this analysis but concludes differently, arguing that, while more choice for alternative foods such as fair trade and organic foods are indeed needed as a way to 'battle back' against conventional food systems, the focus in the first instance should be on the ways that food multinationals have actually constrained consumer choice within a bounded series of goods designed to make a profit. Finally, Barnett et al. (2011) in their treatise on fair trade as a form of ethical consumption, argue that, in actuality, so-called individual consumer choice and practice is instead thoroughly socialized, much of it through the politicized actions of NGOs and food movement groups themselves. Choice is not an individualized act. Rather, it is an act that has social consequences through the ability of these 'consumption singularities' to be globalized 'citizenly' acts that have implications for poor farmers through the mechanisms of fair trade movements and markets.

All these accounts, however, appear to steer clear of the role of *taste* in the politics of these choices and/or their effects. This is not to say that taste is not a key element in the marketing of these 'quality' foods, far from it. Indeed, a third discourse can be identified in which many AFNs, have successfully deployed 'taste' to make inroads into conventional markets and expanded marketability of their products by telling consumers that they 'taste' better. In the UK, for example, there has been a noticeable shift of focus in the marketing of fair trade coffee; with quality and taste first and foremost and the moral economy of development taking a back seat to the desire to be seen as 'better tasting' (for more, see Goodman et al. 2011). As a manager at an organic, fair trade put it recently, 'I think with [our company], taste is the first thing, and then the fact that it's organic and then the fact it's ethical' (Goodman 2010: 110).

These developments suggest that AFNs can be, and often are, as much about the bodily affects of (good) taste as they are about the minded knowledge of improving the conditions of production. In this case, AFNs are not only working across the mind-body dualism of choice but they are 'engineering' choices in such a way that consumers (or at least those who can afford these quality items) have *no choice* but to purchase them due to their quality and taste. In this way, some AFNs – in addition to the use of labels and information about themselves – are using taste and quality as a set of marketing techniques rather than a site of politics. Ironically, these techniques tread very closely to strategies increasingly deployed by the 'conventional' food industry in their attempts to 'nudge' consumers towards AFNs as a 'way of life' rather than as the articulated expression of individualized choice. Here nudging here takes on a 'visceral' quality that moves beyond the simple provisioning of knowledge and information about what is a 'good' choice or not.

This turn to the role of organoleptic taste – perhaps riding alongside the Bourdieusian sense of class and/or culture-based sense of 'taste' and 'distinction' (Bourdieu 1984) – suggests there is a need for scholars and researchers to develop more and better conceptual tools for understanding food choice, not only in the face of the growth of AFNs, but also in the context of food more generally. One attempt at this, and only briefly mentioned here, has been explored in the work of

Alison and Jessica Hayes-Conroy and others (2008, 2010; Longhurst et al. 2009) in their bid for understanding the 'visceral geographies' of food, food choice and AFNs in particular. Building on Elspeth Probyn's (2001) *Carnal Appetites*, for the Hayes-Conroys, understanding the visceral geographies of food is about engaging with the sensual, lived, 'gut' responses we have to food, part of which means engaging with the importance, ambiguities, complexities and problematic of taste, tasti-ness and disgust. Thus 'studying food [choice] in this way could allow [us] to make a powerful link between the everyday judgements that bodies make (e.g. preferences, cravings) and the ethico-political decision-making that happens in thinking through the consequences of consumption' (2008: 462). Taste, and by proxy, choice is both 'differential' and 'particular' (468) and contextualized, contingent and situated. Thus exploring visceral geographies and the role of taste in AFNs and other food networks becomes one of the ways we can understand the ways in which power 'surrounds and penetrates the human relationship with food' and, indeed, food choice (469).

Conclusion: Chewing Tasty Politics

This chapter has traced the origins of 'choice' as a concept informing food and health policy to early developments in the medical sciences in the late fifteenth century. In particular, an a priori severing of the thinking subject from the material body has delimited theorization of the act of eating to two narrow formulations: as either a product of rational choice (by a subject without a body) or a 'gut-level' conditioned response (by a body that cannot choose). In the space between such contradictory and context-free explanations for individual action, 'choice' has proved an elastic concept that has been stretched to its limits in the justification of policies designed to steer consumer behaviour in desired directions. But desired by whom and for whom? Herein lies the conundrum that lies at the core of food and health policy discourses characterized by an increasing deference to the transnational food industry and its purportedly 'essential' role in food policymaking.

The implications of the under-theorization of choice in relation to body, food and eating are illuminated by a detailed examination of the multiple ways in which choice has been framed in public policy – as the UK case study demonstrates. While the presence of 'choice' was a constant across all the policies reviewed, its use has shifted in a direction that accommodates an increasing role for private sector actors who are both complicit in limiting choices to purportedly bad ones while seen as playing a key role in helping to steer consumers towards good ones. The subtlety of discourses and practices surrounding choice in public policy in the UK can be contrasted with the way in which international nutrition and development programmes set out, explicitly, to engineer choice in low and middle income countries. The impoverished understandings of local context upon which such programmes are based ignore both the socio-economic realities that constrain access to choice as well as the rich bio-cultural diversity that has traditionally characterized foodways in much of the developing world – and are ultimately undermined by globalized programmes founded on reductionist thinking originating in a different place and time.

The location of 'choice' in discourses of AFNs muddies the waters yet further, mirroring, to a great extent, its multiple and contradictory uses in 'conventional' food systems. These dynamics highlight the need for more and better conceptual tools to understand choice: within a framework that incorporates a Bourdieusian sense of class and culture, 'taste' and 'distinction' (Kvalvaag 2012). The organoleptic 'turn to taste' of recent scholarship on AFNs, led by Allison and Jessica Hayes-Conroy (2008, 2010), therefore represents a welcome point of departure. These studies re-establish the missing link between 'everyday judgements that bodies make' and the political, 'minded' decisions based on careful consideration of the consequences of consumption (Hayes-Conroy 2008: 462). Interestingly, parallel developments in human biology – notably in neurology and epigenetics (Hart 2008) – are also challenging the established dualistic paradigm, suggesting new possibilities for interdisciplinary engagement (Gordon and Lemond 1997; Kvalvaag 2012). Central to these discussions should be a thorough 'chewing over' of the visceralities of food choice and eating, not only in national and international policymaking, but also in the alternative food movements working to create better, and better 'choose-able' food futures.

References

Aas, A.-M. 2008. *Lifestyle Intervention and Insulin in the Treatment of Type 2 Diabetes: The Effect of Different Treatment Modalities on Body Weight and Cardiovascular Risk*. University of Oslo: Faculty of Medicine.

Amerman, E.C. 2010. *Exploring Anatomy & Physiology in the Laboratory*. Morton Publishing Company.

Andersen, L.F. og Drevon C.A. 2007. Ernæringsepidemiologi, in *Mat og Medisin*, edited by Drevon, Blomhoff, and Bjørneboe. Kristiansand: Høyskoleforlaget, 30–39.

Ashby, J.A. 2009. Fostering farmer first methodological innovation: Organisational learning and change in international agricultural research, in *Farmer First Revisited: Innovation for Agricultural Research and Development*, edited by I. Scoones and J. Thompson. London: Practical Action Publishing, 39–45.

Asia Rice Foundation. 2004. *Rice in the Seven Arts*. Los Baños, Philippines: Asia Rice Foundation.

Barnett, C., Cloke, P., Clarke, N. and Malpass, A. 2011. *Globalizing Responsibility: The Political Rationalities of Ethical Consumption*. London: Blackwell.

Bouis, H. 2004. Hidden hunger: The role of nutrition, fortification and biofortification, *World Food Prize International Symposium: From Asia to Africa: Rice, Biofortification and Enhanced Nutrition*, Des Moines, IA. [Online]. Available at: http://www.worldfoodprize.org/documents/filelibrary/images/borlaug_dialogue/2004/transcripts/bouis_transcript_E097EB8C8381E.pdf [accessed: 30 September 2011].

Bourdieu, P. 1984. *Distinction. A Social Critique of the Judgement of Taste*. London: Routledge & Kegan Paul.

Brooks, S. 2010. *Rice Biofortification: Lessons for Global Science and Development*. London, UK: Earthscan.

Brooks, S. 2011. Is international agricultural research a global public good? The case of rice biofortification. *Journal of Peasant Studies* 38, 67–80.

Brooks, S. and Loevinsohn, M.E. 2011. Shaping agricultural innovation systems sensitive to food insecurity and climate change. *Natural Resources Forum* 5(3), 185–200.

Brooks, S., Thompson, J., Odame, H., Kibaara, B., Nderitu, S., Karin, F. and Millstone, E. 2009. *Environmental Change and Maize Innovation in Kenya: Exploring Pathways In and Out of Maize*. STEPS Working Paper 36. Brighton: STEPS Centre.

Cabinet Office. 2008. *Food Matters: Towards a Strategy for the 21st Century*. London: HSMO.

Caraher, M. and Coveney, J. 2007. Public health nutrition and food policy. *Public Health Nutrition* 7(05), 591–8.

Castillo, G.T. 2006. *Rice in Our Life: A Review of Philippine Studies*, Manila, Philippines, Angelo King Institute, De La Salle University and Philippine Rice Research Institute.

CIAT and IFPRI. 2002. *Biofortified Crops for Improved Human Nutrition: A Challenge Programme Proposal presented by CIAT and IFPRI to the CGIAR Science Council*. Washington, D.C. and Cali: International Centre for Tropical Agriculture and International Food Policy Research Institute.

Clarke, J. 2005. New Labour's citizens: Activated, empowered, responsibilized, abandoned? *Critical Social Policy* 25(4), 447.

Clarke, J., Smith, N. and Vidler, E. 2006. The indeterminacy of choice: Political, policy and organisational implications. *Social Policy and Society* 5(03), 327–336.

Cooper, Z., Doll, HA., Hawker, D.M., Byrne, S., Bonner, G., Eeley, E., O'Connor, M.E. and Fairburn, G. 2010. Testing a new cognitive behavioral treatment for obesity: A randomized controlled trial with three-year follow-up. *Behaviour Research and Therapy* 48(8), 706–13.

Coveney, J. 2003. Why food policy is critical to public health. *Critical Public Health* 13(2), 99–105.

De Groote, H., Kimenju, S.C. and Morawetz, U.B. 2011. Estimating consumer willingness to pay for food quality with experimental auctions: The case of yellow versus fortified maize meal in Kenya. *Agricultural Economics*, 42, 1–16.

Delisle, D. 2003. Food diversification strategies are neglected in spite of their potential effectiveness: Why is it so and what can be done? 2nd International Workshop, Food-based Approaches for a Healthy Nutrition, 23–8 November, Ouagadougou.

Department of Health. 1991. *The Health of the Nation: A Consultative Document on Health in England*. London: HSMO.

Department of Health. 2004. *Choosing Health: Making Healthy Choices Easier*. London: HSMO.

Department of Health. 2005. *Choosing a Better Diet: A Food and Health Action Plan*. London: HSMO.

Department of Health. 2010. *Healthy Lives, Healthy People White Paper: Our Strategy for Public Health in England*. London: HSMO.

Department of Health and Social Security. 1976. *Prevention and Health: Everybody's Business*. London: HSMO.

Fairclough, N. 2001. The discourse of new labour: Critical discourse analysis, in *Discourse as Data: A Guide for Analysis*, edited by M. Wetherell, S. Taylor and S.J. Yates. London: Sage, 229–66.

Food Standards Agency. 1998. *A Force for Change*. London: HSMO.

Frei, M. and Becker, K. 2004. Agro-biodiversity in subsistence-oriented farming systems in a Philippine upland region: Nutritional considerations. *Biodiversity and Conservation* 13, 1591–610.

Gillespie, S., McLachlan, M. and Shrimpton, R. 2004. *Combating Nutrition: Time to Act*. Washington, DC: World Bank.

Goodman, D., DuPuis, E.M. and Goodman, M. 2012. *Alternative Food Networks: Knowledge, Practice and Politics*. London: Routledge.

Goodman, M. 2011. Towards visceral entanglements: Knowing and growing the economic geographies of food, in *The Sage Handbook of Economic Geography*, edited by A. Leyshon, R. Lee, L. McDowell and P. Sunley. London: Sage, 242–57.

Goodman, M. 2010. The mirror of consumption: Celebritization, developmental consumption and the shifting cultural politics of fair trade. *Geoforum* 41, 104–16.

Gordon, E.W. and Lemond, M.P. 1997. An interactionist perspective on the genetics of intelligence, in *Intelligence, Heredity and Environment*, edited by R.J. Sternberg and E. Grigorenko. New York: Cambridge University Press, 323–40.

Greener, I. 2009. Towards a history of choice in UK health policy. *Sociology of Health and Illness* 31(3), 309–24.

Gunnerius, W. 2003. *Aktør, handling og struktur. Grunnlagsproblemer I samfunnsvitenskapene*. Oslo: Tano Aschehoug.

Guthman, J. 2007. The Polyanyian way?: Voluntary food labels and neoliberal governance. *Antipode* 39, 456–78.

Guthman, J. 2008a. Bringing good food to others: Investigating the subjects of alternative food practice. *Cultural Geographies* 15, 431–47.

Guthman, J. 2008b. 'If they only knew': Color blindness and universalism in California alternative food institutions. *Professional Geographer* 60(3), 387–97.
Haddad, H. and Gillespie, S. (eds) 2003. *The Double Burden of Malnutrition in Asia: Causes, Consequences, and Solutions*. CA: Sage.
Hart, S. 2008. *Brain, Attachment, Personality. An Introduction to Neuroaffective Development*. Copenhagen: Karnac.
Hausman, D.M. and Welch, B. 2010. Debate: To nudge or not to nudge. *Journal of Political Philosophy* 18, 123–36.
Haraway, D. 1997. *Modest Witness@Second Millennium. The FemaleMan©* meets *OncoMouse*™. London: Routledge.
Hayes-Conroy, A. and Hayes-Conroy, J. 2008. Taking back taste: Feminism, food and visceral politics. *Gender, Place and Culture* 15(5), 461–73.
Hayes-Conroy, A. and Hayes-Conroy, J. 2010. Visceral difference: Variations in feeling (slow) food. *Environment and Planning A* 42, 2956–71.
Hayes-Conroy, J. and Hayes-Conroy, A. 2010. Visceral geographies: Mattering, relating and defying. *Geography Compass* 4(9), 1273–82.
Howson, A. 2004. *The Body in Society. An Introduction*. Cambridge: Polity Press.
Hunter, D.J. 2011. Is the big society a big con? *Journal of Public Health* 33, 13–14.
Independent. 2009. Harriet Lamb: On the road with the queen of fairtrade (28 February) [Online]. Available at: http://www.lalettredelacheteur.com/LDAENG/archives/539 [accessed: 24 May 2012].
Johns, T. and Sthapit, B.R. 2004. Biocultural diversity in the sustainability of developing-country food systems. *Food and Nutrition Bulletin* 25, 143–55.
Kvalvaag, H. 2012. *Hvorfor er livsstilsendring vanskelig? Kropp, mat og livsstil som praksis og (ernærings-) vitenskap*. Trondheim: Norwegian University of Science and Technology.
Lang, T. 2010. From 'value-for-money' to 'values-for-money'? Ethical food and policy in Europe. *Environment and Planning A* 42, 1814–32.
Latham, M. 2010. The great vitamin A fiasco. *World Nutrition* 1, 12–45.
Longhurst, R., Johnston, L. and Ho, E. 2009. A visceral approach: Cooking 'at home' with migrant women in Hamilton, New Zealand. *Transactions of the Institute of British Geographers* 34, 333–45.
Meenakshi, J.V., Banerji, A., Manyong, V., Tomlins, K., Hamukwala, P., Zulu, R. and Mungoma, C. 2010. *Consumer Acceptance of Provitamin A Orange Maize in Rural Zambia*. Working Paper 4. Washington, DC: HarvestPlus.
MI 2001. *The Micronutrient Initiative 1990–2000: A Decade of Progress, a Lifetime of Hope*. Ottawa: Miconutrient Initiative.
Montano, D., Kasprzyk, D. and Taplin, S. 1995. The theory of reasoned action and the theory of planned behaviour, in *Health Behaviour and Health Education: Theory, Research and Practice*, edited by K. Glanz, B. Rimer and K. Viswanath San Francisco: Jossey Bass, 85–112.

Morris, S., Cogill, B. and Uauy, R. 2008. For the Maternal and Child Undernutrition Study Group. Effective international action against undernutrition: Why has it proven so difficult and what can be done to accelerate progress? *Lancet* 371, 608–21.

Murcott, A. 1994. *Food and Nutrition in Post-War Britain, in Understanding Post-War British Society*, edited by P. Catterall. London: Psychology Press, 155–78.

Pacey, A. and Payne, P. (eds) 1981. *Agricultural Development and Nutrition.* Rome and New York: FAO/UNICEF.

Patel, R. 2007. *Stuffed and Starved: From Farm to Fork, the Hidden Battle for the World Food System*. London: Portobello Books.

Porter, R. (ed.) 1996. *The Cambridge Illustrated History of Medicine*. Cambridge: Cambridge University Press.

Probyn, E. 2000. *Carnal Appetites: Food, Sex, Identities*. London: Routledge.

Schön, D.A. and Rein, M. 1994. *Frame Reflection: Towards the Resolution of Intractable Policy Controversies*. London, Basic Books/Harper Collins.

Scottish Office. 1996. *Eating for Health: A Diet Action Plan for Scotland.* Edinburgh.

Shaw, S. 2010. Reaching the parts that other theories and methods can't reach: How and why a policy-as-discourse approach can inform health-related policy. *Health* 14(2), 196.

Shier, D., Butler, J. and Lewis, R. 2008. *Hole's Essentials of Human Anatomy and Physiology*. New York: McGraw-Hill.

Sommer, A. and Davidson, F.R. 2002. Assessment and vontrol of Vitamin A deficiency: The annecy accords. *Journal of Nutrition, Proceedings of the XX International Vitamin A Consultative Group Meeting* 2845s–2850s.

Stein, A.J., Meenakshi, J.V., Qaim, M., Nestel, P., Sachdev, H.P.S. and Bhutta, Z.A. 2005. *Analysing the Health Benefits of Biofortified Staple Crops by Means for the Disability-Adjusted Life Years Approach A Handbook Focusing on Iron, Zinc and Vitamin A.* Washington, DC and Cali, International Centre for Tropical Agriculture and International Food Policy Research Institute.

Stirling, A. 2007. A general framework for analysing diversity in science, technology and society. *Journal of the Royal Society Interface* 4, 707.

Teixeria, P.J., Patrick, H. and Mata, J. 2011. Why we eat what we eat: The role of autonomous motivation in eating behaviour regulation. *Nutrition Bulletin* 36(1), 102–7.

Thaler, R.H. and Sunstein, C.R. 2008. *Nudge: Improving Decisions about Health, Wealth, and Happiness*. New Haven: Yale University Press.

Chapter 8
'It is the Bacillus that Makes Our Milk': Ethnocentric Perceptions of Yogurt in Postsocialist Bulgaria

Maria Yotova

Introduction

At the public celebration of the 40th anniversary of licensing activity in November 2007, the President of ELBY, a successor to the monopolist socialist enterprise and the only state-owned company in the Bulgarian dairy sector, expressed his gratitude to the chairman of the Japanese yogurt market leader, Meiji, for building 'a positive image of Bulgaria' in Japan and for his constant efforts 'to popularize Bulgarian culture'. In his review of the company's licensing activity, ELBY's President paid special attention to the success of 'Bulgarian yogurt' in Japan and the long-term partnership between the two companies. Returning the compliment, in his congratulatory address, Meiji's Chairman explained the hardships the company had overcome to establish 'Meiji Bulgaria Yogurt' as a leading brand in the Japanese market. In conclusion, he highlighted that the two companies had a common mission; to discover new strains of *Lactobacillus bulgaricus* that thrive in 'Bulgaria, the homeland of yogurt', and to develop new milk-fermented products that contribute to human health.

This event was the biggest organized by ELBY since the collapse of the socialist system. It was widely covered by the mass media and attracted much public attention. The popularity of Bulgarian yogurt in Japan is a favourite topic of the Bulgarian mass media. Whichever aspect of Japanese life, economy or culture is introduced, the coverage consistently starts with the success story of Bulgarian yogurt in Japan. Now, it is taken for granted that Bulgaria is the homeland of yogurt. Most people know that Bulgarian yogurt is prepared with the help of the 'unique' bacterium *Lactobacillus bulgaricus*, and believe that only the plain, set type of yogurt is the 'genuine', health-giving product. In this chapter I explore what lies behind this preoccupation with yogurt in present-day Bulgaria, and examine how its valorization, as a national product, is facilitated by its global marketization, specifically in Japan. I ask what are people actually ingesting when they eat yogurt; a specific bacterium, authenticity, a nationalist tradition, or a health-giving substance?

The diversity of yogurt brands available in Bulgaria, most of which are produced by 'traditional technologies' with milk from 'ecologically clean' regions containing 'original Bulgarian' yogurt-cultures, can be seen during any supermarket visit. Such insistences on 'traditional' and 'locally produced' food, conspicuous in advertising and public space since the opening of Bulgaria to the global economy, are a sign of the general globalization that swept through the postsocialist world after the fall of the Berlin Wall. This preoccupation with 'traditional' or 'national' food, and particularly with such dairy products as cheese and yogurt in present-day Bulgaria should not, however, necessarily be read as a resistance to external global forces of change (cf. Watson 1998; Wilk 1999; Wilson 2006). Perceptions of yogurt as a 'typically Bulgarian' and 'natural' food were formed through the frameworks of mass production and mass consumption under the industrialization doctrine of the communist regime, and are part of larger processes of food standardization and internationalization (see also Jung 2009). It is exactly the 'unique technology' and 'pure cultures' of *Lactobacillus bulgaricus*, used in yogurt mass production during socialism, that provided a springboard for the development of the Japanese top brand 'Meiji Bulgaria Yogurt'. Today, the international valorization of Bulgarian yogurt is reflected back and is adopted as a source of national pride.

Building upon the more established notion that the re-evaluation of traditional food is a reaction against standard, mass-produced industrial food in increasingly industrialized international markets (Lysaght 1994; Sarasua and Scholliers 2005; Terrio 1996), I argue that international consumption and discourses of industrialization are appropriated as a form of national self-definition. In other words, this paper is an ethnographic exploration of the ways in which global processes and industrialization create a mirror through which nationalism is reflected and created.

Bulgarian Yogurt: A Regional Food or National Culture?

People's foodways are not limited by national borders, just as they may not be extended to an entire country. In this sense, we should talk of regional rather than national cuisine (Mintz 1996: 114). Yogurt, just as any other food, is not limited to geography and nationhood; it is a common product for many dairy cultures around the world. Yet, sour milk, as yogurt is called in Bulgaria, is popularly defined as a product specific to the country, superior in taste and health qualities to other fermented milks. Loaded with strong patriotic sentiments, it has come to represent the national tradition.

A number of anthropologists have made an important set of distinctions between, on the one hand, the symbolic capital shared among a group of people that is defined as national and patriotic and, on the other hand, the knowledge and experiences which happen to be contained within national boundaries (Löfgren 1989). It might be tempting to approach food in such separate dimensions:

either as a daily practice or a set of discourses; an aspect of the private (individual) or the public (collective) domain; or as an indigenous part of the local culture or an import from abroad. These are not two separate strata of culture, however, and we should not treat one as the authentic, traditional practice and the other as some artificial product of modernity (Wilk 2002: 70). In spite of the tensions that may exist between practice and performance, local and foreign, domestic and public, it is the very interaction between them that gives birth to what is often called national culture. Moreover, cultural forms perceived as national or traditional are inevitably shaped by larger historical processes, global trade, industrialization and urbanization, and advancements in science and technology. As Sidney Mintz has shown for sugar in modern history, consumer practices and the role of food in daily life are always subject to broader societal processes of economic-political and symbolic value-creation (Mintz 1985). This emphasis on process in recent studies of food is based on the anthropological proposition that the production of national/local culture is a continuous process of creativity and adjustment (cf. Appadurai 1996: 185; Pilcher 2002).

In the Bulgarian case, developments in microbiology from the beginning of the twentieth century in Europe, industrialization during the socialist period and international value-creation in Japan have played a crucial role in the making of Bulgarian yogurt, both in terms of daily practice and as a public symbol, and at the individual level and as a collective tradition. These are reflected in present day consumer attitudes towards yogurt; in notions of what a shop-bought yogurt should taste and look like; in consumer choices and concerns about the health effects of milk fermented products; and in personal narratives and public discourses of 'traditional food'. It is by passing through these processes of industrialization and internalization that yogurt has become nationalized.

The dynamics of the relationship between the daily consumption and production of yogurt, and its symbolic role in public discourses have changed over time in Bulgaria. During socialism when yogurt shifted from the domestic to the public sphere of production it was defined as the 'people's food' and was promoted as a basic food in state nutrition policies. Specialists developed a complete production and supply chain of dairy products, adapting the traditional production technologies to large-scale industrial production. Strict state standards were introduced that provided detailed instructions for every aspect of the manufacturing process.[1] As farmers were transformed into factory workers and villagers into town dwellers under a series of socialist reforms, demand for shop-bought dairy products increased significantly. Yogurt consumption was further boosted by national nutrition policies and health promotional programs. For example, it was included in school and hospital meals, or provided free of charge for workers engaged in heavy industry. Even though compromises in the quality were made almost on a daily basis, basic dairy products were some of the most

1 For more details on standardization and consumers' attitudes towards standardized food in socialist and postsocialist Bulgaria see Jung (2009).

easily accessible products to citizens. A teacher in her 50s from Sofia told me how her mother raised her primarily on yogurt because 'there was nothing else to eat' in the capital. She envied her classmates who had relatives in the country because they had access to a variety of homemade foods. Yogurt was a valuable source of protein especially for those who were not connected with the village, a class marker of the working-class people.

Since the fall of the socialist regime, both the agents and audience for public aspects of national food have changed dramatically. Marketization meant minimized state intervention in food production and consumption, rocketing prices, closure of factories, income disparities and class stratification. As Gerald Creed (2002) has described for rural Bulgaria, the painful reforms and economic instability of the 1990s meant that Western goods were out of reach for many, exposing the capitalist ideal of material abundance and a comfortable life for all as a fantasy. Daily practices and public images of food were now to be shaped by global and domestic companies, consumer associations and NGOs, local producers and tourists. At the same time, individuals were to learn to make their 'wise' choices, adjusting global tastes to their own needs and preferences.

As one of the largest food producers in the Soviet bloc, Bulgarians were taught to be proud of their fertile land and food industry. Since the end of socialism, however, most Bulgarian meat and dairy products could not fully meet the high regulatory standards set by the European Union, and the European market has been practically closed for local producers. 'We always bring up the rear' (*vinagi sme na opashkata*) is a common phrase used to express the shared perception of Bulgaria's weak position in the global economy. At the same time, media exposures of dubious production practices in the dairy sector have fuelled general distrust in food companies. Faced with neoliberal capitalism after 45 years of strict state control, consumers are concerned more than ever with issues of food quality, safety and authenticity. For some, the great diversity of brands and tastes on the shelves of shops and supermarkets is a sign of the normalization of the country (Jung 2009); for others it has become a cause of confusion and frustration (Lankauskas 2002). Yet, what distinguishes Bulgarian yogurt from other foods is that it has been recognized for its health effects in Japan and thus, has become a symbol of national success on the global scene.

Having in mind the importance of the Japanese market for the international valorization of Bulgarian yogurt, I first turn my attention to the way its taste and health qualities are understood and valorized by the Japanese consumers.

'Bulgaria, the Holy Land of Yogurt': Valorization of Yogurt in Japan

At the dawn of the twentieth century Russian scientist Elie Metchnikoff (1845–1916), a 1908 Nobel Prize winner, developed the theory that ageing is caused by toxic bacteria in the gut. He proposed that lactic acid bacteria could neutralize these toxins and thus slow the body's ageing process. He distinguished

Lactobacillus bulgaricus, isolated from homemade Bulgarian yogurt, as most effective in this respect, and recommended its daily consumption (Metchnikoff 1908: 179). Metchnikoff's thesis inspired a number of Bulgarian researchers to study the microflora and nutritional values of yogurt. They shared the position that the Bulgarian product was superior in taste and health qualities to other fermented milks, especially those that were heat treated and produced by European corporations (e.g. Atanassov and Masharov 1981; Kondratenko 1985, etc).[2] Drawing on this research, food technologists responsible for the industrialization of the dairy sector developed 'original' technologies and 'pure' yogurt cultures, which, in turn, were exported to various countries.

The marketing of Bulgarian yogurt was especially successful in Japan. At that time sweetened, heat treated fermented milk with a jelly-like texture was the only type of yogurt on the market. Motivated by Metchnikoff's theory of longevity, the Japanese dairy company Meiji developed a totally different product – plain yogurt with living *Lactobacillus bulgaricus*.[3] Due to a successful branding strategy focused on Bulgarian dairy traditions, 'Meiji Bulgaria Yogurt' gradually rose to be Japan's top-brand yogurt.[4] The product, and its marketing, fully matched the spirit of the health trends of the 1990s, and came to represent some of the trend's core values – natural food, healthy lifestyle, healing and spirituality – and the yogurt's potential to alleviate busy and stressful lifestyles has been at the centre of all its commercials since the second half of the 1980s. One of these states: 'Dr. Elie Metchnikoff. His interest in the connection between food and health led him to the study of Bulgarian yogurt. Its power supports our life today – Meiji Bulgaria Yogurt'.

In Japan today yogurt is synonymous with 'Meiji Bulgaria Yogurt', and the country name Bulgaria has come to stand for yogurt. For some consumers it

2 The basis for distinction between fresh milk yogurt and heat treated fermented milk is the existence of living lactic acid bacteria. Fresh milk yogurt, the only type produced in Bulgaria during socialism, was supposed to contain one million living *Lactobacillus bulgaricus*.

3 The origin of Bulgarian yogurt in Japan is closely related with the Osaka Expo 1970. Meiji's company history tells us that several company employees were sent to the Expo to taste the product served at the Bulgarian pavilion. A year later, they were ready to launch Japan's first plain yogurt. Drawing on Metchnikoff's research, they intended to name the new product 'Bulgaria'. However, Meiji got an unexpected refusal from the Bulgarian side: 'Yogurt is the heart of the Bulgarian people. We can't lend the country's name to a product made by another people'. As a result, in 1971 Meiji launched their product under the name of 'Meiji Plain Yogurt'. Accustomed to the sweet dessert-type of yogurt, however, Japanese consumers rejected the new sour tasting product. Hardly selling 300 packs of yogurt per day, Meiji had no other choice but to renew the negotiations with the Bulgarian side. Two years later, the company received the desired permissions and their new product was crowned with the name 'Meiji Bulgaria Yogurt' (Meiji Dairies Co. 1987).

4 With the acquirement of the FOSHU (Foods for Specified Health Use) Government certificate in 1996, 'Meiji Bulgaria Yogurt' was officially recognized as 'functional food'.

tastes good; for others it is too sour. Yet, even those who do not really like its sour taste make efforts to include it in their daily diet. For example, Akira-san, a 76-year-old man from a prefecture near Tokyo, shared with me that, although he did not like the sour taste of the Bulgarian brand, his doctor had recommended daily consumption of the yogurt for his gastritis. Following this recommendation, Akira-san's wife had been serving him the product every morning and, consequently, 'Meiji Bulgaria Yogurt' and a piece of bread replaced their typical breakfast of rice and miso soup.

The yogurt's sourness may be distasteful to some people but the thesis that it supports human health is of particular value for many Japanese consumers. As Akira-san told me, it was better to take yogurt than a stomach medicine. Others distinguished the product's sourness as 'the natural' taste of yogurt, and used its presence or absence to judge the expected health effects of other brands. Now 'Meiji Bulgaria Yogurt' stands for health and a natural way of life, the most cherished ideals associated with personal happiness and a good life in Japan. The product's name itself has become a 'guarantee' of health and authenticity. Consumers are constantly reminded that Meiji's product is Japan's first plain yogurt ('the legitimate yogurt', as the package says) and that it is licensed by Bulgaria. The package also tells us that the product contains *Lactobacillus bulgaricus* (strain No. 81), directly imported from Bulgaria, 'the homeland of yogurt'. Meiji's advertisements show idyllic pastoral scenery from Bulgaria, big families, and healthy elderly people who lead a peaceful and happy life in harmony with nature. Beautiful pictures from Bulgaria decorate the walls of the only Bulgarian restaurant and Yogurt Bar in Japan, both managed by Meiji. A life-size mannequin of the Bulgarian sumo wrestler Koto Oshu welcomes clients at the entrance.[5] In one of the product's commercials he states convincingly: 'Yogurt? Of course, I eat yogurt every day. It is a gift from heaven'. Trying to convey this sacredness, another advertisement narrates: 'Here the wind is different; the water is different; the light is different. In this land delicious yogurt was born'.

This is how sophisticated marketing forces have transformed Bulgarian sour milk into a culturally meaningful product for the Japanese consumer. Upgrading the 'people's food' with new values, images and meanings, Meiji has not only made a profitable commodity, but also created a beautiful picture of *yoguruto no seichi Burugaria* (Bulgaria, the holy land of yogurt), which is integral to the story told by companies, politicians and the mass media in postsocialist Bulgaria.

---

5 Being healthy (i.e. eating health food including yogurt) is one of the most important criteria for personal happiness. Its personification is Koto Oshu, who is considered to be 'handsome', 'good-humoured' and 'successful'.

Yogurt in Postsocialist Bulgaria: A Reflexive Sacredness

During my fieldwork in the north-eastern village of Getsovo, famous for yogurt production since before the socialist period, I was to hear almost the same story many times; how 'powerful' the local bacterium was, and how it would make delicious and healthy yogurt. For example, Olga, a 74-year-old lady and a leader of the local folklore group, explained: 'They say there is a special herb in the meadows nearby. The sheep and cows eat it, and then you get this natural flavour. The rest is in the *bacillus*. You can't see it, but it is here, in the nature around us'. The members of the local folklore group, whose age varies between 60 and 84, have even created a song that praises the magic herbs and pasture, as well as the industriousness of the local women as 'masters of yogurt'. Its refrain states that 'the milk of Getsovo' (that is the local yogurt) will restore the sick to health, and will make the elderly younger.[6] To know the secrets of milk fermentation is a matter of pride, and hence one of the first culinary skills local women eagerly pass down to their daughters and granddaughters is the knowledge of how to prepare delicious yogurt.

This is probably one of the reasons why almost any woman, even in her teens, can easily explain how to make 'real' yogurt at home. It is estimated that about 20% of the yogurt consumed in Bulgaria is homemade and circulates through informal networks (Marinova 2010), linking rural and urban household economies. In spite of all EU regulations and government efforts to curb the informal distribution of raw milk, which is mainly used for home production of yogurt, one can see farmers from nearby villages selling milk on the corners of streets and in small markets in city residential areas. People believe that Bulgarian yogurt cannot be prepared with milk sold in the supermarkets. It is only the neighbour's cow that can produce the milk for the 'really sour' sour milk, naturally thick and with cream on top, as their grandmothers once prepared.

One day in February 2009, Diana, a 16-year-old high school student with whose family I was staying during my fieldwork, told me *Lactobacillus bulgaricus* was one of the topics of discussion in her geography class. She was surprised to learn that the Bulgarian yogurt cultures were exported even to such a remote country as Japan, and wondered if the Japanese could continuously prepare yogurt at home since the bacterium would change under different climatic conditions in Japan. Diana also told me she was interested to visit the 'only yogurt museum in the world' that had been recently established in a small village, the birthplace of the

6 The song is a part of the Getsovo women's performance at the Yogurt Fair that is held every summer in the nearby town of Razgrad. The event started in 2001 under the initiative of Razgrad Municipality as an attempt to restore the fair, which was held weekly in the region before the socialist era. The main attraction during the three-day fair is the homemade yogurt and traditional meals contest organized in the local ethnographic museum. There is also a bazaar of milk products, demonstrations in ritual yogurt making, and folk dance performances.

Bulgarian scientist who discovered *Lactobacillus bulgaricus*. As 'a great yogurt eater', she herself could easily prepare a jar of yogurt, and preferred to take it for dinner because it was 'a light food, and good for the stomach, too'.

A few days later, Lilly, a self-employed woman in her 50s, once again raised the topic of yogurt. She referred to the recent flood of information in the mass media that warned that there were many 'counterfeit' products on the dairy market and that there was confusion over which product was the 'real' one. Adding fuel to the fire was the comment made by a respected professor in food technologies, during an interview for the Bulgarian National Radio, that *Lactobacillus bulgaricus* had emigrated to Japan and Bulgarian consumers did not consume Bulgarian yogurt any more. Lilly blamed the dairy companies for this situation because 'nowadays they would put anything in the people's food'. She felt especially indignant at Danone, accusing the market leader for changing consumer preferences in favour of much sweeter products than 'the yogurt once prepared by our grandmothers'. She recognized that these new products might meet all EU standards, but she was dissatisfied because of the impositions these standards placed on local farmers, while at the same time allowing the use of additives by manufacturers. She could not understand why 'we' had to obey 'their' standards for the production of 'our' traditional products. Once, she happened to buy yogurt that was rather 'gooey' and 'icky'. Since nobody in her family liked it, she decided to use it for *mekitsa* dough. However, to her disappointment, the dough did not swell enough so, in the end, she had to discard it. Lilly was convinced that the reason for both the strange taste of the product and the failed dough was the foreign standards and yogurt bacteria.

It is not only Lilly who is suspicious of Danone's products. Since the quality of a given company's product is judged by its taste (in terms of sourness and texture), and by its potential to serve as a leaven for homemade yogurt, I would be warned against using Danone's products for home preparation. In this case one needs 'real' yogurt produced by 'traditional' methods. On the Bulgarian National Consumer Association's website (BNAP 2005), one can find the results of a survey on the content of *Lactobacillus bulgaricus* in Bulgaria's top 15 yogurt brands. It shows that Danone's most popular brand 'Na baba' (Grandmother's Yogurt) contains the smallest number of this bacterium. Danone's products are generally said to be filled with many additives; from starch to various flavouring and preservative agents. The same rumour extends to the global maker's brand with *Bifidobacterium*, 'Aktivia', which a man from Sofia defined as 'the fast food' of the yogurt market.

Bifidobacterium is an anaerobic bacterium that can be found in the gastrointestinal tract of mammals and other animals; it does not belong to the group of lactic acid bacteria, and consequently it does not participate in milk fermentation itself. Research has shown that some *Bifidobacterium* strains may exert a range of beneficial health effects, and for that reason these are added to various milk fermented products (Ishibashi and Shimamura 1993; Tamime, Marshall and Robinson 1995). This is the case with Danone's 'Aktivia' brand in Bulgaria. I met several people in Sofia who recognized its benefits. For example, a young woman working as an interpreter in a foreign company expressed her opinion that 'most Bulgarians will never admit it, but Danone produces yogurt of high quality, superior to all domestic products'. She was convinced of the health effects of their *Bifidobacterium* brand, and was regularly taking it for intestinal regulation. Of course, there are many others who do not much care for the 'traditional' taste of yogurt; they are comfortable with the new yogurt market, and are open to new tastes and technologies. Most of these belong to the educated elite or the *nouveau riche*; they have travelled extensively around Europe and welcome Danone's products for their consistent quality.

For the majority, however, it is only yogurt with *Lactobacillus bulgaricus* that could keep them in good health. It is one of 'our' natural foods and its sour taste is perceived as a guarantee that the bacteria is 'alive', while products with *Bifidobacterium* (usually sweetened, in various flavours, and much more expensive) are classified as artificial products 'made for profit' rather than for human health. As in the case of Russian natural food philosophies, such an attitude resonates with cultural concerns for economic morality, or as Cynthia Gabriel has put it, 'healthy food is not-for-profit' (Caldwell 2010; Gabriel 2005: 183). Yogurt made with milk from the neighbour's cow is health giving because it is embedded in social relationships, trust and mastership. I have heard many stories of recovery from various illnesses – from cancer to anorexia – thanks to the local *bacillus*. An elderly woman, who told me the story of her daughter's recovery from breast cancer thanks to homemade, goat-milk yogurt finally concluded: 'It is the *bacillus* that makes our milk, my girl. It is unique. When I was young I didn't eat much yogurt, but since I take it every day, my blood pressure has been normal and I feel so energetic …'.

Such a view of yogurt in Bulgaria is reminiscent of the way the food is imagined in Japan, and reflects recent transnational discourses on the beneficial effects of lactic acid bacteria. During socialism, when yogurt was defined as 'the people's food' and was understood to give nutrition to factory workers and townspeople, there was little thought given to the existence of *Lactobacillus bulgaricus*, its uniqueness or its health effects. There were no public discussions about Bulgarian yogurt traditions, and of course, Japan was never a factor. Since the opening of the country to the global economy, not only new foods and commodities, but also various dietary ideas and philosophies have entered the national market to compete for the ex-socialist consumer. Meanwhile, yogurt consumption has decreased more than twice to 27kg per year, or 70g per day (NSI 2011). Now it may not be consumed as much as previously, when it was one of the few accessible foods on the socialist market, but it has acquired many mythological meanings; it has become healthy, unique, and Bulgarian instead. Its taste should be 'natural' and its preparation methods 'traditional'.

In other words, with the increase of spatial, cognitive, and institutional distances between production and consumption in postsocialist Bulgaria, the culturally constructed stories of commodity flows have acquired a particular intensity (Appadurai 1986: 48). This intensity is obvious in the mythology of Bulgarian yogurt; a mixture of ethnocentric claims enriched with some borrowings from the ideology of transnational food movements. Due to its international valorization, the people's food has become a representation both of national traditions and the health conscious, nature oriented lifestyles associated with Western modernity. Unlike the consumption of vodka at Lithuanian weddings or of 'traditional' Czech food at home, which show how transnational influences may cause a deep conflict of values within the local culture (Lankauskas 2002; Passmore and Passmore 2003), Bulgarian yogurt has the symbolic capacity to unite the nation, not because people are trying to resist global forces in the face of perceived threats to cultural continuity (Watson 1998; Wilk 1999), but precisely because it anchors individual bodies and daily practices to these international scales and transnational values.

In the next section, I will define some specific aspects of this ethnocentric mode of yogurt consumption in Bulgaria, reflecting upon how the national and the transnational are intricately entwined in the concept of Bulgarian yogurt, and its importance as a strategy for redefining the national self in a destabilizing postsocialist transition.

The People's Food Transformed: *Lactobacillus Bulgaricus* in Focus

The perceptions of yogurt that have crystallized in the years since Bulgaria's postsocialist transition have much in common with the nationalist attitudes and practices noted in other anthropological accounts addressing Eastern Europe. For example, in discussing the preoccupation with homemade and natural foods in contemporary Russia, Caldwell shows how it is deeply related to people's understanding of Russianness as a quality deeply rooted in the material landscape (Caldwell 2010: 89). This focus on a 'nature' that has distinctive bio-national qualities can also be seen in the Bulgarian case. In spite of being invisible to the naked eye, *Lactobacillus bulgaricus* has become an important part of a collective consciousness of the national landscape; it is in the soil, on the leaves of plants and trees, or in the morning dew. In the traditional value system, milk fermentation is closely connected with people's understanding of life and wellbeing, with yogurt being perceived as a live food and the leaven itself a symbol of life. In the past, early in the morning on St. George's Day, young women would gather dew and herbs for ritual fermentation (Markova 2006). Today's perceptions of yogurt as a natural food, and the connections people make between its health qualities and sour taste, have their grounds in such pre-socialist practices and values. In this sense, even though socialist industrialization has played a crucial role in the rise of yogurt being central to the national self, these attitudes should not be seen as an attempt to preserve socialist values of sociality and collective responsibility (Caldwell 2002), or as a result of the symbolic devaluation of the 'West' combined with rising nostalgia for the 'East' (Klumbyte 2009). Rather, many of these established practices were threatened under the influence of socialist industrialization (Hirata et al. 2010). As Pilcher has observed in relation to the rise of Mexican tortilla industry, 'the homogenizing effects of national food processing companies may pose as great a threat to local cultures as the more visible cultural imperialism represented by Ronald McDonald' (2002: 223). Interestingly, however, the celebration of yogurt as 'traditional' Bulgarian food is not a reaction against the type of standard, mass-produced industrial product (cf. Lysaght 1994; Sarasua and Scholliers 2005). On the contrary, the taste of industrial yogurt produced to a strict socialist standard has now become a criterion for judging its 'natural' taste. During socialism, ambitious to prove the superiority of the 'original' Bulgarian technology and yogurt cultures, specialists insisted on the 'natural' taste and health qualities of the socialist product (Kondratenko 1985). Except for the 'luxurious' taste of one product with honey ('Medenko') and another with cream and sugar ('Snejanka'), which were hardly ever to be found on the socialist shelves, there was only one dominant taste on the market, and that was the sour taste of the state-subsidized yogurt. Now, many remember with nostalgia its 'natural' taste that is good for both physical and economic health. At the same time, they seem to have forgotten all the skepticism and complaints about factory-produced foods and public food facilities (Jung 2009: 44).

Since the fall of the socialist regime, the advent of the global maker Danone has brought about a variety of flavours and colours on the Bulgarian yogurt market. These inevitably come into conflict with the deeply rooted sensual experiences of yogurt as a product with a 'natural' sour taste and conceptions of what a shop bought product should be like, developed under the socialist system. Ironically, it is exactly the aggressive marketing strategy evolved by the global company that has given rise to the discourse of 'health food' and 'healthy diet', thus paving the way for *Lactobacillus bulgaricus* to appear with fanfare on the national market. And it is some of Danone's most effective advertisement campaigns that, by drawing attention to the importance of lactic acid bacteria for human health, have served as a catalyst for the conception that *Lactobacillus bulgaricus* is a microorganism of special value. Hence, the presence of a global player on the national Bulgarian market has significantly contributed to the recognition of yogurt as 'health food', and provided local people with a powerful resource for self-assertion; it is 'our' bacillus and 'our' yogurt that brings health to the modern world. Being freed from the socialist mission to produce 'the people's food', the invisible bacillus in the form of Bulgarian yogurt has now taken on a new task: to endow the postsocialist consumer with health and a sense of national significance. In this sense, Bulgarian yogurt provides an antidote to the widespread formula of 'we always bring up the rear', and an alternative mode of self-definition ('we, too, have something to be proud of').

This is how the 'local' itself is reinvented through transnational discourses of health and wellbeing. Such a cultural production resembles the process of creolization that Wilk describes for Belizean cuisine, in which different cultural meanings are fused to create new forms of national culture (Wilk 2002, 2006). That the new formula has been integrated quite successfully in the Bulgarian mind is demonstrated by the passionate way people talk about Bulgarian yogurt, from the heated discussions in the mass media, or from the establishment of 'the only yogurt museum in the world'. All these daily conversations, public discussions and exhibitions dedicated to Bulgarian yogurt reveal various aspects of consumer ethnocentric perceptions of yogurt in present-day Bulgaria. One of the most important of these is the wide-spread consciousness that *Lactobacillus bulgaricus* was discovered by a Bulgarian scientist, and that his scientific achievement has significantly contributed not only to the spread of yogurt in the West, but also to its international recognition as a source of life and longevity. Having in mind the overall preoccupation with the native bacterium, it is not surprising that the life of the Bulgarian physician Grigoroff, who discovered its existence in 1905, has drawn much public attention. His achievement is presented as a great contribution to the development of science and to the prosperity of European civilization. In 2009 his discovery was even nominated as one of 'the greatest Bulgarian events of the twentieth century'.[7]

At the same time, Metchnikoff's theory of longevity has become the pillar of the health rhetoric surrounding *Lactobacillus bulgaricus*. His research lies at the root of the ethnocentric attitudes towards yogurt expressed by Bulgarian academics, journalists, and in personal narratives. It is also connected with another important aspect of yogurt perceptions in Bulgaria, namely the widely shared conception that Bulgarian yogurt is the secret of human longevity.[8] No matter if it is a personal story of a miraculous recovery from cancer, a photo of a centenarian in a newspaper, a recipe book, or an academic article on food technologies, the message is that yogurt brings health and longevity to everyone.

Further suggesting the significance of Bulgarian cultural traditions and knowledge to the popularity of yogurt in the Western world, the yogurt museum recently established in Grigoroff's birthplace tells the story of Louis XI of France (1423–1483) whose stomach problems were cured with sheep milk yogurt from Bulgaria. This is just another example that demonstrates the dynamic interplay between the national (traditional or culturally specific) meanings and the transnational (informational, universal) values intricately mixed in the concept of Bulgarian yogurt. It is 'our' tradition and 'our' food, yet at the same time it

7 The video introducing the nominations is quite astonishing. Against the background of Geneva's landscape, where Grigoroff was a student at the time of the discovery, one can see the Japanese 'Meiji Bulgaria Yogurt', from the changes to its packaging to the diversity of products in the brand series, further indicating that the Japanese recognition of the qualities of Bulgarian yogurt is as important as the discovery of the bacterium itself.

8 A joke says that two foreign journalists came to a Bulgarian village to ask for the secret of longevity. An elderly woman directed them to the fields where her 90-year-old husband was working. They found the old man under a tree crying. The journalists asked him why he was crying and he answered: 'Because my father hit me'. Astonished, the foreign journalists found his father working in the vineyard. They asked him about the quarrel, and were even more surprised to hear the reason; the 'youth' had eaten up his grandfather's yogurt.

is assumed to bring health to all people regardless of their national, historical or social background. Bulgarian yogurt might be 'unique' and 'special' in terms of taste and preparation methods, but it has also come to represent health, longevity, and wellbeing as universal human values associated with modern society and the West. As it is both 'the elixir' of the Bulgarian ancestors and a health food that supports modern life, yogurt provides a self-assertive claim persuasive enough to put the national tradition on a par with what is thought to be European modernity. Thus the fusion of 'the universal and the particular' (Robertson 1992: 172) that occurs in the quotidian production and consumption of yogurt gives people a sense of what it is to be Bulgarian in era of dramatic social change and reorientation to the West.

The grounds for this sense of national significance, however, were not acquired through the collision between the 'local' *Lactobacillus bulgaricus* and the 'global' *Bifidobacterium.* It is precisely the international valorization of yogurt in Japan that has created and enriched the beautiful image of 'Bulgaria, the homeland of yogurt', supported with research data on the beneficial effects of *Lactobacillus bulgaricus*. And it is through the Japanese consumption of Bulgarian yogurt that the ethnocentric claims of best health and best taste are continuously confirmed and justified in Bulgaria. In this sense, the disregard for Danone's products should not be read as a resurgence of local practices against perceived threats to cultural homogeneity in the face of globalization (cf. Montagne 2006; Watson 1998). Rather, it is a token of the importance Bulgarian people attach to yogurt as a source of national pride in times of economic instability and social change, and the widespread consciousness that Bulgarian yogurt has been recognized for its taste and health qualities by Japanese consumers. The story of Bulgarian yogurt in Japan, willingly spread by the Bulgarian mass media, is not seen as a successful business strategy deployed by a Japanese private company, but as the very spirit of Bulgarian collective traditions. And it is because Bulgarian yogurt has crossed the borders of the country to become a symbol of health and happiness in one of the world's economic powers that Bulgarians can feel connected to the modern West.

To be recognized as a significant part of Europe has historically been a Bulgarian national ideal. However, the road to Europe has always been difficult and obscured by various economic, political, or ideological factors. Once transformed into a meaningful cultural product for the Japanese consumer, Bulgarian yogurt comes back to its 'homeland' to convince the Bulgarian consumers of its sacredness and their own national significance. Representing cherished national ideals in times of shaking cultural values, yogurt has transformed itself into a cultural emblem; a source of health and nourishment both for the individual bodies and for the national soul.

Concluding Remarks: On the (Inter)National Character of Yogurt in Bulgaria

What is particularly striking about the case of Bulgarian yogurt is that, despite the diversity of agents and consumers, there seems to be a public consensus about what Bulgarian yogurt is and should be. With a few exceptions, even those who did not really like the sour taste of what is thought to be 'typical' yogurt were cautious when talking on the subject. Some would ask me to switch off the recorder – 'this is only between you and me' – others would remind me not to record any negative opinions because 'after all we are Bulgarians' and 'this will spoil our image'. Now, two decades after the collapse of the socialist system, people continue to seek 'real' yogurt on the shelves of supermarkets; they exchange information about various brands, and participate in real-time and virtual discussions, sharing memories and experiences of 'traditional' foods. Postsocialist consumers have also become increasingly health conscious and ensure that the product contains living bacteria. They also look to verify the provenance of the milk, and check the fat content, the expiration date and the standardization label. In other words, they behave like many other consumers around the world, sharing similar anxieties about food safety and quality.

It is because food is ingested and digested by our bodies that it is an important source of personal and public concerns (Nützenadel and Trentmann 2008: 2; see also Abbots this volume). Japanese people take pride in brands like Toshiba and Toyota, but few of them are concerned with, let's say, the origins of the metal in the nuts of the latest model. Food, however, in becoming part of ourselves and our bodies, raises sensitive questions of quality and authenticity. Consequently, it is equally important for Japanese, as well as for Bulgarian consumers, whether the yogurt they ingest is 'real', if its taste is 'natural', or if it contains living *Lactobacillus bulgaricus*. Moreover, as people's 'choices' are inevitably formed through discourse and narratives (Scholliers 2001: 9; see also Brooks et al. this volume), the food they eat every day cannot be separated from their ideas and images of food, from the beautiful Bulgarian scenes in Japanese advertisements to the heated discussions and notions of the 'Bulgarian tradition' in the mass media. The reason for the close connection between the habitus of food and the collective imagination surrounding it is precisely in its capacity to make sense as a signifier, classifier and identity maker (Douglas and Isherwood 1979: 40). As people absorb food inside their bodies, they seize the opportunity to demarcate the outside world, and this 'principle of incorporation' is the basis of collective identity (Fischler 1988: 275).

The classification of yogurt as sour tasting 'health food' attributes a given place in the world to those who eat the product. They are bodily and symbolically linked with other Bulgarians, Japanese and Europeans who supposedly share similar images and ideals of healthiness, happiness and wellbeing. Thus, through the individual consumption and public discourses of 'their' yogurt, Bulgarians get a sense of shared cultural experience not only with other members of the

Bulgarian nation, but also with an imagined international consumer. This feeling of connectedness raises their confidence that the Bulgarian traditions, internationally recognized and appreciated, are of great cultural value to the modern world.

Interestingly, these symbolic and bodily connections are crystallized by discourses of 'unique' industrial technology and health food, international consumption and a significant 'scientific contribution' to the West. And it is because Bulgarian people eat and use yogurt in their daily lives that these discourses of industrialization and internationalization have been appropriated as a form of national self-definition. As the commodity moves through time and space, always enriched by new information, ideas and values, various meanings emerge and fuse together to establish a national myth of international significance. And it is from these close interactions between daily practice and public performance, discourses of industrialization and internalization, nationalistic aspirations and transnational ideas that Bulgarian yogurt was born. The making of Bulgarian yogurt comes to demonstrate once again that common national memories and understandings are sometimes more strongly articulated in shared smells, tastes and visions (Löfgren 1989): in other words, in the non-verbal, even intangible and invisible (like *Lactobacillus bulgaricus*), forms.

Acknowledgements

This paper has benefited significantly from the critical readings of friends and colleagues. I am especially grateful to Chikage Oba, Ayumi Hotta and above all Prof. Hirochika Nakamaki, whose encouragement, guidance and support made this research possible.

References

Appadurai, A. 1986. Introduction: Commodities and the politics of value, in *The Social Life of Things: Commodities in Cultural Perspective*, edited by A. Appadurai. Cambridge: Cambridge University Press, 3–63.

Appadurai, A. 1996. *Modernity at Large: Cultural Dimensions of Globalization*. Minneapolis: University of Minnesota Press.

Atanassov, G. and Masharov, I. 1981. *Mlechnata promishlenost v Bulgaria v minaloto i dnes* (Bulgarian dairy industry in past and present). Sofia: Zemizdat.

BNAP 2005. Sravnitelen test na balgarski kiseli mleka (Comparative test of Bulgarian yogurt brands). [Online: Bulgarian National Association 'Active Consumers']. Available at: http://www.aktivnipotrebiteli.bg/тест/14/ [accessed: 14 December 2009].

Caldwell, M. 2002. The taste of nationalism: Food politics in postsocialist Moscow. *Ethnos* 67(3), 295–319.

Caldwell, M. 2010. *Dacha Idylls: Living Organically in Russia's Countryside*. Berkeley: University of California Press.

Creed, G. 2002. (Consumer) paradise lost: Capitalist dynamics and disenchantment in rural Bulgaria. *Anthropology of East Europe Review* 20(2), 119–25.

Douglas, M. and Isherwood, B. 1979. *The World of Goods: Towards an Anthropology of Consumption*. New York: Basic Books.

Fischler, C. 1988. Food, self and identity. *Social Science Information* 27(2), 275–92.

Gabriel, C. 2005. Healthy Russian food is not-for-profit. *Michigan Discussions in Anthropology* 15(1), 183–222.

Hirata, M., Yotova, M., Uchida, K. and Motosima, H. 2010. Milk processing system in the south-west of Bulgaria. *Milk Science* 59(3), 237–53.

Ishibashi, N. and Shimamura, S. 1993. Bifidobacteria: Research and development in Japan. *Food Technology* 47(6), 126–35.

Jung, Y. 2009. From canned food to canny consumers: Cultural competence in the age of mechanical production, in *Food and Everyday Life in the Postsocialist World*, edited by M. Caldwell. Bloomington and Indianapolis: Indiana University Press, 29–56.

Klumbyte, N. 2009. The geopolitics of taste: The 'Euro' and 'Soviet' sausage industries in Lithuania, in *Food and Everyday Life in the Postsocialist World*, edited by M. Caldwell. Bloomington and Indianapolis: Indiana University Press, 130–53.

Kondratenko, M. 1985. *Bulgarsko kiselo mliako* (Bulgarian yogurt). Sofia: Zemizdat.

Lankauskas, G. 2002. On 'modern' Christians, consumption, and the value of national identity in post-soviet Lithuania. *Ethnos* 67(3), 320–44.

Löfgren, O. 1989. The nationalization of culture. *Ethnologia Europaea* 19(1), 5–24.

Lysaght, P. 1994. *Milk and Milk Products from Medieval to Modern Times*. Edinburgh: Canongate Academic.

Marinova, G. 2010. Potreblenieto na bio kiselo mliako se uvelichava (The consumption of bio yogurt is increasing). *Progressive* 31(1), 40–42 [Online]. Available at: http://www.progressive.bg/htdocs/data/archive/2010-02/progressive_2010-02.pdf [accessed: 27 June 2011].

Markova, M. 2006. Traditsionna tehnologia na bulgarskoto kiselo mliako (Traditional technology of Bulgarian yogurt). *Minalo* 2, 48–56.

Meiji Dairies Co. 1987. *Oishisa to kenko wo motomete – Meiji Nyugyo 70 nenshi* (In search of deliciousness and healthiness: 70 years' history of Meiji Dairies). Tokyo: Meiji Dairies Co.

Metchnikoff, E. 1908. *The Prolongation of Life: Optimistic Studies*. New York and London: G.P. Putnam's Sons.

Mintz, S. 1985. *Sweetness and Power: The Place of Sugar in Modern History*. New York: Viking.

Mintz, S. 1996. *Tasting Food, Tasting Freedom: Excursions into Eating, Culture, and the Past*. Boston: Beacon Press.

Montagne, K. 2006. The quest for quality: Food and the notion of 'trust' in the Gers Area in France, in *Food, Drink and Identity in Europe*, edited by T. Wilson. Amsterdam and New York: Rodopi, 159–77.

NSI 2011. *Households Income and Expenditure*. [Online: National Statistical Institute]. Available at: http://www.nsi.bg/otrasalen.php?otr=44 [accessed: 7 January 2012].

Nützenadel, A. and Trentmann, F. 2008. Introduction: Mapping food and globalization, in *Food and Globalization: Consumption, Markets and Politics in the Modern World*, edited by A. Nützenadel and F. Trentmann. Oxford and New York: Berg, 1–20.

Passmore, B. and Passmore, S. 2003. Taste and transformation: Ethnographic encounters with food in the Czech Republic. *Anthropology of East Europe Review* 21(1), 37–41.

Pilcher, J. 2002. Industrial tortillas and folkloric Pepsi: The nutritional consequences of hybrid cuisines in Mexico, in *Food Nations: Selling Taste in Consumer Societies*, edited by W. Belasco and P. Scranton. New York and London: Routledge, 222–39.

Robertson, R. 1992. *Globalization: Social Theory and Global Culture*. London: Sage.

Sarasua, C. and Scholliers, P. 2005. The rise of food market in European history, in *Land, Shops and Kitchens: Technology and the Food Chain in Twentieth-Century Europe*, edited by C. Sarasua, P. Scholliers and L. Van Molle. Turnhout: Brepols Publishers, 13–29.

Scholliers, P. 2001. Meal, food narratives, and sentiments of belonging in past and present, in *Food, Drink and Identity: Meals, Food Narratives, and Sentiments of Belonging in Past and Present*, edited by P. Scholliers. Oxford and New York: Berg, 3–22.

Tamime, A.Y., Marshall, V.M.E. and Robinson, R.K. 1995. Microbiological and technological aspects of milks fermented by bifidobacteria. *Journal of Dairy Research* 62, 151–87.

Terrio, S. 1996. Crafting grand cru chocolate in contemporary France. *American Anthropologist* 98(1), 67–79.

Watson, J. 1998. *Golden Arches East: McDonald's in East Asia*. Stanford and California: Stanford University Press.

Wilk, R. 1999. 'Real Belizean food': Building local identity in the transnational Caribbean. *American Anthropologist* 101(2), 244–55.

Wilk, R. 2002. Food and nationalism: The origins of 'Belizean' food, in *Food Nations: Selling Taste to Consumer Societies*, edited by W. Belasco and P. Scranton. New York and London: Routledge, 67–89.

Wilk, R. 2006. *Home Cooking in the Global Village: Caribbean Food from Buccaneers to Ecotourists*. Oxford and New York: Berg.

Wilson, T. 2006. *Food, Drink and Identity in Europe*. Amsterdam and New York: Rodopi.

Chapter 9
The Transition to Low Carbon Milk: Dairy Consumption and the Changing Politics of Human-Animal Relations

Jim Ormond

Introduction

Using the case study of a pint of UK milk, this chapter examines how a progressive focus on lowering one's carbon footprint has resulted in co-existent and mutually productive shifts in both agricultural practices and consumers' perspectives on the foods we eat. Drawing on a range of academic and policy-led studies, I explore how the desire for a low carbon pint of milk has resulted in, and is informing, changes to the way in which animals are bred, cared for and eat – and in turn to the food that we consume. This chapter will begin by discussing the landscape within which the transition to low carbon milk is set, including the extent to which consumer desire, corporate branding and market-led governance have integrated environmentally-friendly targets into capitalist food systems. The chapter will then review key approaches that have emerged for the mitigation of greenhouse gas emissions from the dairy sector and consider how these impact upon human-animal, and ultimately production-consumption, relations. Finally, the chapter will offer a summary discussion of the wider potential implications and directions that the transition to low carbon, and more broadly 'ethical' or 'sustainable', food systems may entail. In particular, it will show how these inform existing and future relationships between humans and animals and, consequently, also the food we eat.

For millennia, humans have lived alongside animals, offering feed and refuge in exchange for food, power and companionship. During these centuries of domestication, 'wild' animals have been transformed into creatures of human desire and necessity in terms of their body shapes, feeding patterns and intrinsic behaviours (Rollin 2008). However at the beginning of the twenty-first century, against a backdrop of rising human populations, increasing demands for animal products, ever-more observable limits to planet Earth's finite resources and the looming spectre of global climate change, this ancient relationship between human and animal faces new challenges. Indeed the report *Livestock's Long Shadow*, by the United Nations Food and Agricultural Organization (Steinfeld et al. 2006) has cited the livestock sector as one of the key contributors to the most serious environmental problems facing society, at every scale from local to global.

One of the charges most vociferously levelled at the livestock sector is its role in anthropogenic climate change, in particular through emissions of the greenhouse gases (GHGs) methane (CH_4) and nitrous oxide (N_2O), which have, respectively, global warming potentials 25- and 298-times higher than carbon dioxide (CO_2). Estimates of the GHG emissions from the livestock sector vary considerably, whilst direct emissions account for approximately 9% by including indirect sources, particularly emissions associated with land use change, this figure rises to 18% (Gill et al. 2010). The impact of dairy farming has received particular attention, largely due to the volume of methane emissions released during enteric fermentation. It is estimated that the carbon footprint of the UK dairy sector is 15.5 million tonnes CO_2 per year, approximately 2% of the UK's total emissions (McAllister et al. 2011). This impact is further magnified in light of the prediction that by 2050 the human population will have grown to nine billion, global consumption of meat will have tripled and global milk consumption doubled (FAO 2010). To combat this mounting concern scientists, commentators and politicians have identified a variety of 'solutions', which, as part of a wider proposed transition to a low carbon society, seek to recast humans' relationships with the livestock sector and ultimately with the food that we eat. Proposals range from the abolition of meat consumption, for example, 'go vegetarian save the world', 'low carbon diets', 'meatless Mondays' (Aiking et al. 2006; Davis and Sonesson 2008; Stehfest et al. 2009); to promoting lower carbon alternatives, such as replacing beef and lamb with meat from low carbon animals such as kangaroos (Garnaut 2008) or replacing cows' milk with lower carbon soya milk (*The Ecologist* 2011) or fresh milk with UHT milk (*The Farmers Guardian* 2008); to proposals that seek to reconcile existing production processes with substantial reductions in their greenhouse gas emissions – either through efficiency measures, for example, larger / more productive animals or through more fundamental changes to the way in which the food we eat is produced, for example through genetic engineering.

Milk Production in the UK

To understand the current transition towards, and desire for, a low carbon pint of milk it is worth briefly considering the wider landscape of milk production within the UK – and the actors responsible for the changing nature and spaces of governance of this industry. Specifically three trends, or drivers, can be identified, which have shaped and will continue to shape the production of milk within the UK: firstly a progression towards the market-led governance of food supply chains; secondly a growing consumer awareness and desire for 'responsibly' and 'ethically' produced food and drink; and thirdly a mounting concern and emphasis paid, across both industry and government, to the dangers posed by anthropogenic climatic change.

From a production perspective, the abolition in 1994 of the UK Milk Marketing Board, which guaranteed a minimum price for milk producers, represented a pivotal moment and heralded the entrance of market forces, which have significantly re-shaped the industry, with three trends in particular emerging. Firstly, the number of dairy farms has steadily declined (from 35,741 in 1995 to 15,716 in 2010): in parallel average herd sizes have risen (from 73 in 1995 to 113 in 2010) (DairyUK 2010). Secondly, there has been a concentration of milk processing, with seven processors now accounting for 90% of the UK milk supply – buying from farmers, processing and selling to retailers or directly to consumers (DairyUK 2010). Thirdly, an increasing proportion of UK milk is being purchased and processed through direct contracts between individual farms and large retailers (for example, Tesco[1], Asda[2] and Sainsbury's[3]) and dairy produce manufacturers (for example, Cadbury[4]). In 2010 approximately 2,500 farmers sold their milk through these exclusive, segregated contracts representing a 20% share of the UK market (DairyUK 2010).

The growth of direct contracts is of particular interest as it is representative of a wider shift towards a form of market-based governance within milk supply chains. Alongside direct contracts, and exemplifying the changing relationships between producers and retailers, a number of retailers have formed their own exclusive dairy-farm supplier groups, for example, Tesco Sustainable Dairy Group, Sainsbury's Dairy Development Group and Asda Dairy Link. Within these groups, which are premised upon creating closer relationships between end-retailer and primary producers, farmers typically receive a premium in return for complying with the retailer's specific codes of conduct and procurement requirements. As a result, and through these relationships, retailers and dairy produce manufacturers are able to increasingly influence the conditions in which the milk they sell is produced. For instance the 720 farmers within the Tesco Sustainable Dairy Group have each committed to meet the 17 absolute standards and 15 measures for improvement set by the retailer, which span attributes such as animal health, including lameness; animal welfare, for example, cleanliness; and environmental impact, such as meeting carbon reduction targets (Tesco 2011a). The changing nature of governance structures within milk supply chains can in part be traced to a series of food-safety scares, for example BSE and Salmonella in the final years

1 Tesco is the third-largest retailer in the world and is the grocery market leader in the UK (market share of around 30%). Tesco is the UK's largest milk retailer.

2 Asda, a subsidiary of American retail giant Wal-Mart, is the UK's second largest retail chain.

3 Sainsbury's is the third largest chain of supermarkets in the UK with a market share of 16.5%.

4 Cadbury, owned by Kraft food, is the second largest confectionary company globally.

of the twentieth century, which prompted a re-organization of responsibilities for risk management and food safety policy in which retailers adopted a prominent position.

In parallel to changing producer-retailer relations, the past decade has also witnessed a notable growth in consumer awareness regarding the social and environmental impacts associated with the food we consume. This has informed a growing consumer desire for more 'ethical' or 'environmentally friendly' products and a market in the UK worth £6.49bn (*The Co-Operative's Ethical Consumer Report* 2010). As part of this ethical shift, consumers have also begun to embrace low carbon consumption, with recent research revealing that 45% of shoppers would be prepared to stop buying their favourite brands if they refused to commit to measuring their product's carbon footprint (The Carbon Trust 2011). The emergence of the 'ethical consumer' as a market segment has prompted a raft of initiatives by the UK's main retailers; Sainsbury's now claims to be the world's largest retailer of Fairtrade products, all fish sold by Asda are certified by the Martine Stewardship Council and Tesco has committed to measuring and labelling the carbon footprint of 70,000 of its products.

Finally the UK milk industry, mirroring wider international and industry experiences, is facing mounting public pressure, media attention and scientific evidence regarding its contribution to anthropogenic global climatic change. This pressure has led international and domestic governments to seek pro-active positions towards greenhouse gas emissions, for instance the UK Government has established a legally binding commitment to reduce its GHG emissions by 80% before 2050. In turn, various national measures and targets have been established for the UK dairy sector, including a 20–30% emission reduction target for the UK's dairy farms by 2020 (DSCF 2008). The private sector (both individual corporations and industry bodies) has witnessed similar 'headline-capturing' climate change commitments. For example, Tesco has set the goal to become a zero-carbon business by 2050 and to reduce its supply chain emissions by 30% by 2020 (Tesco 2011b). A crucial factor behind, and facilitating, these carbon reduction targets is how carbon emissions are defined and measured. Specifically over the past five years there has been a switch from sector-based accounting models (which typically attributed vast carbon footprints to energy generation companies) to product carbon footprinting models (which assess the greenhouse gases emitted at each stage of a product's life cycle). Measuring the carbon footprint of a pint of milk this way reveals that on-farm milk production accounts for 73% of total emissions, whilst processing accounts for 9%, distribution 3%, retail 10%, use by consumer 3% and end of life, including disposal and recycling, 2% (Tesco 2008).

The level of detail provided by product carbon footprinting has resulted in, and indeed necessitated, the establishment of mechanisms and tools by retailers and other actors with carbon reduction commitments that engage their supply chains around their collective carbon footprint. This engagement is particularly pronounced for products such as milk, in which up to 90% of the total emissions are outside retailers' direct control. For instance Cadbury have published the *Cadbury Dairy Guide to Low Carbon Dairy Farming*, which provides detailed guidance for their farmers on how changes and modifications to herd health, feeding and breeding can result in emission reductions (Cadbury 2009). Alternatively Sainsbury's has developed a sustainability scorecard to 'look at every aspect of the farm and measure inputs such as electricity, feedstuffs, machinery and fuel use' (Sainsbury's 2009: 60). This scorecard is then used to produce individual farm carbon footprint reports containing specific reduction advice and recommendations. Meanwhile Tesco, in addition to measurement activities and knowledge sharing through the Tesco Sustainable Dairy Group, has established the Tesco Dairy Centre of Excellence with the University of Liverpool to support research on reducing the carbon footprint of its milk supply chain (Tesco 2011a).

To offer a brief summary, the transition to a low carbon pint of milk is set within a landscape which has witnessed significant, indeed fundamental, changes over the past two decades and is marked by new governance structures populated by a variety of actors. Through direct contracts and knowledge brokering relationships, large retailers and dairy produce manufacturers have established complex relationships with their suppliers, exerting control across both the traditional metrics of quality, price and delivery, as well as new requirements relating to how milk is produced. The question remains, however, as to where the desire for a low carbon pint of milk ultimately resides and from where these processes are emanating. In particular, it can be questioned whether low carbon milk represents a genuine consumer desire or is instead a trait attributed to consumers, but mediated and motivated by corporate necessity or requirement. On one hand, proponents argue that public pressure and the rise of ethical consumerism has extended competition between retailers with regards to quantifiable metrics such as cost, quality and convenience and the more complex notions of consumer trust (Pivato et al. 2008). From this perspective, consumer pressure is driving the demand for low carbon milk, as part of a wider consumer requirement for corporate responsibility. In response retailers are seeking to engender consumer trust by integrating sustainability considerations, in this context carbon reduction, across their operations and supply chains.

In contrast, another, more sceptical, perspective can question whether the move towards low carbon milk is not necessarily driven by consumer desire but instead represents a desire which has been generated, or mediated, by corporate requirement (Lyons and Maxwell 2010). Corporate responsibility programmes such as carbon reduction initiatives are thus viewed as 'greenwashing' measures designed to abate negative media coverage, or provide new avenues for brand differentiation in a crowded market place, in part to capture the growing ethical consumer market. Finally, a middle-ground approach suggests that carbon reduction initiatives do not necessarily need to emanate from one particular desire, instead highlighting that reducing a product's carbon footprint represents a win-win situation, resulting in cost savings for consumers and producers and providing market differentiation and enhancing customer loyalty for retailers (Banerjee 2008). For instance working with its Dairy Development Group, Sainsbury's states that it, and its farmers, have reduced their emissions by 6%, saving up to £1.2 million (Sainsbury's 2009).

It is against this backdrop of changes in UK milk production that I now turn to examine how these multidimensional desires for a low carbon pint of milk are influencing humans' relationships with animals and, in turn, our relationship with the food we eat. Specifically, the following section will critically review three central approaches to emerge for the mitigation of GHG emissions from the dairy sector; namely how animals are bred, how animals eat, and how animals are cared for.

How Animals are Bred

Since the eighteenth-century pioneers of Tomkins, Bakewell and Collins, humans have sought to optimize desired genetic qualities in order to breed animals that gain weight faster, mature sooner and grow larger. Whilst techniques may have changed during the last 300 years, the broad goals of selective breeding have remained largely consistent; to identify, select and breed 'improved animals' based on a combination of desired genetic traits, including fertility, health, productivity and longevity. However, recent advances in genetics and breeding techniques have permitted new desirable traits to be considered, for example, modifying crop genetics to increase resistance to insects, or for nutrient enhancement (Ye et al. 2000). Building on these advances, and reflecting the desire for a low carbon society, whether animals can be bred, or genetically engineered, to generate fewer GHG emissions has become a central question. In particular, two routes have been identified; selective breeding to improve productivity and efficiency, and genetic engineering programmes designed to identify and propagate low GHG emission traits. This research, however, is taking place in an era in which there is growing consumer awareness of, and engagement with, food provenance (Dimara and Skuras 2005), often mediated by corporate branding (see, for example, Baker, this volume). This awareness is, I suggest, challenging and exerting influence over the future directions of animal breeding and genetic modification/engineering.

On one level selective breeding programmes, which promote productivity attributes such as increased milk production and improved longevity/health, have long been identified as offering the potential to reduce relative emissions per pint of milk (cf. Beukes 2009; Thompson and Barlow 1986): Hyslop (2003) has shown that bigger continental breeds of cattle produce fewer emissions per unit output than smaller British breeds. Similarly Wall et al. (2008) have reported that increasing the lifespan of a dairy cow from 3.02 to 3.5 lactations can reduce methane emissions by 3%. The potential for GHG reduction through this form of selective breeding, based on productivity, is perhaps best captured by guidance published by Cadbury which states; 'on average, dairy cows each produce around 100kg methane/year and this figure is not greatly affected by yield. If yield per cow is increased and cow numbers reduce correspondingly, then the carbon footprint per litre can be significantly reduced. More milk per cow equals less methane per litre' (2009: 9). Yet this orientation for more efficient milk production, and the emphasis it places on the associated selective breeding programmes, introduces a potential contradiction between consumer desire for low carbon milk and broader public perceptions regarding the intensification of animal/food production. This raises the particular question of whether breeding larger/more efficient dairy cows that produce ever greater volumes of milk can be reconciled with the public sensibilities of 'good' welfare standards: I return to these tensions towards the end of this chapter.

On a different level, alternative breeding programmes have sought to identify specific genetic traits (nutritional and physiological) that can be modified, or replicated (across species), to directly lower the GHG emissions associated with milk production (cf. Hegarty and McEwan 2010; Robertson and Waghorn 2002). The potential for GHG reduction through genetic selection is illustrated by trials conducted at the University of Manitoba, which found that up to 27% of the variation in methane emissions relates to animal-to-animal variation (Boadi and Wittenberg 2002). Whilst genetic engineering is still at an early stage, genome-sequencing projects underway in New Zealand and Australia represent key avenues for understanding the cellular processes responsible for methane production, and in turn potential routes to reduce methane emissions (Attwood et al. 2011; Kebreab et al. 2006). This opportunity for future GHG reductions is, however, set within a highly politicized landscape of public opinion relating to genetic manipulation/engineering and its implementation faces considerable regulatory challenges. In particular, fierce opposition has emerged, especially within Europe, over the potential impacts to food safety and on natural ecosystems associated with genetic modification (Finucane 2002). Indeed a study in 2009 found that only 2% of people in Britain were happy to eat genetically modified food (GM Nation 2003).

Therefore, in response to the societal desire for low carbon milk, selective breeding and genetic engineering are gaining increasing attention. Whilst demonstrating significant reduction potential, both indirectly, by increasing milk productivity and improving herd health, and directly, by manipulating genetic traits responsible for methane production, both approaches have been marked by widespread public concern and regulatory procedures. Questions have been raised by consumers and regulatory bodies alike regarding shifts towards larger, more efficient animals and production practices, amidst concerns over animal welfare and the ways these changes challenge existing perceptions and imaginaries of animal-human relations. Meanwhile advances in genetic understanding that ultimately could be harnessed to develop, or indeed engineer, a low carbon cow face public pressure regarding notions of public safety associated with the potential introduction of 'frankenfood'. Therefore the crucial question, which besets this field, is whether the desire for a low carbon pint of milk, and more broadly a low carbon society, can be reconciled with public sensibilities and perceptions towards animal imagery, welfare and food safety.

How Animals Eat

The principal source of greenhouse gas emissions from the dairy sector is methane, generated as a by-product from animal feeding and digestion processes. This has led to attention (e.g. UK Cabinet Office 2008) being increasingly focused on how animal-eating patterns can be modified to fulfil the human desire for a low carbon pint of milk. Specifically, researchers (e.g. DEFRA in the UK,[5] CSIRO in Australia,[6] Dairy Management Inc.[7] in the US) are exploring how modifying the food that animals eat – for example, reducing fibre levels, managing protein levels and increasing the proportion of non-structural carbohydrates (specifically starches) – can reduce methane emissions. This research has informed changes to dietary management practices on the farm, including adjusting the balance between forage- and concentrate-diets and using various dietary supplements (cf. Wittenberg 2008; Wright 2003). However, as I illustrate below, changes to animal eating practices reveal deep-seated multi-dimensional relationships between the food that animals and the food humans ultimately eat.

One of the key discussions associated with low carbon dietary management relates to the balance between forage- and concentrate-diets. Over the past 50 years, various studies (e.g. Blaxter and Claperton 1965; Johnson and Johnson 1995; Robertson and Waghorn 2002) have concluded that methane emissions from cattle can be reduced by supplementing forage (grass) with concentrate-feed (grain, soy, maize). Broadly this research indicates that whilst concentrate-feed may increase daily methane emissions per animal, overall emissions per kg-feed intake and per-kg product are reduced, thereby representing a net reduction. However the magnitude of these reductions is keenly debated with particular attention called to the wider environmental impacts associated with animal feed production. Specifically, opponents argue that emissions generated by the production of animal feed crops (i.e. from land use change including deforestation and desertification) nullify any potential reductions gained later within the supply chain (Lovett et al. 2006; Phetteplace et al. 2001).

In parallel, research has demonstrated how the greenhouse gas emissions associated with pasture-based diets can be reduced; thereby offering a low-carbon alternative to concentrate-diets. This research has, in particular, focused on changing pasture management strategies (for example, continuous versus

5 The Department for Environment, Food and Rural Affairs as part of its Agriculture and Climate Change field of study has funded various research projects on methane emissions from dairy cows.

6 The Commonwealth Scientific and Industrial Research Organisation is Australia's national science agency and is undertaking an extensive research program focused on developing practical solutions for significantly reducing methane emissions from livestock such as sheep and cattle.

7 Dairy Management Inc. are the promotion and research arm of the US dairy industry.

rotational grazing) and forage species selection including the introduction of certain legumes (Alcock and Hegarty 2006; Boadi and Wittenberg 2002). For instance, McCaughey et al. (1999) concluded that methane production is notably lower for cows grazing alfalfa-grass pastures that grass-only pastures. Meanwhile, in New Zealand, Waghorn et al. (2002) have found that emissions from animals grazing birdsfoot trefoil pasture are substantially lower than those grazing ryegrass white clover pasture.

A second discussion within dietary management has focused on the potential for dietary additives to reduce methane emissions, in particular the role of lipids, such as fatty acids. A prominent avenue of this research has focused on oils extracted from plant sources. For example researchers at France's 'Institut National de Recherche Agronomique' have found that incorporating vegetable oil into dairy cow feed can cut methane emissions by nearly a third. Similarly, other research has concluded that: refined soy oil reduced methane production by 39% (Jordan et al. 2006); sunflower oil resulted in an 11–22% reduction in methanogenesis (McGinn et al. 2004); and linseed oil included in the diets of lactating dairy cows reduced methane emissions by 55.8% (Martin et al. 2008). Moreover, trials of aqueous allicin extract from garlic are linked with a 94% decrease in methane production (Newbold et al. 2010).

The debate between pasture- and concentrate-diets and over the potential introduction of dietary additives is, however, not straightforward, and draws attention to the connections between the food eaten by animals and that which is, in turn, eaten by humans. Consumer tastes, in a physical embodied sense, start to play a significant role, and inform the potential methane reducing measures that can be introduced at the earliest stages of the commodity chain. To elucidate, reports of milk being tainted as a result of animals eating wild garlic and onions have existed since the 1930s (Babcock 1938). Similarly, research reports that grain-fed cattle produce more tender and better-flavoured meat than forage-fed animals (Larick et al. 1987), whilst anecdotal evidence from cheese-makers, compiled over the past century, has stressed the sensory differences between 'winter' cheeses (produced from cows eating a hay and grass silage-based diet) and 'spring/summer cheeses', produced from cows that are turned out in highland pastures (Bosset et al. 1999). In addition, Fisher et al. (2000) have shown that feeding grass and linseed to cattle can have positive effects on the fatty acid profile of their meat resulting in a healthier product. Therefore, while changing animal-eating patterns has demonstrated clear potential in reducing GHGs, the introduction of this knowledge is contained by the tastes, and eating practices, of the consumer. A tension thus emerges between the objectives, and objections, of individual consumers and wider concerns for the future wellbeing of society.

Finally, the desire to reduce methane emissions – both in order to decrease the carbon footprint of milk and also to increase the efficiency of animal production – has led to a range of specific dietary supplements (and hormone replacements, cf. Wittenberg 2008) being developed. The most common of these antimicrobials are ionophores, for example Bovatec™, Rumensin™ and Cattlyst™, which in 2003 had combined yearly sales of more than $150 million, and were estimated to save the cattle industry $1 billion per year (Callaway et al. 2003). Ionophores, which have antibiotic properties, increase the efficiency of feed utilization by altering the composition of an animal's ruminal microbial ecosystems and have long been promoted as dietary supplements to improve productivity and profitability (Russell and Strobel 1989). Crucially, the use of ionophores has also been linked with reducing emissions, for example Monensin™ is reported to decrease methane emissions by 28% (Kinsman et al. 1997). However, over the past 10 years, the use of ionophores as growth proponents has fallen under intense scrutiny, due to the rise of antibiotic resistance within human populations. According to a study by the European Federation of Animal Health (2001), in 1999 6% of antibiotics used in the EU were administrated to farm animals as growth promoters. As a result, in 2006, a European Union-wide ban on the use of antibiotics as growth promoters took effect, although these compounds are still routinely used in a number of non-EU countries, such as China (Sarmah et al. 2006). Notably the re-introduction of additives such as ionophores has been recently cited within national and international climate change strategies as a way to decrease methane emissions from the livestock sector, for example as detailed in the Fourth Intergovernmental Panel on Climate Change report (Smith et al. 2007) and the UK's First Report of the Committee on Climate Change (2008).

Therefore, societal aspirations for a low carbon pint of milk have drawn attention to how, and what, an animal eats, which in turn reveal deep-rooted and multiple connections between what animals and humans eat. This illuminates how human and animal bodies, positioned at polar ends of the commodity chain, become coterminous through their ingestion, or refusal to ingest, certain foods. On one level these connections relate to individual sensory perceptions and preferences, with changes to animal eating patterns (different pastures and different grains) influencing the taste, smell and appearance of dairy produce. Yet on another level these connections refer to wider societal notions of food safety, with the use of new supplements potentially introducing new risks into the human food chain. This leads to a crucial consideration, namely whether consumers will accept changes to the taste and material properties of their food, and adjust their notion of food safety as a trade-off for the wider goal of a low carbon society.

How Animals are Cared For and Dairy Herds Maintained

The final manner in which animal-human relations are being shaped and transformed relates to how on-farm production processes and the maintenance of dairy herds can be linked with, and offer opportunities for, the mitigation and reduction of greenhouse gas emissions. Two avenues of research within this field have been of particular interest; both of which ultimately seek to reduce the relative emissions associated with individual cows and/or dairy herds. Firstly, studies have examined how improving or modifying the reproduction practices of dairy cattle can reduce the 'wastage' emissions associated with undesirable reproduction (male calves or unhealthy offspring), and unproductive periods (intervals between calving). This research has included improving ovulation and conception, through a focus on heat detection and enhancing pre- and post-natal survival. The potential of this research is illustrated in Table 9.1 below. This guidance, provided by Cadbury to their dairy farmers, highlights that by reducing culling rates and calving heifers earlier, fewer young stock are required and the carbon footprint associated with the rearing period is significantly reduced. For instance moving from 30-month calving and a replacement rate of 24% to 22-month calving and a replacement rate of 20% can reduce a 100-cow herd's carbon footprint by more than 120 tonnes/ year.[8] This research is further supported by pilot projects in India that have decreased the interval between births, from 24 months to 12–15 months and reported notable emission reductions (UNFCC fact sheet 271).

Table 9.1 The number of replacement heifers needed to maintain a herd of 100 cows, as featured in *Cadbury Dairy Guide to Low Carbon Dairy Farming*

Cull rate	Age of calving (in months)							
	22	24	26	28	30	32	34	36
16%	32	36	39	41	44	45	50	53
20%	40	44	48	51	54	56	62	66
24%	48	53	57	62	66	70	76	79
28%	56	62	67	72	77	82	87	92

Source: *Cadbury Dairy Guide to Low Carbon Dairy Farming* (2009: 12).

Secondly, attention has explored how improving herd health can reduce 'maintenance' emissions associated with non-productive periods, for instance by reducing lameness or illness and through the removal of non- or less-productive animals (Hunter and Neithe 2009). These studies have shown that poor herd health affects per-unit-of-milk emissions by increasing mortality and decreasing milk production. Recommendations include improving cow cooling during hot weather, grouping animals to minimize behaviour stress, decreasing illness and disease such as mastitis and lameness – for example heat stress is estimated to cost the US dairy industry $1b per year in decreased production and increased deaths (Connor et al. 2011).

---

8 Rearing a heifer to 30 months old on a conventional silage and cake system creates a carbon footprint of 4,766 kg of CO_2. By reducing culling rates and calving heifers earlier, fewer young stock are required and the carbon footprint associated with the rearing period is significantly reduced (Cadbury 2009).

Overall this avenue of research paints a conflicting picture regarding the impact on human-animal relations. On one level the evidence that a healthy and productive herd is a low carbon herd is compatible with the, previously discussed, public concerns for animal welfare. Alternatively reducing calving intervals and the shift towards ever-more production-oriented guidelines introduce potential conflicts with public sensibilities relating to the progressive industrialization of animal-human relationships (see also Baker and Hurn, both this volume). The synergy between low carbon and mass-production, is illustrated by the proposed Nocton 'mega-dairy' (which will be one of the largest dairy herds in Europe[9]) being cited as potentially providing the lowest carbon footprint across all dairy system by the President of the National Farmers Union (*The Farmers Guardian* 2010).

Conclusion

Over the past two decades, amid a series of external pressures and public attention, the UK dairy industry has transformed from a government-led 'command and control' supply chain to a retailer (demand-led) supply chain. Whilst the industry is still subject to national and European regulation and policy, the introduction of direct retailer contracts and exclusive supplier groups has created new, and highly competitive, governance structures and market spaces in which regulation is increasingly being articulated directly by retailers, reflecting (or indeed informing) consumer desires, trends and beliefs. In parallel, aided by the introduction of product carbon footprinting, there has been growing recognition of the dairy sector's role in anthropogenic climatic change, particularly through on-farm animal emissions of methane. Together these two trends have led to, necessitated and facilitated a series of ambitious climate change visions, strategies and targets, articulated by both the public sector (reflecting national strategies) and private retailers (reflecting wider corporate responsibility strategies), which seek to set in motion a transition to a low carbon society. As part of this transition there is a governmental, corporate and consumer aspiration to reduce the carbon footprint of the food, and the focus of this chapter the milk, which we as a society consume.

This societal desire – whether ultimately consumer or corporately inspired – for a low carbon pint of milk has resulted in a series of on-farm initiatives that are transforming animal-human relations; challenging how animals are bred, how animals eat, and how animals are cared for and dairy herds maintained. These initiatives can broadly be split into two categories. On one hand, research has sought to optimize the productivity of dairy herds and thereby minimize the

9 Nocton Dairies is a British company which sought, in 2010, to construct an 8,100-cow dairy at Nocton Heath in Lincolnshire, which would have become the largest in Western Europe. The proposal evoked considerable media interest and public objections.

emissions released per pint of milk through, for example, selective breeding, improving animal health and changing animal-eating patterns. Alternatively more radical approaches have been proposed for lowering on-farm emissions which include genome sequencing as a potential precursor for genetic manipulation, the inclusion of growth additives such as antibiotics and hormones and far-reaching changes to animal reproduction and management.

Whilst displaying notable GHG reduction potential, these initiatives have also revealed multi-dimensional connections between humans, animals and ultimately the food we consume. On one level, reducing the carbon footprint of a pint of milk has illustrated the intrinsic relations between what we eat and what we eat eats. Changing animals' eating patterns to reduce methane emissions can result in different sensory tastes, smells and health properties of dairy produce. Meanwhile, the introduction of artificial dietary additives into animal-eating patterns has resulted in public concerns regarding the safety of low carbon food. On a second level the desire to produce a low carbon cow has revealed and challenged public perception and imaginaries of animal-human relations and food production processes. Whilst a healthy herd is a productive and low carbon herd, the overriding emphasis placed on productivity is fuelling a trend towards larger, more efficient and more intensively reared animals. To many this confirms the wider realization that animal management for food production has irrevocably moved away from the romanticized imaginary of 'wild' animals freely foraging, replaced instead by the reality of selectively bred and bio-engineered animals raised on highly-regulated diets formulated by distant scientists.

Future Directions: Transitions to a Sustainable Society and the Food We Eat

The final section of this chapter draws and reflects upon this brief review of the transition towards a low carbon pint of milk, and seeks to initiate a wider discussion on how multiple notions of sustainable production and consumption challenge and inform the existing and future relationships between humans and animals, and ultimately between humans and the food we eat. As the example of low carbon milk illustrates, the notion of what is sustainable food is far from straightforward. The concept of sustainability is many things to many people, spanning socio-economic factors, animal welfare, biodiversity, food safety, health properties and environmental impact. Furthermore these different elements are often not mutually compatible, as demonstrated by the debate between eco-centric ('green') and anthropocentric ('red') commodity cultures, as illustrated by Bryant and Goodman's (2004) discussion of organic versus Fairtrade foods. As different visions of 'what is a sustainable society' are articulated and promoted, further complexity is added to these debates, thereby introducing new dimensions, and challenges, to the relationships between humans, animals and food. For example, while animal welfare groups have recently succeeded in obtaining new regulation for battery–reared chickens, alternative research has demonstrated that non-organic milk has a lower carbon footprint than organic milk (Williams et al. 2006). Similarly, whilst antibiotics and hormones as growth promoters were originally banned on the grounds of public safety, their re-introduction has been raised due to the significant GHG reduction potential they offer. Therefore as new visions of what is a sustainable society are defined and articulated, so too we as consumers, and as a society, are required to choose between multiple priorities, and between individual and collective desires, aspirations and ambitions. For some, these adaptations and trade-offs may be a small price to pay, for others they may represent a bitter 'taste'.

The multi-faceted nature of sustainability leads directly onto a second question, namely who decides what is sustainable food? As illustrated by this chapter the desire for sustainable food is set within a landscape shaped by changing consumer awareness and desires, corporate branding and marketing strategies, government influence and a variety of other market- and consumer-led pressures. What is, however, clear within this landscape is that the articulation of 'what is sustainable food' is almost exclusively mediated within market-based structures. This confers significant authority to retailers and transnational corporations to define, and market, 'what is sustainable'. Retailers are increasingly seizing this opportunity in an attempt to gain consumer trust, or on a more fundamental level market share, by demonstrating their corporate responsibility through the articulation and promotion of certain discourses of sustainability. For instance when Starbucks began serving 100% Fairtrade coffee in 2009, it 'overnight' increased the volume of Fairtrade coffee sold and consumed in the UK and Ireland by 18% (The Fairtrade Foundation 2009). Likewise, as a result of Tesco's unprecedented campaign to measure and label the carbon footprint of 70,000 of its products, the notion of low carbon consumption has received nation-wide exposure (*The Guardian* 2007). Therefore as a consequence, driven by brand differentiation strategies and/or more honourable notions of corporate societal responsibility, retailers are increasingly determining what food we eat, how it is produced, and how our notions of what is sustainable are defined.

In addition to assuming an omnipotent role in articulating what a sustainable society looks like, retailers, and multi-national corporations, have adopted prominent roles in determining how we, as a society, transition to this vision of sustainability. As I have argued, the transition to a low carbon society has been typically rationalized – particularly by global retailers – through an ecological modernization and technocratic discourse in which carbon reductions can be achieved through efficiency measures and technological advances. Within this vision, the reduction of GHG emission is reconciled with economically viable, market-based solutions, as succinctly outlined in the *Cadbury Dairy Guide to Low Carbon Dairy Farming*; 'efficient farming reduces the carbon footprint of milk production and provides better returns for your business – a 'win-win' situation where the environment benefits and you see improved economic returns' (2009: 18). This vision, and the pervasive influence of corporate actors, has raised two key considerations. On a base level, it is argued that whilst the emphasis placed on 'win-win' or technical solutions may reduce relative emissions, there is the danger this may simply facilitate industry expansion and actually result in an overall increase in global GHG emissions. From this viewpoint it is contended that only a radical change to existing consumption patterns and market-structures will be sufficient in addressing society's currently unsustainable relationship with planet Earth's finite resources. More broadly, concern has been raised regarding the influence corporate funding plays in determining the direction, and priorities, of future research, as illustrated by the emergence of the Tesco Dairy Centre of Excellence in leading low carbon research. Specifically, it is questioned whether these relationships, coupled with the dominant position of retailers within consumer and producer relations, creates a dangerous position in which research priorities are aligned with particular market strategies in which the transition to a sustainable society represents a new market space.

For millennia humans have lived alongside animals. While rising populations and increasing wealth is likely only to further entrench these relationships, the challenges associated with a transition to a sustainable society, and ensuring our mutual survival, dictates that humans must alter how animals are bred, eat and are maintained. However, as society embarks upon this transition, it is crucial that we recognize that these changing relations between humans and animals are deeply political. What is defined as a sustainable society and how we arrive at this destination is a complex political economy, one in which a hybrid of organizations and interests are competing to establish the governance structures, generate the knowledge and capitalize on the new market spaces. One direction or transition pathway, which has gained particular traction, is the growing consensus that to avoid the danger posed by global climate change we must move towards a lower carbon society. At the minimum this transition requires us to reduce the emission intensity of our food – or emissions per pint of milk – if not make far more radical changes to our consumption patterns. This transition will result, and require, continual shifts within human and animal relations that both enact, and are engendered by, alterations in techniques of production, as well as the day-to-day choices about the food we eat, and the food that we eat eats.

Acknowledgements

This chapter was written as part of a PhD funded by the Economic Social Research Council (ESRC-Case Studentship 2010–2013).

References

Aiking, H., de Boer, J. and Vereijken, J. 2006. *Sustainable Protein Production and Consumption: Pigs or Peas?* Dordrecht: Springer.

Alcock, D. and Hegarty, R.S. 2006. Effects of pasture improvement on productivity, gross margin and methane emissions of a grazing sheep enterprise, in *Greenhouse Gases and Animal Agriculture: An Update*, edited by C.R. Soliva, J. Takahashi and M. Kreuzer. The Netherlands: International Congress Series No. 1293, Elsevier, 103–10.

Attwood, G.T., Alternmanna, E. and Kelly, W.J. 2011. Exploring rumen methanogen genomes to identify targets for methane mitigation strategies. *Animal Feed Science and Technology* 166–7, 65–75.

Babcock, C.J. 1938. Feed flavors in milk and milk products. *Journal of Dairy Science* 21, 661–8.

Banerjee, S.B. 2008. *Corporate Social Responsibility: The Good, the Bad and the Ugly*. University of Western Sydney, Australia: Edward Elgar Publishing.

Beukes, P.G. 2009. Modelling the efficacy and profitability of mitigation strategies for greenhouse gas emissions on pastoral dairy farms in New Zealand. Paper to the 18th World IMACS/MODSIM Congress. Cairns, Australia.

Blaxter, K.L. and Clapperton, J.L. 1965. Prediction of the amount of methane produced by ruminants. *British Journal of Nutrition* 19, 511–22.

Boadi, D.A. and Wittenberg, K.M. 2002. Methane production from dairy and beef heifers fed forages differing in nutrient density using the sulphur hexa-flouride tracer gas technique. *Canadian Journal of Animal Science* 82, 201–6.

Bosset, J.O., Jeangros, B., Berger, T., Bütikofer, U., Collomb, M., Gauch, R., Lavanchy, P., Scehovic, J. and Sieber, R. 1999. Comparaison de fromages à pâte dure de type gruyère produits en région de montagne et de plaine. *Revue Suisse d'Agriculture* 31, 17–22.

Bryant, R. and Goodman, M. 2004. Consuming narratives: The political ecology of 'alternative' consumption. *Transactions of the Institute of British Geographers* 29, 344–66.

Cadbury 2009. *Cadbury Dairy Guide to Low Carbon Dairy Farming* [Online]. Available at: http://www.cadbury.co.uk/cadburyandchocolate/OurCommitments/Environmental%20Commitments/Pages/CarbonReduction.aspx [accessed: 2 December 2011].

Callaway, T.R., Edrington, T.S., Rychlik, J.L., Genovese, K.J., Poole, T.L., Jung, Y.S., Bischoff, K.M., Anderson, R.C. and Nisbet, D.J. 2003. Ionophores: Their use as ruminant growth promotants and impact on food safety. *Current Issues in Intestinal Microbiology* 4, 43–51.

Connor, E.E., Hutchison, J.L., Olson, K.M. and Norman, H.D. 2011. Opportunities for improving milk production efficiency in dairy cattle. *Journal of Animal Science* 90(5), 1687–94.

Davis, J. and Sonesson, U. 2008. Environmental potential of grain legumes in meals. Life cycle assessment of meals with varying content of peas. SIK Report 771. Swedish Institute for Food and Biotechnology, Göteborg, Sweden.

Dairy Supply Chain Forum (DSCF). 2008. The Milk Roadmap May [Online]. Available at: http://archive.defra.gov.uk/environment/business/products/roadmaps/documents/milk-roadmap-oneyear090727.pdf [accessed: 2 December 2011].

DairyUK 2010. Dairy White Paper: A report on the UK dairy industry [Online]. Available at: http://www.dairyuk.org/media-area-mainmenu-270 [accessed: 2 December 2011].

Dimara, E. and Skuras, D. 2005. Consumer demand for informative labelling of quality food and drink products: A European Union case study. *Journal of Consumer Marketing* 22(2), 90–100.

European Federation of Animal Health (FEDESA). 2001. Antibiotic use in farm animals does not threaten human health. FEDESA/ FEFANA Press release. 13 July. Brussels, Belgium.

FAOSTAT, 2010. *Food and Agricultural Commodities Production*. Rome, Italy.

Finucane, M.L. 2002. Mad cows, mad corn and mad communities: The role of socio-cultural factors in the perceived risk of genetically-modified food. *Proceedings of the Nutrition Society* 61, 31–7.

Fisher, A.V., Enser, M., Richardson, R.I., Wood, J.D., Nute, G.R., Kurt, L.A., Sinclair, R.G. and Wilkinson, J.M. 2000. Fatty acid composition and eating quality of lambs types derived from four diverse breed × production systems. *Meat Science* 55, 141–7.

Garnaut, R. 2008. The Garnaut Climate Change Review: Final report. Canberra, Commonwealth of Australia. [Online]. Available at: www.garnautreview.org.au [accessed: 2 December 2011].

Gill, M., Smith, P. and Wilkinson, J.M. 2010. Mitigating climate change: The role of domestic livestock. *Animal* 4, 323–33.

GM Nation? 2003. The findings of the public debate. [Online]. Available at: http://www.gmnation.org.uk [accessed: 1 December 2011].

Hegarty, R.S. and McEwan, J.C. 2010. Genetic opportunities to reduce enteric methane emissions from ruminant livestock. Paper to the Proceedings of the 9th World Congress on Genetics Applied to Animal Production, Leipzig, Germany, August 2010.

Hunter, R. and Neithe, G. 2009. Efficiency of feed utilisation and methane emissions for various cattle breeding and finishing systems. *Advanced Animal Nutrition* 17, 75–9.

Hyslop, J. 2003. Simulating the greenhouse gas and ammonia emissions from UK suckler beef systems. Report to the Department for Environment, Food and Rural Affairs.

Johnson, K.A. and Johnson, D.E. 1995. Methane emissions from cattle. *Journal of Animal Science* 73, 2483–92.

Jordan, E., Lovett, D.K., Monahan, F.J., Callan, J., Flynn, B. and O'Mara, F.P. 2006. Effect of refined coconut oil or copra meal on methane output, and on intake and performance of beef heifers. *Journal of Animal Science* 84, 162–70.

Kebreab, E.K., Clark, C., Wagner-Riddle, C. and France, J. 2006. Methane and nitrous oxide emissions from Canadian animal agriculture: A review. *Canadian Journal of Animal Science* 86, 135–58.

Kinsman, R., Sauer, F.D., Jackson H.A. and Wotynetz, M.S. 1995. Methane and carbon dioxide emissions from dairy cows in full lactation monitored over a six-month period. *Journal of Dairy Science* 78, 2760–76.

Larick, D.K., Hedrick, H.B., Bailey, M.E., Williams, J.E., Hancock, D.L., Garner, G.B. and Morrow, R.E. 1987. Flavor constituents of beef as influenced by forage- and grain-feeding. *Journal of Food Science* 52, 245–25.

Lovett, D.K., Shalloo, L., Dillon, P. and O'Mara, F.P. 2006. A systems approach to quantify greenhouse gas fluxes from pastoral dairy production as affected by management regime. *Agricultural Systems* 88, 156–79.

Lyons, T.P. and Maxwell, J.W. 2010. Corporate social responsibility and the environment: A theoretical perspective. *Review of Environmental Economics and Policy* 26(2), 164–81.

Martin, C., Rouel, J., Jouany, J.P., Doreau, M. and Chilliard, Y. 2008. Methane output and diet digestibility in response to feeding dairy cows crude linseed, extruded linseed, or linseed oil. *Journal of Animal Science* 86(10), 2642–50.

McAllister, T.A., Beauchemin, K.A., McGinn, S.M., Xiying H.X. and Robinson, P.H. 2011. Greenhouse gases in animal agriculture – Finding a balance between food production and emissions. *Animal Feed Science and Technology* 166–7, 1–6.

McCaughey, W.P., Wittenberg, K.M. and Corrigan, D. 1999. Impact of pasture type on methane production by lactating beef cows. *Canadian Journal of Animal Science* 79, 221–6.

McGinn, M.S., Beauchemin, K.A., Coates, T. and Colombatto, D. 2004. Methane emissions from beef cattle: Effects of monensin, sunflower oil, enzymes, yeast, and fumaric acid. *Journal of Animal Science* 82(11), 3346–56.

Newbold, J., Kim, E.J. and Scolla, N. 2010. Reducing animal greenhouse gas Emissions IBERS Knowledge-Based Innovations No 3.

Pivato, S., Misani, N. and Tencati, A. 2008. The impact of corporate social responsibility on consumer trust: The case of organic food. *Business Ethics: A European Review* 17, 3–12.

Phetteplace, H.W., Johnson, D.E. and Seidl, A.F. 2001. Greenhouse gas emissions from simulated beef and dairy livestock systems in the United States. *Nutrient Cycling in Agroecosystems* 60, 9–10.

Robertson, L.J. and Waghorn, G.C. 2002. Dairy industry perspectives on methane emissions and production from cattle fed pasture or total mixed rations in New Zealand. *Proceedings of the New Zealand Society of Animal Production*, Abstract No. 55. 62, 213–18.

Rollin, B. 2008. The ethics of agriculture: The end of true husbandry, in *The Future of Animal Farming: Renewing the Ancient Contract*, edited by M.S. Dawkins and R. Bonney, Oxford, UK: Blackwell Publishing, 7–19.

Russell, J.B. and Strobel, H.J. 1989. Effect of ionophores on ruminal fermentation. *Applied Environmental Microbiology* 55, 1–6.

Sarmah, A.K., Meyer, M.T. and Bowall, A.B.A. 2006. A global perspective on the use, sales, exposure pathways occurrence, fate and effects on veterinary antibiotics (VAs) in the environment. *Chemosphere* 65(5) 725–59.

Sainsbury 2009. *Sainsbury Development Dairy Group* [Online]. Available at: http://www.j-sainsbury.co.uk/responsibility/case-studies/2009/sainsburys-dairy-development-group/ [accessed: 2 December 2011].

Smith, P., Martino, D., Cai, Z., Gwary, D., Janzen, H., Kumar, P., McCarl, B., Ogle, S., O'Mara, F., Rice, C., Scholes, B. and Sirotenko, O. 2007. Agriculture, in *Climate Change 2007: Mitigation. Contribution of Working Group III to the Fourth Assessment Report of the Intergovernmental Panel on Climate Change* edited by B. Metz, O.R. Davidson, P.R. Bosch, R. Dave and L.A. Meyer. Cambridge, United Kingdom and New York, NY, USA: Cambridge University Press.

Stehfest, E., Bouwmann, L., van Vuuren, D., den Elzen, M.G.J., Eikhout, B. and Kabat, P. 2009. Climate benefits of changing diet. *Climate Change* 95, 83–102.

Steinfeld, H., Gerber, P., Wassenaar, T., Castel, V., Rosales, M. and de Haan, C. 2006. *Livestock's Long Shadow: Environmental Issues and Options.* Rome, Italy: FAO

Tesco. 2008. Working together for a sustainable future – Emma Jones. [Online]. Available at: http://www.dairyuk.org/component/docman/doc_download/5893-emma-jones-carbon-conference-presentation [accessed: 2 December 2011].

Tesco. 2011a. Why drink Tesco milk? [Online]. Available at: http://www.tescorealfood.com/our-food/milk/why.html [accessed: 2 December 2011].

Tesco. 2011b. Tesco Corporate Responsibility Report 2011. [Online]. Available at: http://www.tescoplc.com/media/60113/tesco_cr_report_2011_final.pdf [accessed: 2 December 2011].

The Carbon Trust. 2011. Low carbon products in demand despite challenging economic climate. [Online]. Available at: http://www.guardian.co.uk/environment/2011/jul/01/carbon-trust-research-footprint-consumer-demand [accessed: 2 December 2011].

The Co-operative. 2010. Ethical Consumer Report 2010. [Online]. Available at: http://www.goodwithmoney.co.uk/report-download [accessed: 1 January 2012].

The Ecologist. 2011. Special report Top 10 … alternatives to cows milk. [Online]. Available at: http://www.theecologist.org/green_green_living/food_and_drink/847876/top_10alternatives_to_cows_milk.html [accessed: 2 December 2011].

The Fairtrade Foundation. 2009. Starbucks serves up its first fairtrade lattes and cappuccinos across the UK and Ireland. [Online]. Available at: http://www.fairtrade.org.uk/press_office/press_releases_and_statements/september_2009/starbucks_serves_up_its_first_fairtrade_lattes_and_cappuccinos.aspx [accessed: 2 December 2011].

The Farmers Guardian. 2008. Dairy roadmap outlines targets for greenhouse emissions. [Online]. Available at: http://www.farmersguardian.com/dairy-road-map-outlines-target-for-greenhouse-gas-cut/17412.article [accessed: 2 December 2011].

The Farmers Guardian. 2010. Nocton could provide lowest carbon footprint. [Online]. Available at: http://www.farmersguardian.com/home/business/business-news/nocton-could-provide-lowest-carbon-footprint-kendall/34646.article [accessed: 22 December 2011].

The Guardian. 2007. Tesco becomes UK's first retailer to display carbon footprint on milk. [Online]. Available at: http://www.guardian.co.uk/environment/2009/aug/17/tesco-milk-carbon-footprint [accessed: 2 December 2011].

Thompson, J. and Barlow, R. 1986. The relationship between feeding and growth parameters and biological efficiency in cattle and sheep. Proc. 3rd World Congress Genet. *Applied Livestock Producers* 2, 271–82.

UK Cabinet Office. 2008. Food matters: Towards a strategy for the 21st century. [Online]. Available at: http://www.foodsecurity.ac.uk/assets/pdfs/cabinet-office-food-matters.pdf [accessed: 2 December 2011].

UK Government. 2008. *Building a Low-Carbon Economy – The UK's Contribution to Tackling Climate Change: The First Report of the Committee on Climate Change*. London: HMSO.

UNFCC fact sheet 271 [Online]. Available at: http://unfccc.int/press/fact_sheets/items/4986.php [accessed: 2 December 2011].

Waghorn, G.C., Tavendale, M.H. and Woodfield, D.R. 2002. Methanogenesis from forages fed to sheep. *Proceedings of New Zealand Grassland Association* 64, 167–71.

Wall, E., Bell, M.J. and Simm, G. 2008. *Developing Breeding Schemes to Assist Mitigation. Livestock and Global Climate Change.* Midlothian: British Society of Animal Science.

Williams, A.G., Audsley, E. and Sandars, D.L. 2006. *Determining the Environmental Burdens and Resource Use in the Production of Agricultural and Horticultural Commodities*. Cranfield University and Defra Research Project ISO205, Bedford.

Wittenberg, K. 2008. *Enteric Methane Emissions and Mitigation Opportunities for Canadian Cattle Production Systems* [Online]. Available at: <http://www.vido.org/beefinfonet/ otherareas/pdf/CcbMethaneemmissionsWittenburg.pdf> [accessed: 2 December 2011].

Wright, T. 2003. *Feeding Dairy Cattle to Reduce Excess Nitrogen Output Cattle Nutritionist/OMAFRA* – factsheet 410/60.

Ye, X., Al-Babili, S., Klöti, A., Zhang, J., Lucca, P., Beyer, P. and Potrykus, I. 2000. Engineering the provitamin A (beta-carotene) biosynthetic pathway into (carotenoid-free) rice endosperm. *Science* 287 (5451), 303–5.

Interlude
Reflections on the Elusiveness of Eating

Anne Murcott

There is a grandfather, now retired, who has lived all his life on the northern coast in rural Ireland. His grandchildren, brought up in Bristol (in south west England) are fortunate enough to be nutritiously fed and are also able to learn to enjoy cuisines from round the world. They have developed a taste for a saffron flavoured fish and shellfish soup (inspired by bouillabaisse) relishing the mussels in particular. On a visit to their grandparents, the children discovered to their great glee that any number of mussels could be had for no more than a trip with a bucket to the shore and a paddle in icy water. But their grandfather retreats from the ensuing feast. Eating mussels takes him back too sharply to a childhood of severe hardship and great deprivation, when survival relied on gathering what could be had from hedgerow, field and shoreline.

Materiality is inescapable in the act of eating: foods and bodies are material objects. It is inescapable in the bodily sensations involved, not just during but also before and after eating. That hunger and the digestive consequences of eating are each the subject of variable, and sometimes quite firm, social conventions governing when, how and among whom they may be made evident, displayed or discussed (cf. Murcott 1993) does not detract from their status as *bodily* sensations – even if these are probably still under-investigated by social anthropologists and sociologists (but for contributions by psychologists, see Blundell, Dalton and Finlayson, forthcoming 2013). All the same, as the French physiologist Jean Tremolières (1970) adroitly pointed out, voluntary ingestion can be thought of as tantamount to a definition of food. 'What does and does not count as food' finds its literal answer in the very act of eating, something that one or other social group would recognize as (in)edible. Here is at least one point at which the material meets and coincides with the social, political and economic. The retired grandfather's definition of mussels as food is in striking contrast to his luckier grandchildren. Both recognize them as food, but the occasion, the circumstances, the ease with which they are acquired, and the alternatives that are, or, worse, are not, available, all become part of the definition.

The realm of the symbolic and the human attribution of meaning are equally inescapable in the act of eating – albeit in a fashion that is analytically detachable from the material – before, during and after eating. That the social, symbolic or political have too often been privileged in anthropological and sociological discussions of eating is a point the editors of this volume propose requires redress, hence their emphasis on the *simultaneity* of the material, political and social.

In consequence, there is an important case for needing empirically to get as near as possible to the act of eating and its close coupling with the meanings attributed to it. Part of the reason for an over-emphasis on the symbolic social and political realms may, however, lie in the difficulty social scientists of many kinds have in approaching, and deciding how to study, food actually going into the mouth. The object of study keeps disappearing.

Three decades and more of writing about eating and carefully avoiding food related puns culminate here in my granting that the one which concludes the previous paragraph has its uses and thus just about earns its keep. For not only does one of the *objects* of study – whatever is put into the mouth – keep disappearing down the gullet, so does the *act* of ingestion on the part of one of the other objects of study – the person eating – keep disappearing. So elusive is eating that, in this particular sense, the act figures not at all in Warde's (forthcoming) chapter entitled simply 'Eating' that usefully reviews a selection of work on the topic.[1]

Putting an item into the mouth, chewing and swallowing is fleeting. Certainly a group of a related succession of such acts can take more or less time – as brief as eating a pear, or a biscuit and cup of coffee, via a daily meal that takes perhaps half an hour or more (Cheng et al. 2007) right up to its being stretched out during a grand, evening-long banquet. Certainly too, it is an integral part of a much wider range of other routine (economic and social consumption) activities – from acquisition (growing, gathering, shopping) to processing, preserving, preparing and transforming (cooking, cooling, assembling) which can of course take far longer than the meal that results. And although each forkful or spoonful carried to the mouth takes its own time it is only a matter of a few seconds. So those activities of spooning or forking represent a very small proportion of the complex of all those others – not just the serving beforehand and of clearing away afterwards – which taken together precede and succeed them. Perhaps it is not surprising that close attention to the fork (or its equivalents of chopsticks, fingers and so on) reaching the mouth and the food arriving inside it is rare. Or, put the other way round, perhaps it is not surprising that so much attention has been devoted to the myriad activities that surround the fork and its brief journey to the mouth.[2] Arguably this adds up to eating's being elusive, which thereby makes its study all the harder.

The reflections that follow are inspired by the three chapters of this section. Its aim is recklessly ambitious, seeking to enlarge on the contribution they make, at the same time as striving to represent the editors' goal of providing a secondary layer of theoretical analysis, whilst also allowing 'the complexities of the central conceptual focus on the encounters between foods and bodies to emerge in clear but nuanced ways'. Each chapter is discussed in turn to reflect upon the elusiveness

---

1 And in the process most importantly calls for more thoroughgoing theoretically informed analyses of eating.

2 Far less social scientific attention has been devoted to the various aftermaths, although new work is starting to appear (cf. Evans 2012; Evans, Campbell and Murcott forthcoming 2013; Alexander, Gille and Gregson forthcoming 2013).

of eating and to ponder on aspects of the way its study may be pursued in future. The discussion is couched in terms of the distinctive sociological tradition of symbolic interactionism (cf. Rock 1979). And before turning to the chapters, a quick reminder of that tradition may be useful. It is one succinctly characterized by Herbert Blumer, who was near enough one of its 'founding fathers':

> The term 'symbolic interaction' refers, of course, to the peculiar and distinctive character of interaction as it takes place between human beings. The peculiarity consists in the fact that human beings interpret or 'define' each other's actions instead of merely reacting to each other's actions. Their 'response' is not made directly to the actions of one another but instead is based on the meaning which they attach to such actions. Thus, human interaction is mediated by the use of symbols, by interpretation or by ascertaining the meaning of one another's actions. This mediation is equivalent to inserting a process of interpretation between stimulus and response in the case of human behavior' (Blumer 1969: 79).

The key point for the present discussion is that the things towards which human beings act and to which they attribute meaning includes not just objects, not just other people, but also a person themselves. This can include a state of being the person experiences, a physical sensation nameable as an itch, a pain or hunger. Via a process Blumer describes as 'self-indication' he presents an extended (faintly autobiographical?) example of someone noting their own hunger who then goes through a long list of possible alternative thoughts – Is it time to eat? Is there enough to eat? Can I afford to eat? Where should I eat? – he arrives at a point where, having decided to go out, he runs into an acquaintance who invites him for a drink with the result that 'the act which started off with a hungry tendency may end up with three hours of beer drinking' (Blumer 1969: 95).

The first chapter to be considered in the light of a symbolic interactionist attitude to eating is Maria Yotova's beautiful ethnography, '*It is the Bacillus that Makes Our Milk*'. At the outset, she asks: what are people actually ingesting when they take yogurt – a specific bacterium, authenticity, a nationalist tradition, or a health-giving substance? It is hard to deny that the answer is all or some of the list. Certainly the yoghurt is being ingested – that takes care of the material. But how are the other things listed to be verified?

First and foremost, for the person actually eating the yogurt, these others are symbolic, immaterial, matters of meaning. The person does not have to know anything about them still to ingest the yogurt, and still to be able to call it yogurt rather than milk or cheese. So under which circumstances does the person add in those other, *metaphorical* meanings to each spoonful of yogurt, and how does one meaning rather than another become attached to the eating? Furthermore, as Yotova so clearly shows, those meanings are not static, but have changed quite dramatically. Where and how does the change take place?

To some extent, the chapter provides a few answers: propaganda either in the form of socialist state 'education' or post socialist advertising. The simple but seemingly effective principle at work appears to be: 'announce something is the case often enough, loudly enough and by public figures who for one reason or another are listened to, and it becomes the case'. This is, however, a little remote from the mundane routine interaction between people eating the yogurt. And in any case, does this type of overarching propaganda always take hold such that in day-to-day interaction in hospitals or schools, and later in homes and among friends, that people have recourse to its content? In this case it does seem to be so, but Yotova is only in a position to report what as an ethnographer she was told – which gets a little closer to occasions of interaction between people and the thing in question.

Closer still is the fieldwork interaction between Yotova and her research informants. They may or may not be eating yogurt together, but the food is the focus of conversation that formed part of her fieldwork relationships. By and large, informants either tacitly or even explicitly were presenting themselves as knowledgeable about the Bulgarian-ness of yogurt, what it should be, how it tastes and that any self-respecting Bulgarian took pride in being of a nation that, is at last, not bringing up the rear. Such accounts would, in addition, seem to need that the informant declared that to be a good Bulgarian, the yogurt not only did Bulgarians good (for health, longevity etc.) but also tasted good. Thus Yotova was able to write of a public consensus about what the yogurt is and ought to be. Even those whose experiential response to the yogurt itself, the material realm, was of distaste were cautious when not wholeheartedly toeing the party line. And then, at last, Yotova is able to offer a glimpse of the fleeting process of the attribution of meaning in social interaction with others – in this case, the researcher herself. 'Some' she writes 'would ask me to switch off the recorder – 'this is only between you and me'; others would remind me not to put down any negative opinions because 'after all we are Bulgarians' and 'this will spoil our image'.

This, however, still leaves the discussion at one remove from the actual occasion of eating. Only reflections of those occasions are found in Yotova's chapter. To study them directly requires some sort of data generation *in situ*. It suggests that perhaps the next analytic tactic is to consider the familiar and fundamental anthropological questions of when, where and with whom this yogurt is eaten. Meanings – as Mintz pointed out in *Sweetness and Power* (1985) – arise in use. Meanings do not arise in use by one person, solo, in some idiosyncratic fashion. They are collectively, not individually, produced and, more to the point, reproduced. Their creation and recreation takes place during social interactions between people. But how does meaning arise when yogurt is being eaten when no-one else is present? Self-indication? How to get at that empirically? Asking someone while they are eating introduces a Heisenberg problem and the risk of a Hawthorne effect: the questioner becomes a party to the eating occasion. Ask someone before or after they have eaten the yogurt and the technical problem of being one remove from the activity is reintroduced. It may be that resorting to the inspection of literary sources is a useful first line of enquiry. Chefs' or food writers' autobiographies, fiction, or diaries, any documents that include a record of a person's reflections on their own eating may provide the source – but pursuing that thought further must be postponed for another discussion.

The question of solo eating and the attribution of meaning on such an occasion is not, however, simply fanciful. It is to some considerable extent a test of Blumer's declaration that meanings arise in the interaction people have with their fellows, and is, in part, at stake in the next chapter to be considered. The authors of *Chewing on Choice* implicitly criticize those seeking to 'educate' the public about healthy eating by analysing a succession of policy documents produced by those 'educators'.[3] In so doing they point out that time and again, the use of the expression 'food choice' entails an assumption of an individual who 'chooses', a person whose 'choice' is shaped by specifiable psychological processes that are amenable to change in a new direction by dint of rational persuasion: show someone the right answer for their health to the question 'what should I eat' and they will readily align their actions accordingly – so this approach goes.

The analysis presented by Brooks et al is welcome. A dissection of the use of terms such as 'food choice' in policy circles is long overdue. It is also a term persistently used and/or criticized by social scientists of varying stripes with very different meanings from one another. For instance:

> … [an] economist's approach to choice is based on the concept of a 'rational' individual attempting to maximise their welfare by choosing the most preferred outcome from a limited set of options (Young, Burton and Dorsett 1998: 82).

> … [(psychologists] define food choice as the selection of foods made by individuals from the range of options available to them. … in everyday life … food choice may include attitudes to different types of food and patterns of purchasing as well as actual intake (Steptoe et al. 1998: 29).

> Much of people's everyday behaviour is predictable to a degree of detail that cannot be attributed to their biological or psychological attributes. Such regularity is the consequence of what many sociologists would describe as the existence of normative order …. Institutionalised norms regulate and steer personal conduct into acceptable, comprehensible and effective activities. As a result of recognising these features of social life, sociology has generally been suspicious of the concept of choice, and has spent much of its history trying to identify the forces, mechanisms and institutional arrangements which restrict the practical range of options open to individuals (Warde and Martens 1998: 129).

Part of the 'moral of the food choice tale' is a reminder to all social scientists that several terms used in the policy community and in public commentary (such

---

3 A term used here as a crude, and somewhat inaccurate, shorthand for members of the policy community who produce the reports analysed from behind the scenes.

as 'convenience' or 'the consumer'[4]) are neither neutral nor somehow 'purely scientific'. On the contrary, just like 'food choice' they are ideologically framed and politicized. The authors show how discussions of healthy eating policies depend on the term 'food choice' with the aim, in effect, of persuading people to attribute distinctive meanings to different types of food and different modes of eating. Here (as in Yotova's chapter) is the attribution of meaning at the level of public information, education and propaganda. In this case, however, the implication is that the lack of success in persuading people to adopt a change in meaning represents a failure to prompt the attribution of a meaning of 'ill advised' sufficiently strongly to deter pursuit of such a diet. It leaves open the question of quite how 'ill advised' is crowded out or 'trumped' by alternative meanings.

Ormond's chapter is a neat discussion of co-existent and mutually-productive shifts in both agricultural practices and consumers' perspectives on the foods we eat. It relies on a case study of milk in the UK and efforts to effect a transition to lowering the carbon footprint of the dairy industry. His discussion notes a change in relations between producers and retailers, with an associated increase in consumer awareness of the environmental consequences of food production and consumption. So the chapter includes consideration of 'consumer desire', as he calls it, for more so-called ethical foodstuffs. It ends, however, on an ironic note: attempts by producers to reduce milk's carbon footprint may represent activities so-called ethical consumers would regard as unacceptable on other, environmental and animal welfare grounds.

For present purposes the 'the moral of the milk tale' is as much methodological as substantive. How is consumer desire to be apprehended and measured? What is the relation between whatever consumer desire is to mean and what actually gets put onto the fork and into the mouth? Substantive questions may, again, draw on Blumer's insights. What are the circumstances in which people learn about milk's carbon footprint, about moves to reduce it and the associated changes in animal husbandry? From whom do people learn about it all? If it is via the mass media, then we must ask under what circumstances and in the company of which others do people read, see and hear? What are the interactional components of people's use of mass media sources? If people learn from others, then the occasions and interactions in which this happens need to be the locus of empirical investigation. Does what has been learned (and continues to be learned, perhaps) play any part in the actual activities of getting food? This question returns to the methodological: how may what people have in mind when the acquire food be dependably studied?

The previous paragraph's list of questions is a suitable ending to these reflections that in any case can be no more than a very small start. The elusiveness of acts of eating themselves deserve attention although they are liable to prove hard to adequately capture social scientifically. At the same time, not only the sources of meanings attributed to those moments of eating but the manner in which some meanings rather than others become attributed also have to be captured.

4 See Trentmann (2007) for a long overdue history of the word consumer.

Doing that will be just as hard. But capture them we must if we are to ensure a means of addressing the simultaneity of the social, the political *and* the material in eating.

References

Alexander, C., Gregson, N. and Gille, Z. Forthcoming 2013. Food waste, in *The Handbook of Food Research*, edited by A. Murcott, W. Belasco and P. Jackson. London: Bloomsbury.

Blumer, H. 1969. *Symbolic Interactionism.* Englewood Cliffs, NJ: Prentice Hall.

Blundell, J., Dalton, M. and Finlayson, G. Forthcoming 2013. Appetite and satiety – A psychobiological approach, in *The Handbook of Food Research*, edited by A. Murcott, W. Belasco and P. Jackson. London: Bloomsbury.

Cheng, S., Olsen, W., Southerton, D. and Warde, A. 2007. The changing practice of eating: Evidence from UK time diaries, 1975 and 2000. *British Journal of Sociology* 58(1), 39–61.

Claflin, K. Forthcoming 2013. Representations of food production and consumption: Cookbooks as historical sources, in *The Handbook of Food Research*, edited by A. Murcott, W. Belasco and P. Jackson. London: Bloomsbury.

Evans, D. 2012. Beyond the throwaway society: Ordinary domestic practice and a sociological approach to household food waste. *Sociology* 46(1), 41–56.

Evans, D., Campbell, H. and Murcott, A. (eds) Forthcoming 2013. *Putting Waste on the Food Studies Agenda: Production, Politics and Everyday Life.* Sociological Review Monograph, Keele: The Sociological Review.

Mintz, S. 1985. *Sweetness and Power: The Place of Sugar in Modern History.* New York: Viking.

Murcott, A. 1993. Purity and pollution: Body management and the social place of infancy, in *Body Matters: Essays on the Sociology of the Body*, edited by D. Morgan and S. Scott. London: Falmer Press/Taylor and Francis, 122–34.

Rock, P. 1979. *The Making of Symbolic Interactionism.* London: Macmillan.

Steptoe, A., Wardle, J., Lipsey, Z., Oliver, G., Pollard, T.M. and Davies, G. 1998. The effects of life stress on food choice, in *The Nation's Diet: The Social Science of Food Choice*, edited by A. Murcott. London: Longman, 29–42.

Trentmann, F. 2006. The modern genealogy of the consumer: Meanings, identities and political synapses, in *Consuming Cultures, Global Perspectives: Historical Trajectories, Transnational Exchanges*, edited by J. Brewer and F. Trentmann. Oxford: Berg, 19–69.

Tremolières, J. 1970. A behavioural approach to organoleptic properties of food. *Proceedings of the Nutrition Society* 29(2), 280–96.

Young, T., Burton, M. and Dorsett, R. 1998. Consumer theory and food choice in economics, with an example, in *The Nation's Diet: The Social Science of Food Choice*, edited by A. Murcott. London: Longman, 81–94.

Warde, A. Forthcoming. Eating, in *The Oxford Handbook of the History of Consumption*, edited by F. Trentmann. Oxford: Oxford University Press.

Warde, A. and Martens, L. 1998. A sociological approach to food choice: The case of eating out, in *The Nation's Diet: The Social Science of Food Choice*, edited by A. Murcott. London: Longman, 129–46.

PART IV
Entanglements and Mobilizations: The Multiple Sites of Eating Encounters

The chapters in this section have a united interest in how foods and the discourses that frame them move and are mobilized across geographical, corporeal and conceptual boundaries, and in how boundaries themselves may be continually re-sited by taking food into the mouth. Accompanied by an interlude by Elspeth Probyn, the chapters map eating's many environments. They reflect on those in which eating takes place, both inside and outside bodies, as well as on those that shape, and are shaped by, eating bodies. In asking what and 'where' is consumed, the chapters engage with, but think beyond, the material properties of foods to reflect on the imaginings, landscapes and discourses that both constitute and fragment them. In turn, by following foods as they move across conceptual, biological and geographical spaces, this section demonstrates that eating is a multi-dimensional act that simultaneously entangles many actors; just a few of those drawn into view in the following chapters are metabolic enzymes; sheep bodies; coffee producers; fish oil and nutritional policies. By placing these alongside one another in text we argue for their simultaneity in the act of eating and suggest that eating is always about multiplicity and, perhaps, transgression. In differing but complementary ways, the chapters in this final section therefore illustrate how the depths of the digestive system constitute a site of multiple encounters and even of collisions. As such, they ask who and what eats, within what value systems, and in what ways. Seeking to move beyond simple discussions of individual agency and responsibility, the contributors to this section demonstrate that in eating, foods, bodies, social environments and value systems are overlaid; they are triangulated in complex and contingent ways that cannot be encapsulated by linear or narrow framings of what eating is and does. At the end of the volume, thus, these chapters illuminate an argument that has resonated throughout; they demonstrate that eating is a continually-shifting assemblage of visceral and political acts performed by actors both within and outside bodies. In these, the biological and symbolic dimensions of foods become coterminous with biological and social bodies at multiple temporal, geographical and conceptual moments of encounter.

Chapter 10
Confessions of a Vegan Anthropologist: Exploring the Trans-Biopolitics of Eating in the Field

Samantha Hurn

Introduction

Food and commensality are important aspects of most ethnographic fieldwork, and of creating and solidifying social relationships not just between our ethnographic subjects, but also between anthropologists and informants. However, food can also be a marker of difference. This chapter explores the difficulties that may arise when anthropologists have very different ideas about food and eating to those of their informants, through a focus on my own experiences of being vegan whilst conducting fieldwork in a rural farming community where meat constitutes a way of life.

It has been suggested that many anthropologists experience some form of 'culture shock' in relation to eating in the field. Indeed, food was mentioned as a potential cause of anxiety leading to 'culture shock' by Oberg when he coined the term (Oberg 1960). With the benefit of hindsight I can recognize that being vegan created certain obstacles during fieldwork that I had (perhaps naively) assumed would be easy to negotiate; the reality proved to be otherwise. However, what my informants saw as my dietary peculiarities provided me with insights into particular issues that I might otherwise have overlooked, and these will be explored here via the post-Foucauldian concept of 'trans-biopolitics' as proposed by Blue and Rock (2011). Following Franklin (2006) and Haraway (1997, 2004, 2008) amongst others, Blue and Rock define trans-biopolitics as 'the classification and evaluation of life as it unfolds in complex, technologically-mediated networks with global reach' (2011: 2). While their paper was primarily concerned with zoonotic disease, it will be argued in this chapter that the concept is as useful, if not more so, when applied to relationships between the human and nonhuman animals involved in contemporary industrialized food networks. Probyn's summation of the Deleuzian conception of desire makes such connections explicit; she states: 'a body, moved by desire, propels itself into networks and milieux of bodies and things. In turn, the milieu must be conceived of as a dynamic arena of social action' (1996: 49). Trans-biopolitics brings other animals into this 'dynamic arena of social action' based on the human desire for meat, and facilitates the exploration of inter-species power relations and entanglements that are integral to processes of food production and consumption.

It goes without saying that dietary preference can be a hugely important marker of identity. While people choose to become vegan for a whole host of reasons, including health, veganism is, rightly or wrongly, associated in popular consciousness with extreme ideological views in relation to animal rights and environmentalism more generally. There are a few published ethnographic accounts of veganism, from Clark's (2008) consideration of veganism as a means of social differentiation within punk sub-culture, to Potts and Parry's (2010) recent work exploring the links between ethical food consumption and sexuality. Some of Potts and Parry's vegan informants claimed that they could only form close intimate bonds with fellow vegans, viewing the consumption of animal products as so polluting as to render off-limits carnivores, omnivores and even vegetarians who ate dairy products (see also Probyn 2000). Such separatist attitudes, while certainly not widespread amongst vegans, have been seized upon by the global media and feed popular perceptions of veganism as a radical lifestyle choice and marker of 'otherness' (Clark 2008: 416; see also Cole and Morgan 2011a).

As an anthropologist who seeks embodied understanding I participated as fully as I could in the lives of my informants during fieldwork. However, as a vegan I did not and could not share their desire to consume flesh. As a result I was one step removed from the network of relations that culminated in consumption. My lack of carnivorous desire, not to mention my gender, nationality and academic motivation placed me in a vulnerable position vis-a-vis my informants. By not eating meat I was thought of as 'abnormal' and this uncomfortable situation brought the hierarchies that exist between the different bodies involved in agricultural production within this particular ethnographic context sharply into focus.

Situating the Ethnographer and the Field

Given my own veganism, the decision to conduct ethnographic fieldwork studying foxhunting and farming might seem surprising. In many respects however, my interest in animal welfare was what decided me on that particular path: I wanted to understand how and why people could participate in activities that involved the suffering and deaths of other living beings. I chose a farmers' hunt in rural Wales in a bid to see whether there might be a genuine argument for hunting as a form of pest control. The particular farmers' hunt whose members were the focus of this research is a functional institution and the hunt events (referred to as 'meets') lack much of the pomp and ceremony associated with mounted foxhunting in other parts of the United Kingdom. The majority of hunt staff and subscribers are either farmers themselves or are involved in the agricultural industry in some way (for example as vets, livestock auctioneers and transporters). I have argued elsewhere that the mounted foxhunting enacted in this particular context shares many similarities with the subsistence hunting found in other parts of the world and which is widely documented in the ethnographic record (see Hurn 2009, 2012 and forthcoming a and b). This is largely because the focus is almost exclusively on fox dispersal and control as opposed to on the riding which takes on more significance for foxhunts elsewhere (see Fukuda 1997; Marvin 2003, 2005). The community within which my ethnography is situated is very rural. Farming (mostly sheep and some cattle and mixed agriculture) is a primary source of income and there is a strong nationalist movement that prioritizes Welsh language and traditions as well as creating, at times, a hostile reception for outsiders who move to the area. These outsiders often hold radically different views relating to, amongst other things, food and animal welfare. As in other rural areas in the UK, this creates tensions in the fabric of rural communities (see Woods 1998).

I have lived and worked in my 'field' for over 12 years, having moved to Wales in 2000 for what was supposed to be one year of fieldwork. Whilst I was employed by the University of Wales from 2003–2012, I continued to maintain an active and consistent presence in the field; I lived for most of that period on a working farm belonging to one of my informants, and helped out when required with manual tasks such as shearing and milking. I still habitually attend events including sheepdog trials, livestock auctions, hunt meets and agricultural shows and regularly meet for tea or lunch and a chat with informants. This latter is to keep up with events and developments and also to maintain the reciprocal relationships that were formed during my initial fieldwork period and which have become binding. Indeed, this sort of commensality was and is integral to daily social interactions, but especially to the hunting experience, as will now be described through a series of fieldwork vignettes.

Commensality and Reciprocity on the Hunting Field

In relation to hunting, one of the ways in which ties of reciprocity were demonstrated between hunt members was through the hosted meet. The hunt staff and followers would arrive at a hosted meet well in advance of the off and, leaving their horses in the vehicles that had transported them to the venue or, in the case of those who had arrived on foot, tying horses to a suitable hitching post, they would venture inside to eat, drink and be merry, taking advantage of their host's hospitality. There were two categories of host: the member-farmer and the public house or social club. The pubs which hosted hunts were typically small, backwater establishments integral to the social and economic life of the farming community and the hunt followers liked to loiter in the bar for as long as possible before venturing out into the invariably cold and wet Welsh weather. In the hour or so before the off large quantities of alcohol were consumed, bringing in much needed revenue to these isolated establishments as well as instilling Dutch courage in the staff and followers.

The meets hosted by farmers and hunt members were very different, although these too were based on hospitality and reciprocity. As noted above, the primary function of mounted foxhunting in the area was fox dispersal and control. Sheep farming was the dominant agricultural industry, and consequently fox predation was a genuine 'problem'. Without exception, the private hosts of the mounted foxhunts were dependent on sheep farming for their livelihood. On entering the host venue, followers would be treated to the 'stirrup cup'; they would be given a drink of their choice (hot alcoholic punch was always available, but also tea, coffee, soft drinks and usually whisky or some other spirit), and let loose on the buffet provided. The spread invariably consisted of sandwiches (ham, cheese and egg being the staples); sausages or sausage rolls; cold meats and cheese; roast potatoes (occasionally); bread and butter; sponge cake; *bara brith* (a traditional Welsh fruit cake made with tea); Welsh cakes (like flat scones) and biscuits. Followers would eat their fill, setting themselves up for a long day in the saddle. The few followers who, for whatever reason, could not leave their horses unattended, would not be forgotten, and food and drink would be taken out to them by a member of the host's family.

These sporadic bouts of conspicuous hospitality were integral to understanding my informants' mounted foxhunting rituals. Indeed, many (over half) of these meets were held at cross roads, bridges, or other liminal landmarks and the 'stirrup cup' was therefore forsaken on these occasions (although from time to time individuals would take it upon themselves to bring refreshments which they distributed to other followers). This clearly demonstrated to me the functionality of the hunt in addition to its social aspects; the hunt is there to do a job of work. Although they are infrequent, hosted meets allow the farmers who benefit from the service provided by the hunt to make some form of repayment to the collective of followers. They not only allow them to ride over their land but, more importantly, provide them with sustenance. Indeed, in addition to the stirrup cup at the start of a meet, hosted meets would always conclude with additional food and drink. Partaking in a communal meal at the end of a day's hunting further served to reaffirm hunt followers' shared values and interests, in this case in sheep, the animals on which most are dependent for survival and in whose honour the whole ritual process is conducted.

Edible Bodies: Symbolic and Physical Consumption

On returning at the end of the hunting day, followers would put their horses away in the lorries or trailers that had brought them to the meet and then the human participants would sit down together to share a communal meal. This meal was always a steaming bowl of lamb *cawl* (a traditional Welsh stew of meat and seasonal vegetables) accompanied by bread, butter and chunks of cheese and as it was served, consumed, and washed down with a glass of something alcoholic, everyone would discuss the day's proceedings and share their experiences as a group. The significance of *cawl* cannot be over-emphasized. The use of lamb in *cawl* is a comparatively recent development, a result of the shift from dairy to sheep-based agriculture in the 1940s. This development also ties in historically with the formation of the hunt itself, which was established in the 1950s after the dissolution of the gentry estates. In the early to mid-twentieth century many tenant farmers became their own bosses and therefore were directly inconvenienced by the foxes who had been imported into the area for the recreational pursuits of their former landlords. As they became responsible for their own livestock, the attitudes of former tenants towards foxes became increasingly acrimonious (see also Hurn 2009, forthcoming b), leading to the establishment of the hunting ritual as a form of sacrifice. In this case, the sacrifice was not conducted with sheep as the physical victims of the sacrifice but rather to propitiate those sheep whose bodies had been unceremoniously consumed by uninvited and predatory transgressors (see also Marvin 2000). In relation to sacrifice more widely, Bloch has commented that 'the emphasis upon the communal meal by Detienne leads to a neglect of issues to do with the self-identification of the sacrificer with the victim' (1992: 30). In this specific ethnographic context, while it is the fox who is killed during the mounted foxhunt, the communal meal is *lamb* and not fox. However my informants identify very strongly with their sheep who might more reasonably be thought of as the sacrificial victim(s) than the fox who is actually killed in the ritual process. Bloch has also asserted that while Turner (1982) and van Gennep (1960) recognize the violence inherent in the act of separation that occurs during the first of the three phases which constitute a rite of passage, 'they completely miss the significance of the much more dramatic violence of the return to the mundane' (1991: 6). Instead, Bloch argues for the recognition of two underlying acts of violence: Firstly, the initial conquest of an inherent weakness within the individual or society (the act of separation, in this case the act of fox predation) and, secondly, a 'rebounding violence', which involves the incorporation of 'vitality obtained from *outside* beings, usually animals' (ibid.: 5, emphasis in original) and typically manifests in blood sacrifice whose victims replenish the initiands during the communal meal. Both of these acts of violence are integral to an understanding of the specific mounted foxhunting ritual under discussion, and the importance of a meat-based diet.

In the context of this working class farmers' hunt, the first violent act, that of fox predation and the death of livestock, is given as the rationale for the sacrificial ritual in the first place. The initiands are separated from the mundane by the donning of hunting dress and through their participation in the hunt itself. On the sound of the huntsman's horn followers are plunged into a period of liminality which continues as they ride over the countryside in pursuit of their quarry, the fox. It is only when they arrive back at the hosting venue after the fox has been killed that they are released from this area of ambiguity and reincorporated back into profane society.It is here that the second act, this time of 'rebounding violence', and the subsequent incorporation of vitality come to fruition. The whole basis of the ritual in this context is to both justify and perpetuate meat as a way of life (Hurn forthcoming a and b). The fox is killed to enable lambs to grow free from predation, to replenish the flock whose members sustain their human caretakers via their deaths at the abattoir. This is an inherently secular sacrifice which lacks any of the reverence or ritual of the foxhunt. Yet by participating in the mounted foxhunting ritual the hunt followers can, in some way, propitiate the sheep on whose lives and deaths they depend. Indeed, as already noted, Detienne and Vernant (1989) were concerned with 'the communal meal which follows the sacrifice and the distribution of the portions of the animal … as a means of *obectifying the social relations of the community*' (Miller 1998: 80, my emphasis). If all of the hunt's followers partake of the *cawl* provided at the end of a hosted meet, engaging in a 'rebounding violence' whereby the vitality of the sacrificial animal is consumed, they are demonstrating their solidarity as a social group and sharing the responsibility for the kills of both fox *and* sheep. This further explains why 'hunters' in this particular context especially are expected to be meat eaters. Because so many local people either make their living or supplement it from buying and selling animals classified as livestock, eating meat, and lamb in particular, has enormous cultural significance.

You Are What You Eat: Vegetarianism and Veganism as Deviance

Outside of ritual occasions, amongst hunt followers there was a tendency towards eating 'meat and two veg' meals. This culinary preference was especially noticeable at hunt functions which were, without exception, 'roasts', typically beef, lamb or poultry preceded by *cawl* and followed by a 'sweet' such as gateaux, trifle or occasionally pavlova. As I prepared myself for entry into the field, I had contemplated how my veganism might be interpreted and dealt with by hunt supporters (see Cassidy 2002 and Sutton 1997 whose vegetarianism featured in their respective fields). I had naively thought I would be able to manage things so that I never had to eat in front of others but in practice it was not that simple. Communal eating and drinking was, as I have already noted, integral to the 'hunting' experience, especially during formal functions such as the Huntsman's Dinner, and the Hunt Ball. It was also impossible to refuse milky drinks and home-made cakes offered when I went to informants' homes to conduct interviews. Because I had been a vegetarian for over 20 years, and a vegan for three prior to my entry into the field, I felt unable to eat meat but resigned myself to the fact that I could not refuse all animal products without causing offence. As a result I accepted meat-free items, which I would eat if there was no other option or, when I could do so without being noticed, I disposed of the offending items at the earliest available opportunity. At hunt meets my horses developed quite a taste for cake and the hounds always flocked round me in the hope of being passed a surreptitious sausage roll or ham sandwich – sacrilegious behaviour as hounds' diets were strictly controlled, especially on hunting days when their hunger was a driving force behind the chase (see Mullin 2007).

Some informants with whom I developed closer bonds realized my dietary peculiarities but accepted them grudgingly because I was 'different from normal veggies'. I felt flattered, believing that I had been accepted. My years of perseverance had paid off, or so I thought. Speaking Welsh, staying out in the torrential rain when other hunt followers had headed for home, working as a farm labourer or unpaid baby-sitter when needed, training unruly horses whom no one else would ride: Maybe I had challenged some of the pre-existing prejudices that many hunt followers held about vegetarians? Unfortunately I was deluding myself, although I only discovered this uncomfortable truth during a conversation with one informant, Lisa.[1] We were discussing my return to health following a serious riding accident sustained whilst conducting participant observation on the hunting field. I exclaimed how relieved I was to get back to a more active life and commented that I had put on two stone in weight since leaving hospital. Lisa agreed that it was good I had put on weight and revealed that she was fed up of fielding questions from other hunt members who were worried that I had an eating disorder. I was horrified and at the next available opportunity 'confronted' some of them about this revelation. It turned out that they had assumed I was obsessively worried about my figure. As one of the less tactful of them explained, 'we thought you were trying to stay slim to attract a husband'. While this last remark was made partly in jest, the fact remained that my eating habits were always noted and commented upon. When I explained that I was actually vegan, my own confession was met with shocked silence. Then, in the knowledge that I was lactose intolerant, Eileen exclaimed 'oh so are you allergic to meat too? Oh you poor thing, how awful for you'. I was not given a chance to explain myself as she and the other followers present started discussing how tragic it must be to 'have to be vegetarian'. From then on, everyone knew about 'poor Sam *bach*' (a term of endearment meaning 'small') and her unnatural inability to digest meat. My attempts at explanation consistently fell on deaf ears.

At one of the pre-season dinners I left the table to make a phone call just as the waitress arrived to take our table's order and, on my return, I was informed by Dan that his wife Bella had asked the waitress if I could have melon instead of the *cawl* that everyone else seemed to be having as a starter. 'Don't know why she bothered *bach*, there ain't no fuckin' meat in this *cawl* in any case!' he concluded. Another member of the hunt staff, Merf, who was sitting to my left interjected, 'so what do you live on then eh? Fuckin' air? Ain't nothin' of yer. You need to get some fuckin' meat in yer, init, a young girl like you'. This innuendo was met with raucous laughter from the other men sitting within ear shot (see for example Adams 1990 and Fiddes 1991 for discussions on the association between meat, sex and male virility). When the main course arrived, I pushed the meat to the edge of my plate and tried to scrape off the gravy as inconspicuously as possible to salvage the vegetables. Bella's husband noticed and without a word relieved me of the slabs of roast lamb, heaping them onto his already-overflowing plate. When Bella and Eileen realized that I had been given a plate with meat on, they called the waitress over and asked if I could have an empty plate so I could stock up on

1 All names are pseudonyms.

vegetables from the communal platters in the centre of the table. This became standard practice at hunt social functions and, at the same event the following year, they even ventured to ask the waitress if there was a vegetarian lasagne in the freezer they could heat up for me. I was consistently mortified that my informants drew such attention to what was seen to be my plight, but was hushed and told I needed 'feeding up' as they piled my plate high with vegetables, not allowing any of the men to have seconds until I had finished eating. They also found my embarrassment, and indeed the whole situation, somewhat amusing; 'What must they think? A vegetarian at a hunt dinner! Whatever next?' and it became a long-running joke that followers would compete to sit next to me so as to get my 'portion' of meat (again, the innuendos would be churned out each time. See S. Jones 2003).

Thus, in spite of their support, and my protestations to the contrary, my 'fussy' eating habits continued to be interpreted as either an eating disorder or some other medical condition. For example, the first meet I attended following my riding accident went from Eileen's farm and I was informed on arrival that she'd made me a vegetable *cawl*: 'There's no meat in it at all. And you're not to leave the house until you've had some!' At the next meet, this time an unhosted event which saw followers congregate at a crossroads in the middle of nowhere, Bella ordered me to dismount and I was taken to the boot of her car and forced to eat a peanut butter sandwich and drink a plastic tumbler full of lemonade from the spread she had laid out on her parcel shelf! At subsequent meets Bella took it upon herself to pack jam or banana sandwiches for me to eat ('just in case'), which she would produce from her saddle bags at opportune moments. 'We don't want you passing out and falling off again!' she explained, 'it's taken you long enough to get back in the saddle after the last time!' Even my accident was explained in terms of what was perceived to be my inadequate diet. It was easier for my informants to believe that I was 'ill' than that I chose not to eat animal products of my own free will. And so the belief persisted that vegetarianism and healthy country people do not mix. The desire for meat was normalized and anyone who chose to eschew that norm was considered deviant, with all that that entails. As Eileen confided during an interview, 'I've always been brought up to think that you have to eat meat to be healthy. If I didn't feed my kids meat I'd feel like I was *neglecting* them'.

Numerous scholars have noted that consuming meat is widely thought to be essential to human health (for example, Peace 2008). Meat-eating also symbolizes human 'culture' through the domination and control of animals and the environment (see Hurn 2012). As Joy (2011) argues, 'carnism' is a dominant belief system within which it is considered appropriate and ethically unproblematic to eat meat. She adds the suffix '-ism' to denote that, for an omnivorous species such as humans, this emphasis on meat is a choice not a biological necessity. However, because the majority of my hunting informants are involved in some way or another in livestock production, meat-eating is regarded as an absolute necessity – one which ensures not just physical but also socio-economic health.

The act of eating meat, especially meat which they themselves have raised, places my human farming informants at the top of their food chain.

While I have argued that sheep constitute totems of sorts, their relationship with their human caretakers is primarily functional and persists because of a desire for meat at a local and global level. Notwithstanding the few individuals who are occasionally spared as pets, the bodies of sheep are protected and allowed to live only for as long as they are productive. For the majority of males, that can be from six to 18 months while for most females and those individual males whose bodies are required for sexual reproduction, that may extend to four, five or even six years (way below the 'natural' lifespan of domestic sheep who can live into their late teens and early 20s). It should be noted that foxes are not the only nonhumans to take ovine lives (see, for example, Hurn 2009). Yet they are the only species to be dealt with ritualistically. Their role as scapegoat and not totem is further evidenced in the manner of their death and especially the distribution of their flesh. As noted above, it is lamb that forms the basis of the obligatory *cawl* consumed at the communal meal. Fox meat is, as Wilde famously noted 'uneatable' (2007: 11). It is the hounds, the liminal servants of the hunt staff, who act as sacrificial knife and then devour the body of their quarry. The fox becomes the sacrificial substitute who pays a heavy price for 'his' own desire for forbidden flesh. Indeed, the 'meat is natural' discourse certainly rings true for foxes, whose evolutionary history and contemporary physiology leave little room for doubting their carnivorous nature. Their treatment at the hands of the omnivorous primates who also desire meat needs to be considered alongside the development of large-scale livestock production; a system which simply does not allow for the propitiation of the individual lives who are killed to satiate that desire.

From Biopolitics to Trans-Biopolitics

It is here, after recognizing the link between desire, sacrifice and consumption that we can begin to explore the trans-biopolitics of eating in this particular 'dynamic arena of social action' (Probyn 1996: 49). Foucault's concept of 'biopolitics' is a means of conceptualizing power relations as they manifest in and on the physical bodies of social actors through the state control of bio-power – the biological functions of individuals, such as sexuality, reproduction and so on. In his lecture series, *Society must be defended* (2003), Foucault observed a paradigm shift in sovereign power in the nineteenth century which saw 'the biological come under State control … [accompanied by] a certain tendency that leads to what might be termed State control of the biological' (ibid.: 240). Foucault clarified the position as follows: 'I wouldn't say exactly that sovereignty's old right – to take life or let live – was replaced, but it came to be complemented by a new right which does not erase the old right but which does penetrate it, permeate it. This is the right, or rather precisely the opposite right. It is the power to "make" live and "let" die' (ibid.: 241). However this power to 'make' live and 'let' die leads to a paradox: how to maximize life and yet also take it? For Foucault, racism provided the answer when considered as a 'means of introducing … a fundamental division between those who must live and those who must die' (Foucault cited in Stoler 1995). For Foucault, racism facilitated the creation of hierarchy between self and other, us and them (see also Sartre 1995), which has enabled acts of violence such as slavery and genocide to be committed in other historical and ethnographic contexts.

In a similar vein, Ryder defined 'speciesism' as 'the widespread discrimination that is practised by man [sic] against other species' (1975: 16) and Singer, who popularized the term, defined it as 'the idea that it is justifiable to give preference to beings simply on the grounds that they are members of the species *Homo sapiens*' (2006: 3, see also 2009). Many contemporary scholars in the field of human-animal studies argue that the perpetuation of exploitative and discriminatory practices such as the farming of livestock indicates the 'ongoing normalization of speciesism' (Cole and Morgan 2011b: 145). However, one might take some small comfort from Pleasants's assertion that the industrialization of slavery in the eighteenth century [an industry grounded in racism] and the intensification of 'agribusiness' [an industry grounded in speciesism] in the twentieth century have facilitated the realization that these practices are immoral and unjust (2008: 208–9), realizations which are necessary preconditions for discriminatory, normative practices to be challenged. Racism, as Foucault conceived of it, 'fragments the biological field, it establishes a break … inside the biological continuum of human beings by defining a hierarchy of races, a set of subdivisions in which races are classified as "good", fit and superior. More importantly, it establishes a *positive* relation between the right to kill and the assurance of life' (Stoler 1995: 84), which suggests that 'the more you kill [and] … let die, the more you will live' (Foucault in ibid.: 84). Such a sentiment lies at the heart of 'carnism' and the desire for meat.

The global agricultural industry is immensely powerful (Peace 2008), and this power is based on the normative assumption of 'carnism' – that meat eating is both natural and justifiable because of perceived differences between humans and other animals (Hurn 2012; Joy 2011). That is not to say that all meat eaters consciously harbour speciesist principles or that individual farmers carry significant political and economic clout. On the contrary, my ethnographic experience with farmers at the coal face (to mix some Welsh metaphors) suggests that many have complex and contradictory ideas about the animals in their care (see also Baker, this volume). They are also marginalized and not in control, driven by global forces and almost entirely dependent on government subsidies. Yet in this, as in other agricultural contexts, countless nonhuman lives are brought to an abrupt end so as to sustain and preserve the lives of others and it is the termination of the lives of the animals whose actions are seen to undermine human agency and control which is the focus of the sacrificial ritual because these animals serve as useful scapegoats for the propitiation of guilt or rather the shame which accompanies excessive desire (see Probyn 2005).

Foucault has been widely criticized by post-humanist scholars for his anthropocentric focus (see Haraway 2008; Blue and Rock 2011). Yet there is plenty of scope for the application of Foucault's work to the field of human-animal studies or anthrozoology. In relation to racism for example, it might be ventured that this hierarchical biological continuum extends beyond the human species. Humans share biological continuity with other animals, which makes the exclusion of nonhuman species, especially 'companion species' (Haraway 2003), from discussions of biopower surprising. Indeed, in what Bulliet (2005) refers to as post-domesticity, a state of affairs characterized by the alienation of the majority of consumers from processes of production, some nonhuman animals such as pets are embraced as honorary humans, kin even, while others, especially those enrolled in intensive agricultural production systems are regarded as inferior species. As a result, a strictly Foucauldian biopolitics appears ill-equipped when it comes to accounting for many contemporary power relations, especially inter-species relationships which are inherently paradoxical.

Blue and Rock advanced the concept of trans-biopolitics 'to account for the manifold interconnections among animal bodies as they emerge and unfold in complex industrial and technological systems with global reach' (2011: 355). Drawing on other areas of anthropological investigation (such as trans-nationalism) which allow for a multitude of voices to be heard and which link and analyse 'transformations in the nation-state in light of cross-border flows of humans, enabled and facilitated by global technological and economic processes' (ibid.: 3), trans-biopolitics facilitates a consideration of multi-species relations. Indeed, 'the prefix *trans* signals the transgression of boundaries, borders, and barriers, enabled and augmented by modern scientific and technological systems. Thus, trans-biopolitics implies more mobility in the creation and exercise of power than that designated by biopolitics' (Blue and Rock 2011: 2). This breaking down of borders and the inclusion of nonhuman voices and bodies in previously anthropocentric theoretical discussion are of fundamental import in a world where, through their lives and particularly their deaths, nonhuman others are inextricably implicated in every aspect of human existence (see Hurn 2012).

Animals classified as livestock exist to feed and clothe a human population whose members, as noted above, *desire* to eat meat. As a vegan I have, in the past, taken the moral high ground, seeing my own lifestyle choice as inherently more 'ethical' than that of those who eat meat (see Probyn 2000), and especially of those who produce it. Yet the process of fieldwork enabled me to reconsider both my position and the processes of meat production engaged in by my informants. Unlike many consumers, including those who are vegetarian or vegan on moral grounds, my farming informants are, as I have noted above, deeply interconnected in manifold ways not just with humans but also with other animals. In relation to foxhunting, it is through their social relations with sheep that foxes become actively embroiled in human social relations and, as the eaters of forbidden flesh, their bodies become coterminous with human bodies, facilitating their substitution as human scapegoats in the sacrificial act (Hurn 2009, forthcoming a and b; see also Marvin 2000).

Embodied Knowledge and Power

Farming, or large scale animal management, is a truly embodied livelihood which dictates the lifestyles not just of the individuals concerned, but also of their families and other members of the community. But, more importantly, as a result of co-existing so closely with large animals, each individual farmer carries the scars of their inter-dependency on their own bodies; broken bones, calloused hands, aching backs and weathered skin. Moreover, animals are not inanimate passive objects; rather, as with the fox whose actions initiate the mounted hunting ritual, they frequently exert influence on the humans with whom they interact. Elsewhere I have discussed the relationships between some of my human informants and their horses as a form of symbiosis (Hurn 2008a, 2008b, 2011). However, when it comes to animals involved in food production the balance of power shifts somewhat. Yes, the lives of farmers are still inextricably bound up with their livestock, but the nonhumans in question have less opportunity to exert influence; the end point will usually see the humans concerned regaining the upper hand through the assertion of physical control, often by force (see O. Jones 2003). The use of force in human-animal interactions provides another window into how the humans involved think about both themselves and the animals in question and this domination of animals is facilitated by the same sort of anthropocentricity and speciesism that facilitates 'carnism' (Joy 2011).

Conversely, a significant factor often influencing an individual's decision to become vegan is how they perceive other animals. Unlike some other political orientations, being vegan is often an embodied ethical or moral decision. The notion of consuming the flesh of another animal, or consuming products such as milk or honey which may have caused suffering and/or death is off-putting because of recognition of the continuity between humans and other animals. In many cases this is compounded by embodied experiences and inter-subjective interactions with individual animals through the course of individual lives (Hurn 2012, forthcoming b; Milton 2002, 2005). Despite my earlier vignettes suggesting that all farmers were rampant carnivores, the reality is actually quite different. Over half of my farming informants had at least one child who was vegetarian and one farmer was himself vegetarian (but this was not common knowledge within the community). Many would not eat their own animals, while others would only eat their own animals 'because they had a good life', as one informant put it (see also Serpell 1999). Such contradictions suggest that psychological devices are at work in the food choices of individuals for whom meat is a way of life, but one that requires a close association with the animals who become food.

One final vignette from my fieldwork illustrates the contradictions, conflicting emotions and the balance of power involved in raising animals for meat.

I arrived late to one of the hunt dinners and had to take the only remaining seat. I found myself next to a hunt supporter whom I had always been rather wary of. He followed the hunt religiously and was renowned for being outspoken, with staunchly nationalistic (i.e. anti-English) political views. I was, therefore, rather apprehensive about making small talk while pushing slices of uneaten meat around my plate. However, the frank exchange which ensued took me completely by surprise; he said:

> You're a vegetarian aren't you? Don't look so embarrassed! I don't blame you, not eating meat, not really. I don't care – more for me. Do you mind? [as he took the meat from my plate]. It even bothers me, you know, taking the sheep to slaughter. I don't hang around. Just drop them off and leave. They know they're going to die, every one of them. They know and it scares the shit out of them. You can see it in their faces, and in their calls – you can hear the panic. I don't enjoy that. But that's how things are. I've got to eat. They're not stupid, sheep, they're very perceptive. People don't like to think so but they are, they know what's going on, and I have to live with that. But they're not the *same* as humans, we're different. It's a different sort of intelligence. Like foxes, now they're too clever for their own good …

This conversation was important on many levels, and it resonates with many of the central arguments of this chapter as a whole. Firstly, it gave me a concrete indication that there was some sense of guilt or shame associated with sheep farming. Secondly, it drew my attention to perceived hierarchical *differences* between species, prioritizing the human in a manner consistent with speciesism (Ryder 1975; Singer 2006, 2009). And thirdly, it suggested that there was a link between the guilt of speciesist animal exploitation and foxhunting for those who were sufficiently affected by it. Guilt or shame, the emotional residue left behind after social or moral transgression, has to be dissipated in some way, and the mounted foxhunting ritual explored in this chapter can be seen to provide such a means of absolution. By sacrificing the fox in such a heavily ritualized manner (when many other farmers simply shoot, trap, poison or gas foxes on their land), those farmers who participated in the mounted foxhunt were acting to propitiate both the sheep and lambs who had died while they should have been under human protective custody, and the hundreds and thousands of sheep whose lives were terminated every week at the local abattoir to be shipped off in cellophane packaging to feed a national and international desire for meat.

Conclusion

To come full circle and draw the chapter to a close, I will return to veganism as a secular lifestyle choice. In the contemporary world, elective veganism and (to a lesser extent) vegetarianism are social movements often (but not always) rooted in an ethical concern for the wellbeing of nonhuman animals and/or environmental sustainability. While most contemporary post-domestic consumers are considerably removed from processes of agricultural production, the 'technologically-mediated networks with global reach' referred to by Blue and Rock (2011: 354) facilitate an awareness of the plight of the animals who then become food. Indeed, while they may choose to ignore or 'forget' it, thanks to television and the Internet the majority of consumers are well aware of the often brutal exploitation of animals which is part and parcel of industrialized livestock production; 'carnism' (Joy 2011), the desire for meat and an entrenched specieisism allow for such realities to be overlooked. However, as Pleasants has noted (2008), becoming vegetarian (or indeed vegan) is predicated on the existence of a socio-cultural structure that makes such a dietary preference a possibility. As my own ethnographic experiences within the farming community suggest, even those individuals involved in meat production are not immune to feelings of guilt, shame and crises of conscience when it comes to making a living from the deaths of other animals. Many find ways of asserting their moral agency – 'neutralizing' their actions (Matza and Sykes 1961) – by, for example, taking animals to slaughter themselves, only eating their own meat, singling out individuals for preferential treatment or, as explored in this chapter, participating in the propitiatory ritual provided by the mounted foxhunt. Nonetheless, such actions are also grounded within a speciesist ontology whereby 'carnism' and the desire for other animal products provide ultimate justification for exploitation. Consequently these neutralizations arguably do not improve the lives of the animals concerned, nor do they destabilize power inequalities between human producers, consumers, and nonhuman 'products' in any meaningful way.

References

Adams, C.J. 1990. *The Sexual Politics of Meat: A Feminist-Vegetarian Critical Theory*. Oxford: Polity Press.

Bloch, M. 1991. *Prey into Hunter. The Politics of Religious Experience*. Cambridge: Cambridge University Press.

Blue, G. and Rock, M. 2011. Trans-biopolitics: Complexity in interspecies relations. *Health: An Interdisciplinary Journal for the Social Study of Health, Illness and Medicine* 15(4), 353–68.

Bulliet, R.W. 2005. *Hunters, Herders, and Hamburgers: The Past and Future of Human–Animal Relationships*. New York: Columbia University Press.

Cassidy, R. 2002. *The Sport of Kings. Kinship, Class and Thoroughbred Breeding in Newmarket*. Cambridge: Cambridge University Press.

Clark, D. 2008. The raw and the rotten: Punk cuisine, in *Food and Culture: A Reader, Second Edition*, edited by C. Counihan and P. Van Esterik. London: Routledge, 411–22.

Cole, M. and Morgan, K. 2011a. Vegaphobia: Derogatory discourses of veganism and the reproduction of speciesism in UK national newspapers. *The British Journal of Sociology* 62, 134–53.
Cole, M. and Morgan, K. 2011b. Veganism contra speciesism: Beyond debate. *The Brock Review* 12(1), 144–63.
Detienne, M. and Vernant, J.P. 1989. *The Cuisine of Sacrifice Among the Greeks*. Chicago: Chicago University Press.
Fiddes, N. 1991. *Meat: A Natural Symbol*. London: Routledge.
Foucault, M. 2003. *Society Must Be Defended: Lectures at the College De France, 1975–1976*. New York: Picador.
Franklin, S. 2006. The Cyborg embryo: Our path to transbiology. *Theory, Culture & Society* 23(12), 167–87.
Fukuda, K. 1997. Different views of animals and cruelty to animals: Cases in fox-hunting and pet-keeping in Britain. *Anthropology Today* 13(5), 2–6.
Haraway, D. 1997. *Modest.Witness@Second.Millenium.FemaleMan©.Meets. Oncomouse™*. Feminism and Technoscience. New York: Routledge.
Haraway, D. 2003. *The Companion Species Manifesto: Dogs, People and Significant Otherness*. Chicago, IL: Prickly Paradigm Press.
Haraway, D. 2004. A manifesto for Cyborgs: Science, technology and socialist feminism in the 1980s, in *The Haraway Reader*, edited by D. Haraway. London: Routledge.
Haraway, D. 2008. *When Species Meet*. Minneapolis: University of Minnesota Press.
Hurn, S. 2008a. What's love got to do with it? The interplay of sex and gender in the commercial breeding of Welsh cobs. *Society & Animals* 16(1), 23–44.
Hurn, S. 2008b. The 'Cardinauts' of the western coast of Wales: Exchanging and exhibiting horses in the pursuit of fame. *Journal of Material Culture* 13(3), 335–55.
Hurn, S. 2009. Here be dragons? No, big cats! Predator symbolism in rural west Wales. *Anthropology Today* 25(1), 6–11.
Hurn, S. 2010. What's in a name? Anthrozoology, human–animal studies, animal studies or…? *Anthropology Today* 26(3), 27–8.
Hurn, S. 2011. Dressing down: Clothing animals, disguising animality. Special Issue: Les Apparences de l'Homme. *Civilisations* 59(2), 123–38.
Hurn, S. 2012. *Humans and Other Animals. Human-Animal Interactions in Cross-Cultural Perspective*. London: Pluto Press.
Hurn, S. forthcoming a. *Human-Animal Farm*. London: Ashgate.
Hurn, S. Forthcoming b. Post-domestic sacrifice: Exploring the present and future of gifts for the gods, in *Worlds of Sacrifice: Exploring the Past and Present of Gifts for the Gods*, edited by C. Murray. Buffalo: IEMA Monograph Series.
Jones, O. 2003. 'The restraint of beasts': Rurality, animality, actor network theory and dwelling, in *Country Visions*, edited by P. Cloke. Harlow: Pearson, 283–303.

Jones, S. 2003. Supporting the team, sustaining the community: Gender and rugby in a former mining village, in *Welsh Communities: New Ethnographic Perspectives*, edited by C.A. Davies and S. Jones. Cardiff: University of Wales Press, 27–48.

Joy, M. 2011. *Why We Love Dogs, Eat Pigs, and Wear Cows: An Introduction to Carnism.* Newburyport, MA: Red Wheel Weiser.

Marvin, G. 2000. The problem of foxes: Legitimate and illegitimate killing in the English countryside, in *Natural Enemies: People–Wildlife Conflicts in Anthropological Perspective*, edited by J. Knight. London: Routledge, 189–212.

Marvin, G. 2003. A passionate pursuit: Foxhunting as performance. *Sociological Review* 52(2), 46–60.

Marvin, G. 2005. Sensing nature: Encountering the world in hunting. *Etnofoor* 18(1), 15–26.

Matza, D. and Sykes, G. 1961. Juvenile delinquency and subterranean values. *American Sociological Review* 26(5), 712–19.

Miller, D. 1998. *A Theory of Shopping*. Ithaca, NY: Cornell University Press.

Milton, K. 2002. *Loving Nature: Towards an Ecology of Emotion*. London: Routledge.

Milton, K. 2005. Anthropomorphism or egomorphism? The perception of non-human persons by human ones, in *Animals in Person: Cultural Perspectives on Human–Animal Intimacies*, edited by J. Knight. New York: Berg, 255–71.

Mullin, M.H. 2007. Feeding the animals, in *Where the Wild Things Are Now: Domestication Reconsidered*, edited by R. Cassidy and M. Mullin. Oxford: Berg, 277–304.

Oberg, K. 1960. Cultural shock: Adjustment to new cultural environments. *Practical Anthropology* 7, 177–82.

Peace, A. 2008. Meat in the genes. *Anthropology Today* 24(3), 5–10.

Pleasants, N. 2008. Structure and moral agency in the antislavery and animal liberation movements, in *Eating and Believing: Interdisciplinary Perspectives on Vegetarianism and Theology*, edited by D. Grumett and R. Muers. London: T&T Clark, 198–216.

Potts, A. and Parry, J. 2010. Vegan sexuality: Challenging heteronormative masculinity through meat-free sex. *Feminism and Psychology* 20(53), 53–72.

Probyn, E. 1996. *Outside Belongings: Disciplines, Nations and the Place of Sex*. New York and London: Routledge.

Probyn, E. 2000. *Carnal Appetites: FoodSexIdentities*. London: Routledge.

Probyn, E. 2005. *Blush: Faces of Shame*. Minneapolis: University of Minnesota Press.

Ryder, R. 1975. *Victims of Science: The Use of Animals in Research*. London: Davis-Poynter.

Ryder, R. 2000. *Animal Revolution: Changing Attitudes Towards Speciesism*. Oxford: Berg.

Sartre, J.-P. 1995. *Anti-Semite and Jew: An Exploration of the Etiology of Hate*. New York: Schocken Books, New Edition.

Serpell, J.A. 1999. Sheep in wolves' clothing? Attitudes to animals among farmers and scientists, in *Attitudes to Animals: Views in Animal Welfare*, edited by F.L. Dolins. Cambridge: Cambridge University Press, 26–33.

Singer, P. 2006. *In Defense of Animals: The Second Wave*. Chichester: John Wiley & Sons.

Singer, P. 2009. *Animal Liberation, Updated Edition*. New York: Harper Collins.

Stoler, A.L. 1995. *Race and the Education of Desire: Foucault's History of Sexuality and the Colonial Order of Things*. Durham, NC: Duke University Press.

Sutton, D.E. 1997. The vegetarian anthropologist. *Anthropology Today* 13(1), 5–7.

Turner, V. 1982. *From Ritual to Theatre: The Human Seriousness of Play*. New York: PAJ Publications.

Van Gennep, A. 1960. *The Rites of Passage*. London: Routledge & Keegan Paul.

Wilde, O. 2007. *A Woman of No Importance*. London: Penguin Classics.

Woods, M. 1998. Researching rural conflicts: Hunting, local politics and actor networks. *Journal of Rural Studies* 14(3), 321–40.

Chapter 11
Metabolism as Strategy: Agency, Evolution and Biological Hinterlands

Rachael Kendrick

Introduction

In sociocultural discussions of eating, metabolism is often either absent or framed as a dull biological hinterland to the interesting business of taste, meaning, ideology and so on. In this chapter I offer a definition of metabolism and use this to show how thinking metabolically affords traction on particular questions related to eating, health and the politics of survival – questions about the relations between fat, eating, metabolic illness and agency. Thinking metabolically, I suggest, allows us to attend to the materiality of epidemic obesity. A discussion of metabolism is important to this book for two reasons. Firstly, the concept of metabolism advances the volume's aim of exploring the networks that are made and unmade by eating, and it is my aim to usefully trace out the contours of this concept and its implications for studies of food and eating. Secondly, questions of metabolism and its relationship to illness are becoming important to researchers of embodiment, health and culture, particularly those concerned with obesity, even if obesity is frequently rendered as a representational or cultural issue rather than a material or metabolic one. This chapter is drawn from a broader materialist analysis of epidemic obesity, an analysis that attends particularly to the behaviours of specific bodily tissues, such as fat, the brain, the stomach and bowel. As such, the chapter is less concerned with the moment of eating itself; rather it explores what happens afterwards, when food descends into visceral depths beyond conscious awareness and control to become part of the body's biochemical hustle and flow, which includes the growth and decline of fat tissue. In so doing, the chapter offers an alternative perspective that troubles how these bodily visceral and biochemical behaviours are often reduced to a crude economic rubric of 'calories in, calories out', where fat is rendered as a kind of carnal balance sheet – visual evidence of caloric imbalance (see Aphramor et al. and Yates-Doerr, both this volume).

The first half of this chapter engages with several theoretical and ontological concerns. Specifically, I advance a definition of the metabolic body as a bounded, self-organizing system that uses material flows and transformations to maintain the distinction between system and environment. I am particularly interested in

Maturana and Varela's concept of structural coupling (1980) as a way to overcome two paradoxes, the first the question of how the living system as a whole can be simultaneously bounded and relational, as it appears in the eating encounter and, second, the problem of how the human body can at once be a bounded unity and also contain multiple agencies and modes of being-in-the-world, which is also highlighted by the act of eating. These concerns emerged from what was initially intended to be a brief engagement with the most prominent evolutionary 'just-so' story about obesity, James Neel's thrifty gene hypothesis (1962). The thrifty gene hypothesis (TGH) is a familiar explanation for the aetiology of obesity and human fatness: Neel proposes that the twin human capacities to accumulate body fat and become diabetic are the result of genetic mechanisms intended to preserve reproductive capacity in times of famine. The TGH, with its market-based logic of surplus, investment and strategy, has retained an aura of common sense and remains a frequently-invoked explanation for the relationship between eating, obesity, diabetes and techno-economic development, despite intense criticism from both within and without population genetics. My interest in Neel's theory is not to do with his narrative, with the meanings he ascribes to the human body, or with his use of mechanistic concepts and metaphors such as strategy and investment to describe bodily functioning. What interests me is the presence of metabolism in this account, and the relationship between environmental conditions and phylogeny. Here, the human body has evolved to be metabolically responsive to the environment in specific ways.

The second half of this chapter, then, is concerned with evolutionary accounts of the metabolic body. It considers the human body as a product of evolution, but reads the TGH against another account of evolution and metabolism, the selfish brain hypothesis. I read these accounts against each other to tease out some of the ways in which the body has become responsive to the environment, and to plot out how fat has been figured as relating to the environment and technology in two different ways: as both a product of eating behaviour *and* a mediator of eating behaviour. In one, fat is an *effect* of the technological stabilization of energetic environments, but in the other fat is a *precondition* for technological relations to the environment; eating, thus, is differently positioned in relation to fat and implicates different modes of agency. To map out this positioning, it is my intention to take account of the many agencies in the eating encounter, agencies present on a range of scales and temporalities within the human body. As Myra Hird (2009), for example, demonstrates in her study of the micro-ontologies of microbial life, much of the human body is quite literally composed of nonhuman matter, in the form of the colonies of bacteria on our skin and mucous membranes on whom we, as humans, are metabolically dependent. As such, eating is a social act, a conscious act, but it is also a physiochemical act, which is both performed by, but also arises from, a body whose actions exceed human consciousness and, indeed, exceed the human.

Re-creating Internal Environments

What happens when you eat? You take food into your mouth, chew, swallow, and that food descends into the bowels where it is squeezed, ground, dissolved by acids, dismantled by enzymes, massaged by cilia, and then elements of your worked-upon meal enter your bloodstream. Why? Because the body may only remain alive if it enters into constant material exchange with its environment, by inhaling and ingesting air, water and food and exhaling and excreting the by-products of this metabolism. This is a simple observation, so simple it might appear to be at best a critically uninteresting aspect of the eating encounter. Yet, by focusing on how the internal bodily environment is sustained by metabolic flows, we can map out an ontology of the metabolic body, which shows it to be at once stable and autonomous, yet also dependent and relational. At the heart of these two concepts is the problem of entropy, broadly defined as the tendency of matter to lose heat, become disorganized and cease moving over time.

Erwin Schrodinger (1992), in his collection of essays on molecular biology, defines metabolism as a set of physiochemical processes by which living systems resist succumbing to a state of permanent equilibrium with their surroundings. When a non-living system is placed into a stable environment movement comes to a halt as a result of friction; all reactions that can occur take place then cease as stable compounds are formed, and a permanent state is reached. Living systems are however, to make explicit a point implicit throughout this chapter, partially open to their environment. They engage in energy exchange with the surrounding environment in order to maintain their structural and functional organization and thus, internal entropy. As Schrodinger describes it:

> Every process, event, happening – call it what you will, in a word, everything that is going on in nature means an increase of the entropy of the part of the world where it is going on. Thus a living organism continually increases its entropy – or, as you may say, produces positive entropy, which is death. It can only keep aloof from it, i.e. alive, by continually drawing from its environment negative entropy … What an organism feeds upon is negative entropy. Or, to put it less paradoxically the essential thing in metabolism is that the organism succeeds in freeing itself from the entropy that it cannot help producing while alive (Schrodinger 1992: 71).

Living systems, then, are positioned precariously between the open dynamism of life and the closed stability of death. From this perspective, metabolism is a conservative process through which the dynamic living system resists death, or entropy, by remaining partially open to its environment, engaging in an energetic exchange where material substrates from the environment are brought in and catabolized, used to maintain the self-sustaining functions of the organism. The internal environment may only sustain itself and maintain its life-giving equilibrium through a metabolic relation with the world outside of the living system. This metabolic relation, of course, is mediated by the eating encounter.

Also thinking about metabolism, but in the context of artificial life, Margaret Boden (1999) argues that metabolism is such a defining, essential, and absolutely material feature of life that anything abstracted or removed from such material exchanges cannot be considered in any way 'living'. She defines metabolism as a state of energy dependency where the living system uses, collects, spends, stores and budgets energy. In order for an artificial life form to be considered 'alive', Boden argues, it must operate in a similar state of energy dependency, resolved using similar forms of material exchange. Accordingly, she defines three 'senses' of metabolism that must be present in an artificial life form in order for it to be considered sufficiently metabolic. Firstly, Boden argues, to be metabolic is to be, as mentioned previously, in a state of energy dependency, and therefore reliant on a set of internal processes to incorporate external energy sources into the system. Secondly, individual energy 'packets' power the functions of the system: energy is finite, and if internal energy sources are used up and no more can be obtained from the environment, the organism ceases its energy-using activities and dies. Many of these energy-using functions are directed at maintaining the internal environment of the system, aside from other functions like movement, growth and reproduction. As Boden explains:

> Additional, purely internal energy exchanges are required as the collected energy is first converted into substances suitable for storage and then, on the breakdown of these substances, released for use. Very likely, these processes will produce waste materials, which have to be neutralized and/or excreted by still other processes. In short, metabolism necessarily involves a nice equilibrium between anabolism and catabolism, requiring a complex biochemistry to effect these functions (Boden 1999: 238).

To be metabolic, then, is also to produce waste. Thirdly, metabolism is a process of material self-organization, where energetic processes organize the matter of which the living system is made. It is this third 'sense', in particular, that locates life solidly in the material world, and it is for this reason, Boden contends, that virtual life forms that are, say, contained in an energy-stable environment, such as creatures made of software 'living' in a computer, cannot be regarded as 'alive'. Boden explains that:

> Metabolism concerns the role of matter/energy in organisms considered as physically existing things. It is not an abstract functionalist concept, divorced from specific material realities. By contrast, the other features typically mentioned in definitions of life – self-organization, emergence, autonomy, growth, development, reproduction, adaptation, responsiveness and (sometimes) evolution – can arguably be glossed in functionalist, informational terms (ibid.: 231).

Boden's aim in introducing metabolism into discussions of artificial life is to reinsert the material into these debates, arguing that the biochemical complexity and specificity of metabolism must be included in definitions of life. As Zaretsky and Latelier (2002) note, Boden's concept of metabolism as a form of material self-organization, and her insistence that the materiality of metabolism is a defining feature of life, is broadly equivalent to the definition of life given by Maturana and Varela (1980) as 'autopoiesis in physical space' (ibid.: 265). Autopoiesis, which may be broadly defined as 'self-creation' or 'self-production', is a term employed by Maturana and Varela to convey what is, for them, a critical feature of living organisms: their autonomy, and their self-regulating organization. For both Boden and for Maturana and Varela, the parts that make up the organism – the various muscle fibres, digestive membranes, neural tissues and bones – are, in a sense, irrelevant compared to the relations they generate to constitute a system or unity. What matters in the living system, then, are the relations between lens of the eye and larynx that mutually allow both to persist. The internal environment is produced by these relations, not the form or structure of each organ.

In their work on autopoiesis, Maturana and Varela take the phenomenology of living systems as their initial object of investigation. They conceptualize the organism as an autopoetic system or machine, an autonomous unity that is 'an homeostatic (or rather relations-static) system which has its own organization (defining the network of relations) as the fundamental variable which it remains constant' (1980: 79). Autopoietic machines are self-organizing systems that work in a state of operational closure; they are networks of processes of production and transformation that subordinate all internal changes to their own maintenance. Maturana and Varela define their version of the 'machine' as 'a unity in physical space, defined by its organization, which connotes a non-animistic outlook, and whose dynamisms are apparent' (ibid.: 136). These machines constantly turn over their components; what remains stable and unified is the organization of these components, the relations between them. This is an important point. When Maturana and Varela talk about *boundaries*, what is 'bounded' is the organization of the machine. In the metabolic encounter, for instance, metabolic substrates are drawn into such a bounded machine in order to sustain the distinction between machine and environment; the substance of the machine is changed – its structure – but the organization of the machine is not. I will return to this distinction later; first, I will discuss Maturana and Varela's argument about observation, teleology and structural coupling as it relates to autopoietic machines.

The authors also make an important epistemological point about autopoietic machines, arguing that such machines are, fundamentally, *purposeless*. They remind their reader that '[t]he use to which a machine can be put by man is not a feature of the organization of the machine, but of the domain in which the machine operates, and belongs to our description of the machine in a context wider than the machine itself' (Maturana and Varela 1980: 77–8). In other words, notions such as 'aim,' 'purpose' and 'function' belong to the observer, not to the machine itself. They therefore position teleonomy as an artefact of observation that does not describe or reveal anything about the organization of the machine itself. This forces the focus of their project, of understanding the organization of living systems, on to the autonomous individual. Maturana and Varela's organic machine, then, is a singular, autonomous system, spatially bounded and absolutely separate from the surrounding environment or medium, defined by the self-organization and self-maintenance of relations between its constitutive parts. Maturana and Varela's argument about observation and teleology goes beyond understanding the behaviour of an organic machine. Most notably, they position cognition and perception as embodied, as occurring in the closed internal world of the organism, itself an aggregate of closed systems – an aggregate of autopoietic unities. This understanding of cognition illustrates their concept of structural coupling, the means by which autopoietic unities are responsive to their medium or environment. As they put it, '[t]he cells of multicellular systems normally exist only by taking other cells in close proximity as a medium for realizing their autopoiesis. These systems are the result of a natural drift of lineages in which close proximity has been conserved' (Maturana and Varela 1987: 77). In other words, a multicellular organism is the result of many autopoietic unities coming to share a similar bounded space, generated by the mutual production of an internal environment upon which they all depend.

To borrow from Cary Wolfe in his discussion of second order systems theory, for Maturana and Varela '[t]he full definition of "embodiment," then, is a self-referential, self-organizing and nonrepresentational system whose modes of emergence are made possible by the history of structural coupling between the autopoietic entity and an environment to which it remains closed on the level of organization but open on the level of structure' (Wolfe 1988: 61). Thus, Maturana and Varela distinguish between the organization of a machine, the relations that constitute it as a bounded unity, and its structure. Indeed, it is through this distinction between organization and structure that they are able to generate an account of a system that is at once open to its environment and functionally closed. They define 'organization' as that set of relations between components that define the system, and 'structure' as the components themselves. While the components of an organism – the structure – may change and, indeed, need to change for an organism to stay alive (for instance, the continuous shedding of epithelial cells from the lining of the bowel), the organization remains the same; the bowel remains a bowel. In *Tree of Knowledge* (1987), a toilet may be described as an arrangement of tanks, pipes and levers; if you were to replace the plastic float with a wood float the structure of the arrangement would change, but the organization of the toilet remains the same. Plastic float or wood float, it is still organized, and functions, as a toilet. To reflect upon my body, the closed operational milieu I, as author of this text, function in, I am now largely composed of matter of an entirely different origin from that which I was when I was born. I have become what I ate, and I continue to slough off skin and hair and other tissues wherever I go. However, I am still organized as a human organism, and for the time being I persist, a bounded living machine distinct from my environment. It is therefore this distinction between organization and structure that accounts for how autopoietic machines may remain open to their environment without losing their unity. Structure eats; organization digests. As Maturana and Varela put it:

> Autopoietic machines [...] can be perturbed by independent events and undergo structural changes which compensate for these perturbations. If the perturbations are repeated, the machine may undergo a repeated series of internal changes which may or may not be identical. Whichever series of changes may take place, however, they are always subordinated to a maintenance of the machine organization, a condition of which is definitory of autopoietic machines (Maturana and Varela 1980: 81).

In other words, environments can shape and modify the structure of autopoietic machines, while their organization as machines remains unchanged. Moreover, Maturana and Varela go further to talk about how autopoietic unities may be structurally coupled to other unities:

> In describing autopoietic unity as having a particular structure, it will become clear to us that interactions (as long as they are recurrent) between unity and environment will consist of reciprocal perturbations. In these interactions, the structure of the environment will only *trigger* structural changes in the autopoietic unities (it does not specify or direct them), and vice versa for the environment. The result will be a history of mutual congruent structural changes as long as the autopoietic unity and its containing environment do not disintegrate: there will be *structural coupling* (Maturana and Varela 1987: 75, emphasis in original).

Thus, in this section I have argued that metabolic systems are at once bounded and relational; metabolic processes are directed towards the maintenance of the distinction between system and environment, a distinction maintained, for example, by the material flows and transformations mediated by eating. In the following section I discuss how changes in the environment trigger changes in the body, such as the activation of stress systems in the brain, the secretion of cortisol and, eventually, the deposition of fat around the viscera. In other words, I explore how the tissues of the body respond to environmental stresses in ways that impact eating and metabolism, troubling 'calories in-calories out' understandings of the aetiology of obesity and the physiochemical bases of hunger and eating. I consider how environments shape metabolic systems over time. It is my concern to render metabolic illnesses as illnesses of environmental contingency, showing how the tissues of the body have their own modes of responsiveness to environmental stress, engendering the metabolic conditions that produce illness; I find support for these claims in a reading of evolutionary accounts of diabetes and reflect on relationships between this condition and the act of eating.

Evolved Metabolic Strategies

In his work on insulin resistance, population geneticist James Neel described diabetes as an illness of ecological contingency in an organism shaped by evolution (1962). Neel observed that diabetes and its key pathology – non-insulin dependent diabetes mellitus (NIDDM), or type-2 diabetes – were highly prevalent in the indigenous populations of the Americas, Greenland, Micronesia and Melanesia (West 1974). Neel argued that individuals in these populations had evolved what he called a 'quick insulin trigger' (Neel 1962) in response to hyperglycaemia, the high levels of blood glucose following a meal. This 'quick insulin trigger' stimulates the uptake of blood glucose by the tissues of the body, allowing the individual to readily store available glucose as glycogen in the liver and eventually as fat. This store of energy in the liver and in the body's fat, Neel argued, not only provides an 'energy buffer' against conditions of famine, it also ensures the maintenance of reproductive capacity. For Neel, 'differential conception rates among women able or not able to conceive under energetically marginal conditions could have an extremely powerful and rapid effect on selection' (Prentice, Henning, and Fulforrd 2008: 1608). The quick insulin trigger then, Neel suggested, is heritable, ensuring its spread through populations regularly affected by scarcity and famine. In times of plenty, however, the quick insulin trigger becomes a physiological liability, prolonging states of hyperglycaemia following a meal, straining the capacity of the pancreas to produce insulin and regulate blood sugar. Eventually, the body's tissues become insensitive to the effects of insulin – a condition called insulin resistance – which is a pathological precursor to the development of diabetes. Neel, in particular, argued that diabetes, insulin resistance and obesity should be considered to be aspects of the same pathological process, writing that 'the overweight individual of 40 or 50 with mild diabetes is not so much diabetic because he is obese, as he is obese because he is of a particular (diabetic) genotype' (Neel 1962: 92).

The TGH was not a long-lived hypothesis, as later physiological research found little experimental evidence for Neel's proposed quick insulin trigger, and Neel subsequently published major revisions to the theory in what Michael Montoya called an act of 'uncanny reflexive modesty' (2007: 104). In particular, Neel gestured towards the frailty of evolutionary explanations for disease, concluding:

> All these speculations may be utterly demolished the moment the precise etiologies of [NIDDM] become known. Until that time, however, devising fanciful hypotheses based on evolutionary principles offers an intellectual sweepstakes in which I invite you all to join (Neel in Montoya 2007: 105).

Despite this admission, populations such as indigenous communities living in reservation-style settings (Richards and Patterson 2006), Hispanics (Haffner et al. 1997), Japanese American immigrants (Fujimoto, Leonotti, and Kinyon 1987), and Polynesians and Pacific Islanders remain key subjects for population geneticists to study the genetic mechanisms that, in their view, underpin the metabolic changes characteristic of late modernity. It is important to emphasize the kind of temporalities involved in the operations of these proposed genetic mechanisms and their response to energetic environments. Neel et al. capture this temporality in a later report on the TGH, writing:

> If our thesis that the three disease entities briefly discussed in this review all in very large measure reflect genetically complex homeostatic systems now pushed to and beyond their limits, a collective term would be useful. With respect of these three entities [of NIDDM, hypertension and obesity], the genes involved are very predominantly fine old genes … honed by millennia of selection for harmonious interactions and appropriate epigenetic relationships, whose proper functioning is overwhelmed by extraneously imposed parameters of really very recent origins (Neel, Weder and Julius 1998: 60–61).

Here, Neel's thrifty genetic mechanisms appear to see 'fine old genes' rendered obsolete by 'extraneously imposed parameters' or environments that *some* populations are adapted to and *others are not.* Here then, metabolic illnesses are diseases of obsolescence that arise in bodies poorly suited to the energetic conditions of modernity, a logic that has generated remarkable (and frequently racist) accounts of the interactions between population, environment and economy, such as Jared Diamond's (2003) perplexing tale of a 'genetic epidemic' of NIDDM in Nauru. Opponents of the TGH argue that NIDDM is too complex to be explained by genetics. Paradies et al. suggest instead that '[t]he fact that over 250 genes have been studied as possible causes for [NIDDM], but together these genes explain less than 1% of the prevalence of diabetes worldwide, should give researchers – and others – pause' (2007: 217). Further, many critics question the legitimacy of using ethnorace as a category of analysis. Indeed, racial difference is pivotal to Neel's analysis of diabetes and obesity, and it is frequently invoked in the work of Neel's intellectual descendants. As one commenter notes, Neel's work teems with:

> Contrasting images of modern civilized man and primitive hunter-gatherers, 'affluence' and 'feast and famine,' 'technologically advanced nations' and 'the Stone Age body'. Neel consistently maintains the focus on difference, if only at the level of imagery. After all, he is a population geneticist, with nothing to contribute as a scientist if there is no difference among populations that might be ascribed to evolutionary advantage (Fee 2006: 2992).

Michael Montoya (2007), in his analysis of the use of ethnic and racial (what he terms 'ethnorace') categories in the study of diabetes, argues that such categories are used by scientists in two ways: either descriptively, as a coherent way to distinguish between (presumably) genetically similar populations, and attributively, to attribute qualities to certain groups.

However, although the Thrifty Gene Hypothesis has been thoroughly discredited within population genetics, as one group of researchers complains, '[d]espite the conceptual difficulties [...] the racialized TGH remains one of the most widely cited hypotheses in all of genetic epidemiology' (Paradies, Montoya, and Fullerton 2007: 212). In arguing for the presence of a discrete 'gene' responsible for such bodily behaviours, Neel's hypothesis and its surprising resilience as an explanation, betrays a bias for genetic causes as the only acceptable 'excuse' for obesity and diabetes, in contrast to stigmatizing explanations of 'over-eating'. Yet, importantly, the TGH also distributes the mechanisms of insulin resistance throughout the body. Neel's proposed 'quick insulin trigger', while directly implicating the gut, liver and pancreas, has implications for processes of glucose metabolism in potentially all tissues by affecting the rate at which glucose is made available to tissues.

Other related evolutionary and physiological explanations of glucose metabolism and insulin resistance move past genetic explanations by attending to the behaviour of three specific sites of the body: the energy-intensive, large human brain, the blood brain barrier, and the short, 'cheap' human gut. I shall refer to these explanations collectively as the selfish brain hypothesis, but it is important to note that the following section discusses two separate explanations, the first an evolutionary theory on how the human body came to have the structure it does, the second a physiological theory on the regulation of energy supply in the organism.

It is common for evolutionary explanations to conceptualize the structure and function of organisms in economic terms of price, profit and loss. For instance, sexual reproduction is frequently figured as an expensive activity, an investment on the part of the organism, one that requires trade-offs in competing functions. Physical anthropologists Leslie Aiello and Paul Wheeler (1995), in their study of how humans and other primates came to develop large brains – a transition called 'encephalization' by evolutionary theorists – developed what they called the expensive tissue hypothesis. According to Aiello and Wheeler, 'the average human has a brain that is 4–6 times the size expected for the average mammal, and the average non-human primate anthropoid has a brain almost twice as large of that of the average mammal' (1995: 200). For Aiello and Wheeler, a central question for studies of encephalization is how primate bodies evolved to manage the energetic cost of a large brain. It is suggested that brain tissue is the most energetically 'expensive' tissue in mammalian bodies, with physiologists estimating that brain metabolism accounts for 20–25% of basal metabolic rate (BMR) in the adult human body (Leonard, Snodgrass, and Robertson 2007). Therefore, it might be expected that humans and other encephalized primates would have a high overall BMR compared to other mammals of similar mass. However, Aiello and Wheeler observe that there 'is no evidence of an increase in basal metabolism [in encephalized primates] sufficient to account for the additional metabolic expenditure of the large brain' (Aiello and Wheeler 1995: 201). The authors speculate that a possible answer to this 'cost question' is 'that the increased energetic demands of a larger brain are compensated for by a reduction in the mass-specific metabolic rate of other tissues' (ibid.: 201).

In looking for this hypothesized reduction in metabolic rate, Aiello and Wheeler compare the energy uses of various organs in the human body. They explain that the splanchnic organs (liver, kidneys, gut), combined with the brain, make up 60–70% of the total BMR budget, despite taking up less than 7% of total body mass. Skeletal muscle, on the other hand, the tissue most commonly figured as energy-consuming in popular understandings of human metabolism, contributes only 14.9% of the BMR, while taking up 41.5% of total body mass in the average human. Furthermore, working from allometric predictions about the composition of placental mammal bodies, the human body has a smaller than expected gut. Aiello and Wheeler assert that the size of the brain then, composed as it is of metabolically expensive tissues, may be negatively correlated with the size of the gut. Extending their cost-analysis method, Aiello and Wheeler predict that:

> A relatively high BMR would require a correspondingly high energy intake, and unless the environmental conditions were unusual, this would not only require devoting a significantly larger percentage of the daily time budget to feeding behaviour but also put the animal in more intense competition for limited food resources. Further, it is unlikely that the size of other metabolically expensive tissues (liver, heart or kidneys) could be altered substantially (Aiello and Wheeler 1995: 205).

Working through the functions of other splanchnic organs as they relate to the function of the enlarged human brain, Aiello and Wheeler argue that the human gut became attenuated over time so as not to compete with the brain for energy. The liver maintains blood glucose levels, both through the metabolism of stored glycogen and other energy sources mobilized from elsewhere in the body; the maintenance of blood glucose within narrow parameters is essential for brain function, as will be discussed further. The kidneys have an unavoidably high metabolic rate due to their key function of ion transport, and so on. Therefore Aiello and Wheeler argue that '[i]f the hypothesis of coevolution is correct, what is essential for understanding how encephalized primates can afford large brains is identifying the factors that allow them to have relatively small guts. The gut is the only one of the expensive metabolic tissues that could vary in size sufficiently to offset the metabolic cost of the encephalized brain' (ibid.: 206).

In Aiello and Wheeler's theory, then, encephalized mammals offset their large brains with an attenuated, less expensive gut, a gut that requires a high-energy, readily digestible diet. This hypothesis, of a gut shortened in favour of the brain, articulates well with accounts of the brain's role in energy management throughout the body. A team of neuroscientists, led by Achim Peters (2004), argues that not only is the brain the biggest consumer of energy in the body, it also controls various 'setpoints' in the allocation of energy and the intake of nutrients. In Peters et al.'s work, the brain has the capacity to allocate energetic resources throughout the body, and in so doing, it prioritizes its own energy needs above those of other bodily tissues. They link the brain's capacity to allocate energy resources to the precarious metabolic position the brain occupies within the body. Firstly, the brain is separated from general circulation by the blood brain barrier. Secondly, although the brain is the largest consumer of energy in the body, unlike other organ systems such as the liver, kidneys and skeletal muscle, it has no capacity to store energy for later use. Thirdly, brain tissue is almost entirely dependent on the metabolism of glucose, while other tissues, such as muscle, can metabolize glucose, fat or proteins; this is partly because few other energy substrates can pass through the blood brain barrier. Fourthly, the brain is able to record information from, as well as control, peripheral organs. Due to its dependence on the metabolic capacities of the rest of the body, and its control and surveillance of the body, the brain ensures its supply of glucose by either impairing glucose uptake in peripheral tissues, through activation of what the authors refer to as the brain's stress system, or by stimulating eating behaviour through hunger. Just as, in the expensive tissue hypothesis, encephalization came at the price of re-ordering metabolically costly tissues, truncating an expensive gut in favour of a developed brain, so in the selfish brain hypothesis the brain ensures its own energy needs at the expense of other bodily systems. The selfish brain hypothesis's contribution, then, is to gather together the mechanisms of control of blood glucose and fat, previously distributed throughout the body, and place the brain at the centre of homeostatic control. This perspective is summarized in an article by Fehm et al. who write:

> The target of the regulatory system is not a constant energy intake or a constant body weight, but rather a constant energy content within the brain. Of course, in a healthy subject the hippocampal outflow is such that energy expenditure and energy uptake are equal, a situation that results in a constant body weight. The situation represents the set-point of the system: the energy needs of the brain are satisfied and peripheral energy stores are stable (Fehm et al. 2006: 134).

If in the TGH the metabolic mechanisms allowing for the development of insulin resistance and obesity were selected because they maximized reproductive success, in the selfish brain/expensive tissue hypothesis these mechanisms were selected because they provided the stable metabolic conditions necessary for the development of a large brain. As one commentator on the expensive tissue hypothesis noted:

> The expensive-tissue hypothesis also differs from [conjectures that encephalization may be accounted for by behavioural factors like tool production, hunting, warfare, language etc.] in that it suggests a physiological/anatomical complex that acted as a *prime releaser* permitting selection for increased brain size rather than speculating about one hypothetical behaviour

> that was the prime target of that selection (Falk in Aiello and Wheeler 1995: 212, emphasis in original).

To take this hypothesis one step further, the evolved arrangement of short gut, energy storing liver, comparatively small muscle mass, plentiful adipose tissue and diverse, energy-rich diet provides the conditions of possibility for the development of human cognitive capacity, and the arrangements of neurological control.

Peters et al. (2004) position the selfish brain hypothesis against two similar theories of homeostatic control, the 'glucostatic' and 'lipostatic' theories of fat and glucose control. The lipostatic theory was supported by the identification of leptin, a hormone secreted by adipose tissue (fat) as a feedback signal to the brain, so that the brain is informed of the status of peripheral energy stores. 'Most researchers', the authors note, 'considered this to be a closed regulatory system in which the absorption of nutrients is the regulator, body mass is the controlled parameter, and leptin is the feedback signal' (ibid.: 145). In the glucostatic theory it is assumed that blood glucose is the regulated parameter in a system that uses endocrine changes (for example, hormones such as insulin, glucagon and cortisol) and feeding behaviours to keep glucose constant. In both, the regulators of glucose and, ultimately, the deposition of body fat are found outside the brain, with '[t]he implicit assumption that an adequate energy supply to the brain automatically results from the constant behavior of the fat reserves and the blood glucose is common to both the glucostatic and the lipostatic theory' (ibid.: 145). In other words, what emerges from these accounts is a body that requires a tightly-controlled internal environment, consisting in part of appropriate concentrations of glucose, and the mechanisms that control these concentrations are found throughout the body – in the endocrine system, in fat, in the blood. These control mechanisms are conceptualized as feedback loops with a regulator, controlled parameter and feedback signal. It is important to note here that, where eating is not directly addressed in work on the TGH, in the selfish brain hypothesis '[t]he brain changes eating behavior so that it can then alleviate the stress system and return it to a state of balance' (ibid.: 147).

The pathological mechanisms of diabetes, then, within the theory of the selfish brain hypothesis, are a disruption of arrangements of neurological control over glucose metabolism and, ultimately, eating, where food intake, mediated by complexities of hunger and the preferential allocation of energy to the brain, regulate the 'set point' for body weight. A slight but chronic activation of the brain's stress systems activates a neurological feedback loop that, in preferentially maintaining the brain's glucose level, leads to a chronic rise in blood sugar and subsequent 'spikes' in insulin secretion, which ultimately lead to insulin resistance and diabetes. This point deserves emphasis: in the TGH diabetes appears as an atavism, as the continued action of metabolic mechanisms rendered obsolete in contemporary energetic environments ('times of plenty'); in the selfish brain hypothesis diabetes appears as a *response to stress*, an attempt to maintain the brain's glucose levels. 'In the selfish brain theory,' researchers explain, 'insulin resistance and an activated sympathetic nervous system represent counter regulatory mechanisms that aim to compensate for a displacement of the set point' (Fehm et al. 2006: 136). This is a significant point of difference between the TGH and the selfish brain hypothesis. Through the concept of a homeostatic 'set point' for body weight that may be disrupted by cumulative stress, the selfish brain hypothesis opens up the possibility that obesity is a bodily response to stressful and insecure, rather than energetically plentiful, environments and this is further supported by research on metabolic syndrome and its relationship to obesity.

While the relationship between diabetes and obesity is complex, the kinds of endocrine and metabolic conditions within the body engendered by diabetes do produce particular kinds of fat (Després 2006; Després and Lemieux 2006; Godoy-Matos et al. 2009). This opens up possibilities of discussing multiple obesities, multiple 'fatnesses', multiple kinds of fat with different actions and behaviours and different relationships to eating. The metabolic syndrome is, broadly, a set of symptoms that indicates an elevated risk of diabetes and cardiovascular disease. These symptoms include abnormalities in the metabolism of fat and glucose, such as high levels of triglycerides and increased fasting blood glucose, hypertension, and abdominal obesity (Black 2003; Rosmond 2005). Key to this is abdominal or central obesity, the deposition of fat around internal organs, as opposed to peripheral obesity, where fat is deposited beneath the skin. Central obesity is particularly related to stress; individuals with central obesity have been found to have excess levels of cortisol, a stress hormone (Rosmond 2005). One group of researchers argues in an article in *Nature* that central obesity is a marker of 'dysfunctional adipose tissue'. They write that:

> There is evidence suggesting that if the extra energy is channelled into insulin-sensitive subcutaneous adipose tissue, the individual, although in positive energy balance, will be protected against the development of the metabolic syndrome. However, in cases in which adipose tissue is absent, deficient or insulin resistant with a limited ability to store the energy excess, the … surplus will be deposited at undesirable sites such as the liver, the heart, the skeletal muscle and in visceral adipose tissue – a phenomenon described as ectopic fat deposition (Després and Lemieux 2006: 882).

Metabolic syndrome is significant here for two reasons: firstly, it positions obesity and metabolic illness as a response to chronic stress; secondly, it distinguishes between central or pathological obesity and other kinds of obesity that do not have such negative impacts on health. Here, *some forms* of obesity are an adaptive response to stress, while others are not. It is worthwhile briefly noting what researchers define as 'stress'; in a collection on stress science, stress is most often defined as a long-term threat to homeostasis, where 'environmental demands tax or exceed the adaptive capacity of an organism, resulting in psychological and biological changes that may place persons at risk for disease' (Cohen, Kessler annd Gordon, in Contrada 2011: 1, emphasis in original). Stress, then, threatens the homeostatic continuity of the organism by wearing away at the capacity to adapt, or by triggering internal changes that cumulatively cause illness. In the context of the selfish brain hypothesis, the chronic activation of the brain's stress systems creates the preconditions for pathological forms of obesity. Large deposits of fat ensure that concentrations of glucose in the brain never drop below a certain level, and they also ameliorate the risks of an evolved short gut that requires a calorie-intensive and difficult-to-find diet. Here, diabetes and insulin resistance appear as a response to stress and precarity, an attempt to conserve a metabolically fragile brain at the expense of other tissues. In other words, the body that may become diabetic, that may become obese, is also a body that may reorder the world around itself. It is a body oriented towards preserving its own internal environment. The selfish brain hypothesis therefore opens up the possibility that obesity and metabolic disease are cumulative responses to stressful, insecure environments; environments that are engendered by the operations of capital. From this perspective, obesity may be read as another instance of, as Judith Butler puts it, the 'different ways that corporeal vulnerability is distributed globally' (2004: 31).

Conclusion

In this chapter, I have argued that the ontological status of the metabolic body is one of simultaneous boundedness and relationality, and that this is, in part at least, mediated by the eating encounter. Yet, this also highlights how eating may be more complex than it appears, with its relationship to bodily fat drawn into question. In Neel's thrifty gene hypothesis, certain strategies of glucose metabolism, specifically the 'quick insulin trigger', are selected over time as a way to ensure reproductive success. Neel (1962) argued that populations that retained the 'quick insulin trigger,' when shunted into new energetic environments by the expansion of global capital and, by extension, the logics of colonization, succumbed to diabetes as a result of the over-activation of their 'quick insulin trigger'. In this way the thrifty gene sexes and races populations through their susceptibility to metabolic illness, illnesses caused by changed social, technological, economic and natural environments to which they are poorly suited. In the selfish brain hypothesis, on the other hand, the capacity for obesity and diabetes are the result of metabolic strategies that create the conditions of possibility for a large brain, a brain that drives endocrine and physiochemical changes that alter eating behaviour. Here, diabetes and insulin resistance are framed as a response to stress and precarity, and signify ways in which to conserve a metabolically fragile brain at the expense of other tissues. Thus, what emerges from this metabolic account is a vision of obesity as a survival mechanism, a way to ensure bodily survival in precarious times driven by the regimes of debt-capitalism. The body as it appears in many critiques of epidemic obesity is largely understood in terms of language and representation. When such bodily phenomena as hunger and eating appear in these accounts they largely do so as the bearers of meaning: hunger is something that is either encouraged or proscribed, or the manifestation of psychic trauma and conflict. By focusing on the metabolic body, the material body may be productively re-inserted into accounts of obesity. In particular, the contentious tissue of body fat is then able to emerge as active and vital, with its own history and agency, as an actor in practice. In the beginning of this chapter I positioned eating as a matter of survival, a material encounter required to maintain the bounded, operationally closed internal environment of the organism. Here, hunger as the need for food, arises from a series of complex, frequently conflicting, physiochemical imperatives. Importantly, what is done with food once it is eaten – how it is metabolized – is also far from straightforward. Together these two temporalities offer a nuanced portrait of eating and the multiple agents it implicates: we eat because we are hungry, we eat to live and sustain ourselves as organizations of living matter, and we eat because the tissues that compose our bodies ask 'us' to. By attending to the agency and behaviour of these tissues, thus, it becomes possible to better understand the complexities of how and why we eat.

References

Aiello, L.C. and Wheeler, P. 1995. The expensive-tissue hypothesis: The brain and the digestive system in human and primate evolution. *Current Anthropology* 36(2), 199–221.

Black, P.H. 2003. The inflammatory response is an integral part of the stress response: Implications for atherosclerosis, insulin resistance, type II diabetes and metabolic syndrome X. *Brain, Behaviour and Immunity* 17, 350–64.

Boden, M.A. 1999. Is metabolism necessary? *British Journal for the Philosophy of Science* 50(2), 231–48.

Contrada, R.J. 2011. Stress, adaptation and health, in *The Handbook of Stress Science: Biology, Psychology, and Health*, edited by R.J. Contrada and A. Baum. New York: Springer Publishing Company, 1–9.

Després, J.-P. and Lemieux, I. 2006. Abdominal obesity and metabolic syndrome. *Nature* 444, 881–7.

Diamond, J. 2003. The double puzzle of diabetes. *Nature* 423, 599–602.

Fee, M. 2006. Racializing narratives: Obesity, diabetes and the 'Aboriginal' thrifty genotype. *Social Science & Medicine* 62(12), 2988–97.

Fehm, H.L., Kern, W., Peters, A., Fliers, E., Hofman, M.A., Swaab, D.F., van Someren, E.J.W., Kalsbeek, A. and Buijs, R.M. 2006. The selfish brain: Competition for energy resources. *Progress in Brain Research* 153, 129–40.

Fujimoto, W.Y., Leonotti, D.L., Kinyoun, J.L., Newell-Morris, L., Shuman, W.P., Stolov, W.C. and Wahl, P.W. 1987. Prevalence of diabetes mellitus and impaired glucose tolerance amongst second-generation Japanese-American men. *Diabetes* 36(6), 721–9.

Godoy-Matos, A.F., Vaisman, F., Pedrosa, A.P., Farias, M.L.F., Mendonça, L.M.C. and Pinheiro, M.F.M.C. 2005. Central-to-peripheral fat ratio, but not peripheral body fat, is related to insulin resistance and androgen markers in polycystic ovary syndrome. *Gynecological Endocrinology* 25(12), 793–8.

Haffner, S.M., Howard, G., Mayer, E., Bergman, R.N., Savage, P.J., Rewers, M., Mykkanen, L., Karter, A.J., Hamman, R. and Saad, M.F. 1997. Insulin sensitivity and acute insulin responses in African Americans, non-Hispanic whites, and Hispanics with NIDDM: The Insulin Resistance Atherosclerosis Study. *Diabetes* 46(1), 63–9.

Hird, M. 2009. *The Origins of Sociable Life: Evolution After Science Studies*. London: Palgrave Macmillan.

Leonard, W.R., Snodgrass, J.J. and Robertson, M.L. 2007. Effects of brain evolution on human nutrition and metabolism. *Annual Review of Nutrition* 27, 311–27.

Maturana, H.R. and Varela, F.J. 1980. *Autopoiesis and Cognition: The Realization of the Living*. Dordrecht, Holland; Boston: D. Reidel Pub. Co.

Maturana, H.R. and Varela, F.J. 1987. *The Tree of Knowledge: The Biological Roots of Human Understanding*. Boston: New Science Library.

Montoya, M.J. 2007. Bioethnic conscription: Genes, race, and Mexicana/o ethnicity in diabetes research. *Cultural Anthropology* 22(1), 94–128.

Neel, J.V. 1962. Diabetes Mellitus: A 'thrifty genotype' rendered detrimental by 'progress'? *Americal Journal of Human Genetics* 14, 353–62.

Neel, J.V., Weder, A.B. and Julius, S. 1998. Type II diabetes, essential hypertension, and obesity as 'syndromes of impaired genetic homeostasis': The 'thrifty genotype' hypothesis enters the 21st century. *Perspectives in Biology and Medicine* 42(1), 44–74.

Paradies, Y.C., Montoya, M.J. and Fullerton, S.M. 2007. Racialized genetics and the study of complex diseases: The thrifty genotype revisited. *Perspectives in Biology and Medicine* 50(2), 203–27.

Peters, A., Schweiger, U., Pellerin, L., Hubold, C., Oltmanns, K.M., Conrad, M., Schultes, B., Born, J. and Fehm, H.L. 2004. The selfish brain: Competition for energy resources. *Neuroscience & Biobehavioral Reviews* 28(2), 143–80.

Prentice, A.M., Hennig, B.J. and Fulford, A.J. 2008. Evolutionary origins of the obesity epidemic: Natural selection of thrifty genes or genetic drift following predation release? *International Journal of Obesity* 32(11), 1607–10.

Richards, T.J. and Patterson, P.M. 2006. Native American obesity: An economic model of the 'thrifty gene' theory. *American Journal of Agricultural Economics* 88(3), 542–60.

Rosmond, R. 2005. Role of stress in the pathogenesis of the metabolic syndrome. *Psychoneuroendocrinology* 30(1), 1–10.

Schrödinger, E. 1992. *What is Life?: The Physical Aspect of the Living Cell; Mind and Matter; Autobiographical Sketches*. Cambridge and New York: Cambridge University Press.

West, K.M. 1974. Diabetes in American Indians and other native populations in the New World. *Diabetes* 23(10), 841–55.

Wolfe, C. 1998. *Critical Environments: Postmodern Theory and the Pragmatics of the Outside, Theory Out of Bounds*. Minneapolis: University of Minnesota Press.

Zaretzky, A.N. and Letelier, J.E. 2002. Metabolic networks from M, R systems and autopoiesis perspective. *Journal of Biological Systems* 10(3), 265–80.

Chapter 12
Ingesting Places: Embodied Geographies of Coffee

Benjamin Coles

Introduction: Placing Coffee

Under the arches of the railway line that feeds into London's Cannon Street station, beneath a slow bend after London Bridge Station that causes innumerable delays for the City's commuters, it is always damp. Drips are common and commonly horrifying. I watch as one falls through the little hole in the lid of my paper coffee cup, and throw it into the bin. Who knows where that water's been, or if it's water at all, and now it's in the coffee I bought from Monmouth for two pounds. Drips don't happen in the famous Seattle-based chain coffee shop down the street and, if they did, I'd probably get a new cup for free. Then again, the chain outlet down the street does not boast the same kind of provenance as Monmouth. Monmouth Coffee is 'hand selected' from 'single estates' and 'craft roasted' by an 'expert team of mad roasters'. It is also next to London's Borough Market, and helps to feed and feed off its 'buzz'. In a place like Borough Market, where customers come to consume affective consumption experience alongside its products, drips happen but they are part of the charm. Borough 'fine foods' Market is located amidst the drips and the damp of these arches. They provide a backdrop for its distinctive visual material culture, a material culture that utilizes geographical imaginaries to variously 'place' the market's products into an ethical register based on shared understandings and knowledges about where food comes from. Some of the geographies of the market, such as those about place and provenance, are explicit. These are made available to consumers through signage and displays that, along with the built environment, assemble into a material semiotic discourse about food and eating. Yet there is an implicit geography at work as well. Its unique environment, along with the wide variety of food, makes the market popular and exciting. On given trading days, Borough Market is filled with people who come to consume its performances of food and geographies, and this makes Monmouth Coffee fit right in.

On my way back to the coffee shop, I notice just how many people carry paper cups with Monmouth's logo on the side. One could possibly make the argument that coffee (or at least caffeine), much of which comes from Monmouth, helps to fuel the frenzy of the market by contributing to its buzz. These cups, with plastic lids and cardboard hand-insulator-sleeve-things are coffee's material culture.

They provide the physical means for coffee to be transported and ingested throughout the market. By the time I make it to the shop, the queue winds down the steps to the pavement and across the street until it resembles a mob and mingles with the throng of Borough Market. Cars edge their way through, while drivers gesticulate. To give me something to do while I wait, I read a copy of Monmouth Coffee Company's newsletter, a file of which I have in my office. Through personalized-sounding updates, 'thick descriptions' (Geertz 1973), of where their coffee comes from and what it tastes like, and biographical profiles of their roasters, buyers and baristas, each newsletter tells me about Monmouth, their various coffees, and their philosophy. One coffee is from Mbura Farm, Kenya, but 'it's not your typical Kenyan'. Rather, 'it's more demanding in its flavours' and 'good for someone who wants to explore coffee a bit further'. Another is from Na Bagak Bagak, Indonesia. This one is 'very distinctive' with a 'heavy body, smoky, peaty, earthy … reminiscent of fresh leafed tobacco'. Yet another is from Finca Las Bubes, Guatemala. It is 'juicy citrus with milk chocolate, medium acidity and smooth body … an entry level coffee for those not familiar with single origin. A good place to start your coffee drinking exploration. Perfect for all day drinking.' Monmouth Coffee presents coffee drinking as a journey of discovery, which implies a kind of travel where drinkers' bodies are on the move, travelling from place to place through coffee.

As a geographer, I am interested in these places and how they relate to each other and to the multiple bodies such a journey of discovery might imply. This is the focus of this chapter; it explores the ways in which place and food (or drink) are interrelated and how this interrelatedness becomes ingested by the body. It employs 'topography,' an ethnographically-inspired form of place-writing, to trace out inter-relations. In this process a kind of topology of food and drink emerges. The discussion draws on some of the theoretical concerns that characterize multi-sited ethnography (Marcus 1995) as well as some of its methodological and discursive devices. Particularly, this chapter takes the form of an essay (Adorno 1958) and uses it to follow the roaming and meanders (Geertz 1973) implicit in any consideration of topology. Likewise, methodologically at least, this chapter borrows the language of Actor-network theory (ANT), particularly as it has been reclaimed by Latour's recent (2005) reflections. This chapter talks of 'assemblages', of 'tracing out' and of making connections, but only so far as these terms, and the ideas they present, work for examining or otherwise interrogating place and place-making. To say that places have topology is to suggest that places interrelate with other places through relations organized across space and time. Topology, in this sense, destabilizes Cartesian geography by suggesting that as bodies move, places fold into each other. Distance and time are often collapsed as non-Cartesian geographies are made. Thus, topography destabilizes linear conceptions of connectivity by considering how places fold into each other through mutually constitutive processes (Coles and Crang 2011). Through these, foodstuffs emerge and so to eat a food is to ingest its topology. Moreover, places are materialized through food. The biophysical elements of place, minerals, water, and climate become the very matter of a foodstuff which, upon ingestion, ultimately becomes the very matter of the body. Yet places are more than their biophysical properties. They are produced by interrelationships between material (biophysical), social and discursive elements within a particular locale, as well as by interconnections with other places. By the time they are eaten, foodstuffs are more than 'mere' things, but complex bundles of all these relations (Ekers, Loftus and Mann 2009). This means that places are embedded into the very materiality of foods, and so when foods are eaten these places become part of the body. Biological processes of body-making happen because of the geographical processes of place-making. To eat a food is to eat its geography, defined as an assemblage of places. Moreover, wrapped up in this geography is a collection of bodies that inhabit, and are ingested in, the different places of coffee.

This chapter first articulates the geographical implications that surround ingestion and interrogates what it means to eat or drink place. Yet, arguing that to drink coffee is to drink a place and its geographies is not as straightforward as it seems. The geography of a foodstuff is more complex than the geography of its commodity chain might suggest (Leslie and Reimer 1999). Secondly, the chapter suggests that foodstuffs like coffee that, for one reason or another, evoke geographical knowledges about their production come from places that are imaginatively as well as materially fashioned. This imaginary emerges as discursive, social and material processes intersect. Fashioning these places, their assemblage and their ingestion are active processes of place-making that imprint the imaginary into the material and the material into the imaginary. The places this chapter follows are those of coffee. While all foodstuffs are implicitly geographic, coffee's geographies, especially the 'speciality'/'specialist' coffee that Monmouth sells, are explicit. Beans are named after the places in which they are grown – Na Bagak Bagak, Indonesia, for instance; taste-based experiences – 'heavy body, smoky, peaty earthy … reminiscent of fresh leafed tobacco' are attached to these places and, more generally, geographical designations and signifiers arrive with coffee as part of its material culture. At Monmouth, customers are invited to 'explore' the taste of place, but a lot has to happen to coffee and a lot has to happen to place before they are allowed to go on this 'journey'. Even then, the journey is biased by a tour guide who socially, discursively, and materially manipulates place and coffee, while crafting and performing its geographies. This chapter therefore follows coffee in order to critically examine the different relations of place that are variously embedded and embodied into the very stuff of coffee. In the process, it interrogates the ways these relations blend geographical imaginations and imaginative geographies to reproduce embodied geographies of coffee. Through this perspective it seeks to understand the geographies of eating (or drinking), where they are fashioned, and what it might mean to our own bodies when we find out. It asks: if the geographies of our foods make us, what are we made of?

Locating Place

Because when we eat we ingest place, it is little wonder that critics of the modern, industrial food system, in which our foods have become homogenized into undifferentiated commodities that simultaneously come from anywhere and nowhere, have labelled it a 'placeless-foodscape' (Holloway and Kneafsey 2000). An alternative to the geographical 'non-place' (Auge 1995) of modern food is one where places are present or at least (re)presented. I argue that no food is 'placeless' and that the 'placeless-foodscape' refers rather to a foodscape where the places that foods come from do not map onto some kind of geographical imaginary about where they ought to come from. This imparts an ethical imperative to the relationship between place and placeless-ness. 'Placeless' places, often suggested as 'typical' to the industrialized world of food, are usually not particularly pleasant or desirable. They might be sites of disease or anxiety such as the offshore 'factory' chicken farm (Wallace 2009) or industrial slaughterhouse (Dunn 2007). They may be indicative of shifts in agriculture at large that have rescaled food production globally, and consolidated it into fewer and fewer producers, marginalizing people and their environments, and rendering places of food banal, boring or homogenous (Boyd and Watts 1997). Because they do not fulfil an imaginary geography of food, and their 'geographical imagination' is limited, these places become 'placeless'; their foods become 'commodities' and their geographies are obscured by the commodity fetish. The placeless foodscape is then critiqued for distancing consumers from producers by disconnecting them from the social relations of production through the commodity fetish. Researchers seek to 'lift the veil of geographical and social ignorance' (Harvey 1990: 423) that underlies the commodity fetish in order to reveal the hidden and often exploitative relations that constitute commodities. The effect of this, however, is to spatialize ethics.

Geographical knowledge leads to an awareness of other people and other places, and the distances over which relationships are constituted. Awareness leads to ethical and moral outcomes (Sack 2003). The proposition of the placeless-foodscape is that 'typical' commoditized food production relies on obscuring relationships and that eating its foods makes (unaware) consumers complicit in the constitution of unfair and potentially immoral geographies that characterize commodity production and capitalism. Making consumers 'geographically aware' however, leads them to places that make foods fit into a vision of 'right' or 'proper' production (and consumption). It is argued that through awareness, relationships between producers and consumers or between one place and another in a commodity system transform 'trade into a vehicle of global solidarity between conscientious consumers and empowered producers' (Trentman 2007: 1081). In other words, good food is tied to good places, and good places to good ethics. If to eat food is to eat place and geography and to be responsible for implicit ethical relations, dubious or otherwise, then it is necessary to interrogate how these places are made and represented and how these geographies are established.

The geo-ethical implications of the commodity fetish, alongside unveiling it, getting behind it or otherwise 'with it' (Cook et al. 2006), inform academic interest in food as well as the ways in which food is thought of, and indeed used, by producers, retailers and consumers. Both seem to be driven by the power of revelatory politics to reclaim food from its placeless landscape and to reconnect relations that have otherwise been broken or displaced through its commoditization. Conceptually, these projects 'follow' food from place to place; they describe, articulate and analyse the commodity relations of food, and variously illuminate its once-hidden geographies. Academic interest is broadly organized around notions of the 'commodity fetish', noted above. Yet food practitioners, such as producers, retailers and consumers have responded to the placeless-foodscape by producing, purveying and seeking-out a range of foodstuffs, production methods and specialized locales that celebrate the geographies of food. These deploy a variety of material-semiotic devices such as labelling, shop displays or, as in the case of the foods in Borough Market, an entire food retail environment, all of which fashion food into the material embodiment of their geographies. As opposed to a factory production, meat becomes 'farm-raised' (or 'free-range'), bread 'artisanal', and fruit and vegetables 'organic' or 'farm-assured'. And, regardless of whether these products are local in any legal sense, the places where they are made and geographies in which they circulate somehow all become 'local' as well. From this, an ethical narrative that subverts the trope of placeless food is variously woven into an imaginative geography and commodity lore about where food comes from, how it ought to be produced and how it, and its places, ought to be consumed.

For something like coffee, whose geographies are only at least partially revealed yet openly deployed, this raises a question about the commodity itself: Is it the bean? Is it the place? To reflect on this, I turn to the work of Cook et al. (2004, 2006, 2008). Working with ideas put forward by Appadurai (1986) and Kopytoff (1986), as well as the moral imperative set out by Harvey (1990) to unveil the commodity fetish, Ian Cook (2004) 'follows the thing' to examine the different 'commodity contexts' through which values are ascribed to objects. Along the way, he interrogates the spatial dynamics of these values as objects move from context to context, or indeed through time and space. This perspective problematizes the ways in which commodity fetishes and commodity 'double fetishes' obscure as well as inform their own geographies (see Cook and Crang 1996). Extending these arguments with reflection on topology and topography illustrates how relations of place and place-making are at the centre of object/commodity contexts, and indeed *are* these contexts. Values are ascribed to objects in place as well as ascribed to place. What this means is that for some commodities, such as coffee, or 'specialist' food more generally, part of the value(s) that makes it a commodity lie in how some aspects of its production are celebrated and others ignored or even obscured (see also Baker and O'Connor, this volume). Through this, some places are made visible and others invisible. For coffee these places of visibility are places where coffee is grown and in cases like that of Monmouth, roasted. These places of celebration however are themselves produced within an imaginary and discursive field that creates geographical narratives of coffee. This goes beyond Cook and Crang's (1996) notion of 'double' commodity fetish because the exchange value of coffee relates to the bio-physicality of its cultivation as well as to how these physical processes are managed, translated and ultimately performed, to customers and to various producing and consuming bodies along coffee's already-contentious 'commodity chain'. The value of coffee is physically and affectively embodied when coffee is ingested, and is reproduced as these bodies move through space. Rather than being fixed, these embodied commodity processes are constituted through flows and circuits within porous and leaky spaces and are ultimately held in tension by and between other places. Commodities are therefore topological, and paying attention to this topography teases out ways in which bodies are made to be (in)visible within these geographies, as well as ways in which bodily traces of place – visible or otherwise – are embedded into the materiality of the bean. If Cook et al.'s (op cit.) project is to 'follow things' in order to better understand their commodity contexts, then here I am following the places where things and their 'contexts' are constituted.

Narrating Place

On my way to get a new coffee, I have ended up at Monmouth's warehouse and roasterie located in Bermondsey talking to AJ. She tells me that 'green coffee is valueless. It takes roasting … burning it to get anything out of it … roasting enhances characteristics already there in the bean, so you need to start with good beans'. The company's chief roaster, Angela, agrees. She tells me that roasting 'is about releasing potential … you can't bring out good flavours from bad beans … hide them, perhaps, but you can't add anything that isn't already there'. Good beans, at least for Monmouth and other dealers in 'specialist' coffees, come from locales whose biophysical properties are capable of producing desirable flavours and aromas in coffee. Sourcing 'high quality coffee beans [from these locales] is a crucial activity in achieving the strategic objectives of speciality coffee firms' (Donnet, Weatherspoon and Hoehn 2007: 3). Monmouth Coffee Company sells top-quality coffee beans, and it secures these beans by travelling to their sites of origin, and sampling and tasting prospective products. When suitable beans are located, it develops long-term contracts with producers to ensure their supply. AJ tells me that she is interested in sourcing coffee that she thinks her customers would like. 'When I taste', she says, 'I generally have flavours in mind for each coffee that reflect how varietals from different regions should taste. You don't compare different regions, or different varietals, but the same from different farms, or different producers, but always from the same general region with similar taste profiles'.

Over the years, the coffee industry – a collective or producers, exporters, importers, roasters and retailers – has, by consensus, developed formalized schema through which coffee growing regions are categorized. Producing a taste profile entails a series of tasting rituals where members of the industry gather to 'cup' – meaning to taste through proscribed procedures – coffees from a particular region. Whilst tasting, individual tasters detail the aromas and flavours of a coffee and enter them on a proforma. Afterwards proformas are compared and a profile is compiled from the collective palate of the group. Regional taste profiling is an iterative process where numerous coffees from the same region are tasted and over time generic characteristics are ascertained and documented. As a result, specific regions have taste profiles that dictate a set of flavours and textures associated with them. Indonesian coffees, for instance, are said to be 'smoky', 'peaty' and 'earthy', with a 'heavy body' mouth feel. Coffees from individual producers in a region are therefore evaluated based on how their coffee matches up to the regional profile as well as to each other. The tastes of places are ultimately 'stored' in the collective body of expert tasters who collaborate to produce socially and affectively imagined geographical knowledges about coffee origins. The geography of place-making, however, is even more complex. It requires the assemblage and mediation of different bodies in different times and places. In AJ's industry shorthand, 'a place should taste right', and part of her job as a taster for the company is to 'look for flavours that [she] knows her customers will like … I'm buying for them. If I were buying for myself, we'd only sell like ten kinds of coffee. I always have customers in mind'. Not only should a particular place taste a particular way, but also the taste characteristics of place are defined by experts speculating on how their customers might *expect* to experience place through coffee. The ingesting body of the consumer is present, albeit mediated by proxy, and as these bodies come together, places are interrogated, evaluated, compared and ultimately produced. As these geographies are reproduced, and commoditized, they are recycled within the materiality of particular coffee beans.

Monmouth Coffee Company ensures that it has a steady supply of coffees that will appeal to customers' tastes by establishing long-term contracts with producers. This is partly a strategic response to competition in the industry. AJ comments that 'the coffee world is not all that big. Everyone knows everyone, and everyone knows what their coffee's worth'. Additionally, this is a response to historic industry practices where short-term contracts, along with dubious social-economic dealings, were used to speculate on open commodity exchanges; these practices ultimately prompted, with good cause, the emergence of Fair Trade, and have forced those within the industry to rethink their corporate social responsibility (Raynolds 2002; Raynolds, Murray and Taylor 2004). Primarily, however, Monmouth relies on long-term contracts to maintain control over its supply chains and the fashioning of place within its coffees. For Monmouth, long-term contracts secure access to consistently high-quality coffee that they know (or at least think) they can sell without having to rely on agents and coffee auctions, or having to continually worry about their supplies. These contracts also allow for information exchange between growers and importer/roaster/retailers like Monmouth. Monmouth's team of tasters and roasters is able to provide feedback on particular crops and the way they taste. Soil nutrients, moisture, and sun/shade – the biophysical properties of place – all impact how coffee tastes after it is roasted. Changes in any one of these can be sensed on the palate and the company, seeking consistent products, must somehow manage this. Monmouth manages these dynamics through their relative position of power within their own supply chain. Their roasters can feed back quantitative and qualitative information about coffee – such as how it tastes, how it roasts, and its chemical profiles, all of which are monitored during roasting – to growers who can manipulate their crops by adjusting agricultural inputs. For instance 'adding potassium', AJ tells me, 'enhances citrus notes'. Alongside the biophysical side of coffee production, where the materiality of place is carefully managed, Monmouth's roasters are able to detect flaws in coffee's processing:

> Not picking the ripest fruits, allowing it to not dry [and therefore rot] and any number of things that might be okay for lower quality coffee is unacceptable, and we can tell when we roast. Angela will feed these back so that growers can sort themselves out…and we can help so everyone wins. We're able do this because we have long-term contracts with our growers … though we've also ended contracts [for non-compliance].

The way that coffee tastes is partially determined by where it grows and how it was processed. This sensory potential is stored in the bean's bio-chemical materiality, which is translated socially and discursively as embodied geographical knowledges and imaginations are shared and transferred. Transferring geographical knowledges means translating place. For coffee the relations of place are locked into its materially, and this materiality undergoes different transformations that variously express its origins.

The rituals of tasting that lead to a regional taste profile and the physical manipulation of growers' farms and estates are active processes of place making that totalize place into an imperial gaze (Duncan and Ley 1993) and normalize and objectify it into a hegemonic Cartesian discourse. These rituals nod towards a scientific process of tasting that disciplines the complexities and variations of place so that it fits into a generic and categorical regional profile. This process 'flattens' the variations that make places unique and agglomerates them into a generic understanding of what a particular place should be like. The disciplining of place is further mediated through the bodies of tasters who have trained themselves to taste places in particular ways. Despite the 'science' or indeed because of it, place is materialized *as* coffee beans and both become subjective social constructions that, to borrow language from actor-network theory (ANT), emerge not only through material, social and discursive assemblages (Thrift 1996) but also through an embodied assemblage that draws elements from other places and other bodies into the 'network'.

Translating Place

Each material process that coffee beans at Monmouth undergoes, including picking, pulping, drying, roasting, grinding, and brewing, is designed to bring place to the palate; AJ goes so far as to suggest that the quality of their product is defined by the 'way taste is translated down to the customer ... from the estates [where coffee is grown] to the shop counter where you can sample what we're talking about'. For Monmouth Coffee, translating taste means translating place. I'm still in Bermondsey; I'm still talking to AJ and Angela, and we're tasting coffee in a room adjacent to two roasting machines. Roasting coffee oxidizes the chemical compounds of the coffee beans. In this process, different flavours emerge from the raw beans, depending on their biophysical properties, which in the end are dependent on how they are grown. Roasting at Monmouth, however, is a craft practised by skilled roasters and machine operators who are taught to roast beans in accordance with how those at Monmouth think their customers want to drink and experience place. The roasters mediate between places of production and places of ingestion (see Smith-Maguire 2010), and they fulfill an important role in translating places of coffee to customers and translating customer imaginations through tasters to growers. Roasters utilize heat and time to highlight and suppress particular elements of coffee's physical geography and, in the process, translate place by transcribing it onto the bean. These places are mediated by a variety of bodies: expert tasters like AJ who are judging customer demands; expert tasting panels who construct place profiles; (in)expert customers who lead, as well as are led by, the production of these 'geographies'; and growers who are subjected to the tastes of others, and who reproduce coffee and place through their labour.

Monmouth has two roasting machines. One is large in order to handle the company's high-volume needs. It is programmable to prescribed settings, like temperature and duration; these settings have been determined by scientific processes of trial and error that have, over time, established a series of roasting profiles to match the regional taste profiles of coffees or, in Angela's words, 'the right roast for the right place'. Roasting machines work by forcing hot air through a rotating drum containing the coffee beans. The beans are tested periodically during roasting to ensure that the machine is doing what it is meant to. And, the machine produces a computer-generated readout that records roasting times, temperatures and basic chemical information. These machines are newish to Monmouth, having replaced a set of old flame-roasters, which were considerably more 'hands-on' for the roaster. One reason for the investment in the new machines is that the business has grown and their commercial customers have asked for a more consistent product. Alongside their ability to produce analytical readouts through which roasting profiles can be developed, stored and reused, the computer-programmable roasting machines standardize the variables of roasting, which are coffee quantity, air, heat and time. As such, they have created a method that, as Angela puts it, 'allows a bunch of non-scientists to be more scientific in standardizing our products'. A peculiar tension has emerged as Monmouth's business has grown.

The company's discourse stresses the value of individuality within its coffees, and uses place-narratives to differentiate itself from the 'placeless-foodscape'. Yet in practice, these same places are shaped into homogenous representations of individuality through spatially distant homogenizing processes, such as tasting panels and automated roasters. At the same time, these very same experts recognize that their products are comprised of heterogeneous geographic processes and seek to impart this geography to their customers.

Buyers like AJ want to offer different flavours to their customers, so they work closely with the roasters to manipulate the roasting process to 'maximize the experience for the customers'. 'Roasting is cool because it brings out the flavours that I've tasted when at origin … it turns a raw product into something else, but there are parameters … parameters between green and burned'. Roasting also converts raw, green, 'valueless' coffee into a commodity, the value of which is created by highlighting its individual characteristics and the ways in which these relate to their places of origin. Roasting coffee is one component of this process. Roasted coffee is graded based on colour: light, medium and dark, and these correspond to time spent in the machine. Darker roasted coffee means that more volatile chemicals have been oxidized and more sugars converted to carbon. This means that dark roasted coffees lose their place-based individuality but gain a generic consistency that (can) hide flaws. Assuming they are not burned away, typical coffee flavours are described as 'fruit', 'citrus', 'chocolate', 'caramel', 'tobacco', 'smoked' with characteristics of 'elegant', 'smooth', 'bright', 'funky', or 'wild'. These are the characteristics that lead to a particular coffee's profile and what roasters seek to express. Roasters at Monmouth work at the direction of AJ who, having tasted all coffees on the farm, seeks to impart her own interpretations of place onto her customers. She tells me that 'typically for a new coffee I have the roasters try a variety of different things … first I ask for a medium and a dark. Then if I want to make changes I'll ask for more a more medium than medium but less dark than dark, or darker than light but not medium … they usually know what I mean and translate this into something that makes sense'. Yet despite these attempts at consistency, the machines and the profiles, coffee is a natural product (in one way or another) and has variations in its composition. Roasters work with its natural variations and make adjustments to roasting profiles. In Angela's words, they 'craft roast coffees to get the best from them', imparting their own interpretations of AJ's demands and their own interpretations of what they think particular coffees should taste like. AJ tells me that 'I can tell who in the company's roasted a coffee, Angela or Jason, or their protégées … and I can certainly tell our coffee from other roasters, even if they're the same lot of beans from the same place'. As a craft, roasters at Monmouth burn their own ideas of place, standardized by their own scientific practices, into the materiality of the bean. Monmouth's roasters celebrate other places, but unlike other 'craftsmen' (Sennett 2008) the body of the roaster is expected to hide behind the machine.

Monmouth's roasters strive to maintain the particular place-based characteristics of the coffees that the company sources, but as coffee becomes a commodity, and as place becomes commoditized, maintaining these characteristics becomes one of several priorities. The value of Monmouth's coffees is in their individuality, but this value can only be realized as individuality is reproduced time and time again. The imagined geographies that are materialized into each coffee bean's physical properties are shipped across the world to roasters who seek to realize these geographies by physically transforming the coffee bean to match. The roasters, however, receive their instructions from the buyers who have already had a hand in shaping how the places of coffees' origins can be realized. Roasters strive to reproduce these places and they report any failures to the buyers who entreat producers to make changes through material manipulations in growing and processing. By the end of roasting, not only has coffee emerged as a commodity, but also the places of coffee, how it should taste, and how place ought to be, are physically burned into the bean. Place is imprinted into the material stuff of coffee through this violent transformation that subjugates place and remakes it in coffee's geographical image.

Brewing Place

It is a 20-minute walk from Monmouth's roasterie and warehouse back to their shop in Borough Market. Judging by the traffic it is a 20-minute drive as well. In the time it takes for roasted coffee to travel from Bermondsey and arrive at the shop in Borough, place has emerged from coffee, ready to be sold, bought and ingested. I am back in the queue waiting to buy a new coffee to replace the one lost to Borough Market's drips, and I'm keen now to buy some beans too. The shop on Park Street is divided into two parts. The left is for customers who only want filter coffees, lattes, espressos, and pastries. In the centre of the room there is a large communal table for sitting. There is also a table near the back that displays pastries. Above this are shelves displaying a variety of coffee-making implements – filters, grinders, stovetop 'mocha' pots – all of which are for sale. This coffee shop does not have the comfortable seating that many international chains advertise. It is no 'third space' (Oldenburg 1999) popularized by Howard Schultz, founder of Starbucks. Indeed, Monmouth's seating comprises old wooden school and church chairs as well as slightly wobbly stools on which to perch. Its space is cramped, often hot in the summer and cold in the winter. This is a place, according to a voice that wishes to remain anonymous, 'where you don't sit and linger. You come in, drink your coffee on a hard wooden chair and fuck-off again … and that's how [they've] designed it. At Monmouth, it's all about coffee'. An apparatus that facilitates making Monmouth's 'signature drink', a 'single origin filter coffee', is located on a countertop next to the till. It comprises a rack that holds the filters and a shelf below that holds the cups. Behind this countertop there is a handwritten chalk sign along the wall. It lists the coffees and drinks available, as well as prices. Next to the sign is a pictorial 'commodity chain' – complete with requisite photograph of happy coffee plantation workers – which illustrates various stages of coffee production, from planting on the farms to roasting, and ending with a steaming mug of coffee. Everything Monmouth thinks I need to know about coffee's geography is right there.

The other side of the shop, where I'm queuing, is dedicated to bean sales, although if they are buying beans, customers may also buy a drink. I'm here to buy beans and to buy a coffee because I'm curious to find out how places are translated to customers before they experience the coffee, and because the queue is much shorter. In front of me are bins of coffee, each with a handwritten origin label. A brass scale for measuring beans is located directly behind. Even though the queue is shorter, I still have to wait my turn. As I'm waiting, I flip back to the newsletter and read a brief account of a recent trip to origin in Colombia:

> Looking out of the window on the flight from Bogota to Neiva we could see that the rainy season had arrived in Colombia. The waters of the Magdalena River were overflowing with muddy water from the rain in the hills above. Arriving in Neiva from Bogota is like getting out a fridge and into a frying pan. It is hot and humid and it is always a relief to get onto the road for the three and a half hour drive to Pitalito. … The group has established its own cupping laboratory … [and] the farmers are learning about their own coffees and how changes in the picking and processing affect the flavour of the final cup (Tasting Notes provided by Monmouth Coffee Company, July 2008).

By the time I finish reading this I feel like I have been on a trip, and I am surprised to find myself at the front of the queue talking to the barista. I know from the company's owner and founder, Anita, that the hardest part of the business is:

> Transferring all of the stuff we learn [about coffee's origin, its production, its profiles, etc] as managers to our staff … It's not just a question of giving [them] a handout … it takes so much more for that information to become absorbed in not just an intellectual way but to have real knowledge of place … the best way … [to learn] … is by going out to origin, so all the stuff we tell them [the staff] … you get there and actually see it happen, every time we take somebody out they say 'okay, I understand now'.

I tell the barista behind the counter that I'd like to buy some coffee beans, as well as a filter coffee, and (taking on coffee lingo) I explain that I'm trying to learn more about coffee and the ways that origin can be tasted in the cup. Coincidently, she asks if I've ever had 'the Cooperativa Quebradon, Huila' from Colombia. I respond, 'no, but I've just been there'. She laughs and tells me, 'it's one of my favourites and works well as filter or in a press pot. Both bring out its citrus … This one's bright, lemony chocolate and with nice fruit'. She indicates towards the tasting notes so that I can read what she has just told me. I also ask for a filter coffee. The filter coffee is, according to Anita, 'Monmouth's signature drink'. It is made by pouring hot water though ground coffee and through a paper filter fitted to a ceramic cone suspended above a cup. This is a gentle method for brewing coffee that preserves the volatile chemicals that comprise coffee's flavour and link it to its origin. Monmouth brews and sells filter coffees on the premises in full view of the customers and it sells the equipment – ceramic cones and paper filters – so that its customers can make it at home. Anita tells me that the filter coffee is the best expression of origin, which is:

> What we're all about … we were the only people that I was aware of that was making a single cup with a ceramic cone … it's always been located at the front of the shop so customers could see it. We didn't want it relegated to the back … over the years it became our signature. And now it's not just coffee. Someone will say I'll have a country: A Colombian … imagine that.

At Monmouth, taste and place are preconfigured for customers before they reach the front of the queue. Monmouth's filter brewing method is the primary way in which it translates the places of coffee to customers. Its entire system of taste profiling, signature roasting and the mutually-construed processes that lead to the physical manipulation of coffee's places of origin ultimately come down to flavours and tastes that are accessible within their 'signature' filter coffee. The semiotic devices of the shop, such as the posters, the signage and the 'field notes' bracket the experience of drinking the coffees and help to provide a geographical imagination with which customers can develop their own. The filter coffee itself, at least according to those in the company, is the 'best way to translate coffee's terroir[s] to the customers'. However, different brewing methods will produce different flavours in a cup of coffee, so Monmouth's promotion of a single method, along with their provision of the means to make a 'Monmouth signature' filter coffee at home, is a way to insure against irregularities. As a measure of quality control, everyone in the company tastes Monmouth's product in the same way through the filter cup system, ensuring that the same kinds of flavours, and indeed the same kinds of places, are transmitted. The filter coffee is one final mode of standardization for the company. It is one that focuses Monmouth's entire place-making efforts, all of its imaginative geographies, and all of the customers' own imaginations and expectations of coffee's geography into an eight-ounce cup of coffee. It seems that the questions that began this journey across South London and beyond have shifted from a concern with which places I am ingesting when I drink my Monmouth coffee, to whose places I am drinking and, finally, to whose palate I am experiencing.

Conclusion: Drinking the Body-Place

The very act of ingesting food mobilizes spatially vast, typically invisible, geographical relationships. That these relationships fashion places is one thing. That these relationships dictate to places how they are meant to be fashioned is another. And that these relations then dictate the fashioning of places through the body suggests a kind of consumptive bio-power where bodies are variously mobilized and disciplined to fit into a geographical imaginary. Many of the foodstuffs we eat and drink are not only produced somewhere else by others, but are also the product of multiple spatio-temporal relations. Up until the point at which coffee is ready to be sipped, glugged, or otherwise 'thrown down', its places, bodies, geographies and commodifications – alongside the ways in which these all fold into some type of topology – form a story of geographical potential. The moment that coffee passes the lips, these are all realized; they are ingested and become part of the drinker's body. And, since coffee is mediated by a constellation of bodies, to drink coffee is to drink a lot of bodies. Watts (2004), when examining the differences between one type of agrarian meat production and another (chickens and pigs), queries whether the biology of a commodity makes 'a damn bit of difference' He suggests that all commodity systems rely on processes of displacement and distancing to make their constitution palatable. I argue, instead, that biology does make a difference because eating – the visceral reproduction of the body and its biology – is constituted by geographies of embodiment. Ingesting food is profoundly intimate and profoundly local. It signifies a moment when the objective outside becomes the subjective inside, and all of the things that go along with food, such as tastes, feelings, affects, materialities and socialities are realized within the body. This biological intimacy is what differentiates food from (most) other objects and other commodities because its value is at least partially dependent on what happens to it when it enters the body, and also, the bodies of others. To eat a food is to eat its geography but for coffee this is more complex; coffee is traded based on how it tastes and, although this is dependent on its places of origin, how it tastes also depends on tasters themselves and their embodied ability to communicate their own experiences when tasting. The experience of ingesting is thereby shared, sought out, and otherwise commoditized to be reproduced. The story of, particularly specialist, coffee, such as that sold by Monmouth, is therefore one of bodies as well as places on the move. Yet attention to these bodies is largely missing from discussions of a foodstuff that celebrates its places of origin to its consumers and imbibers. In this chapter I have suggested that this biology matters because its geography matters, and its geographies matter because its biologies matter.

References

Adorno, T. 1958. The essay as form. Translated by Bob Hullot-Kentor and Frederic Will. *New German Critique* 32(2), 151–71.

Appadurai, A. 1986. Introduction, in *The Social Life of Things: Commodities in a Cultural Perspective*, edited by A. Appadurai. New York: Cambridge University Press, 1–63.

Auge, M. 1995. *Non-Places: Introduction to an Anthropology of Supermodernity*. London: Verso.

Boyd, W. and Watts, M. 1997. Agro-industrial just-in-time: The chicken industry and post-war American capitalism, in *Globalising Food: Agrarian Questions and Global Restructuring*, edited by D. Goodman and M. Watts. London: Routledge, 139–65.

Coles, B. and Crang, P. 2011. Placing alternative consumption: Commodity fetishism in Borough Market, London, in *Ethical Consumption: A Critical Introduction*, edited by T. Lewis and E. Potter. London: Routledge, 87–102.

Cook, I. et al. 2004. Follow the thing: Papaya. *Antipode* 36(4), 642–64.

Cook, I. et al. 2006. Geographies of food: Following. *Progress in Human Geography* 30(5), 655–66.

Cook, I. et al. 2008. Geographies of food: Mixing. *Progress in Human Geography* 32(6), 821–33.

Cook, I. and Crang, P. 1996. The world on a plate: Culinary culture, displacement and geographical knowledge. *Journal of Material Culture* 1(2), 131–53.

Donnet, L.M., Weatherspoon, D.D. and Hoehn, J.P. 2007. What adds value in specialty coffee? Managerial implications from hedonic price analysis of Central and South American e-auctions. *International Food and Agribusiness Management Review* 10(3), 1–18.

Duncan, J. and Ley, D. 1993. *Place/Culture/Representation*. London: Routledge.

Dunn, E. 2007. Escherichia coli, corporate discipline, and the failure of audit. *Space and Polity* 11(1), 35–53.

Ekers, M., Loftus, A. and Mann, G. 2009. Gramsci lives! *Geoforum* 40(3), 287–91.

Geertz, C. 1973. *The Interpretation of Cultures: Selected Essays*. New York: Basic Books.

Harvey, D. 1990. Between space and time: Reflections on the geographical imagination. *Annals of the Association of American Geographers* 80(3), 418–34.

Holloway, L. and Kneafsey, M. 2000. Reading the space of the farmers' market: A case study from the United Kingdom. *Sociologia Ruralis* 40(3), 285–99.

Kopytoff, I. 1986. The cultural biography of things: Commoditization as process, in *The Social Life of Things: Commodities in Cultural Perspective*, edited by A. Appadurai. New York: Cambridge University Press, 64–91.

Latour, B. 2005. *Reassembling the Social: An Introduction to Actor-Network-Theory*. Oxford: Oxford University Press.

Leslie, D. and Reimer, S. 1999. Spatializing commodity chains. *Progress in Human Geography* 23(4), 401–20.

Marcus, G. 1995. Ethnography in/of the world system: The emergence of multi-sited ethnography. *Annual Review of Anthropology* 24(1), 95–117.

Oldenburg, R. 1999. *The Great Good Places: Cafes, Coffee Shops, Bookstores, Bars, Hair Salons, and Other Hangouts at the Heart of Communities*. New York: Marlow and Company.

Raynolds, L.T. 2002. Consumer/producer links in fair trade coffee networks. *Sociologia Ruralis* 42, 404–24.

Raynolds, L.T., Murray, D. and Taylor, P.L. 2004. Fair trade coffee: Building producer capacity via global networks. *Journal of International Development* 16(8), 1109–21.

Sack, R.D. 2003. *A Geographical Guide to the Real and the Good.* London: Routledge.

Sennett, R. 2008. *The Craftsman.* New Haven CT: Yale University Press.

Smith-Maguire, J. 2010. Provenance and the liminality of consumption: The case of wine promoters. *Marketing Theory* 10(3), 269–82.

Thrift, N. 1996. *Spatial Formations*. London: Sage.

Trentman, F. 2007. Before 'fair trade': Empire, free trade, and the moral economies of food in the modern world. *Environment and Planning D* 25(8), 1079–102.

Wallace, R. 2009. Breeding influenza: The political virology of offshore farming. *Antipode* 41(5), 916–51.

Watts, M. 2004. Are hogs like chickens? Enclosure and mechanization in two 'white meat' filieres, in *Geographies of Commodity Chains*, edited by A. Hughes and S. Reimer. London: Routledge, 39–62.

Chapter 13
Complex Carbohydrates: On the Relevance of Ethnography in Nutrition Education

Emily Yates-Doerr

Introduction

Twelve months into ethnographic fieldwork in Guatemala, I returned to New York for the week of Christmas. Before my departure I made an offer to a friend whose younger brother had, several years earlier, travelled from their rural K'iche' community to the United States in search of work and had been relying on intermittent and unpredictable construction jobs in Manhattan ever since. If my friend wanted to give his brother a gift, I would deliver it. It was no trouble; my trip would take me less than a day and my own bags were empty. On the day I was to leave I had heard nothing from my friend and assumed he had dismissed my offer. But late into the evening, just before departing for the bus that would descend from Guatemala's verdant highlands to the international airport in the capital five hours away, he arrived at my home with what he told me was the 'perfect gift'. I opened the bundle he handed me to find ten loaves of *pan dulce* (sweet bread). Each one was carefully stored in its own plastic wrapping, having been baked by his mother that afternoon.

Shortly after landing in New York, I met the brother beneath icy skyscraper scaffolding in midtown Manhattan. He greeted me shyly but when I handed him the package an eager smile spread across his face. He seemed to be no longer aware of me as he tore through the plastic, split open the crust of one of the loaves and, transfixed, buried his face against the soft texture of the bread, eyes closed, inhaling deeply. Some moments later, he glanced up: 'It has been years since I have tasted this – not just bread from my town, but fresh bread from my town, bread that has been formed by my mother's hands. This is the best gift I could have received.'

Energetic Equivalencies

At the time, I was studying a process that scientists call 'the nutrition transition', which refers to a growth in rates of dietary-related chronic diseases in regions around the world that have been long dominated by infectious illnesses and underweight malnutrition. Against a background of increasing concern over obesity in Guatemala (Groeneveld, Solomons and Doak 2007), bread was a main topic of conversation in the clinics and classes where I carried out fieldwork. Countless doctors advised their patients to avoid the kind of bread my friend's mother had baked. This bread was too sweet; when ingested this sweetness would transform into fats and be stored by a body that would become too large if this was done too often. They explained that this bread was made of a substance called carbohydrates, which are 'our source of energy' until they make us overweight.

To minimize consumption of carbohydrates, doctors commonly advised their patients that sweet breads should be substituted with whole-wheat bread, which had fibre that would slow the rate of sugar absorption, create satiety, and help with weight loss. Whereas bread – a Catholic staple – had augmented the regional mainstay of tortillas for centuries, and family-run *panaderías* selling bread cheaply could be found on nearly every city block, whole-wheat bread was sold only in supermarkets or chain bakeries that produced it off-site and sold it at a premium.

'Yes, this is more expensive', doctors would acknowledge, 'but the price you pay now will be offset by how much you will save in the future'. The calculations they expected their patients to make were not only those of nutrients – more fibre! Fewer carbs! They also expected patients to weigh the pleasure of the moment against the prospect of future health as though these were objects that could be clearly added to, and subtracted from, one another. The following examples of nutritional advice, drawn from my 16 months of intensive fieldwork in nutrition and health clinics in the highlands of Guatemala (Jan 2008–May 2009), reflect typical ways in which doctors would discuss eating bread with their patients:

Example One:

> Doctor: For the moment try not to eat things with fats. Instead you should eat fruits, but not too much fruit. If you eat too much you will become sick because even though fruit is good if you eat too much it will make your blood sugar levels rise. Do you eat whole-wheat bread? Are you used to eating this?
>
> Patient: Which kind?
>
> Doctor: Whole-wheat.
>
> Patient: Regularly? No I don't eat this.
>
> Doctor: You eat French bread instead? Well the truth is that you should be eating whole-wheat bread. Do you not eat it because you don't like it?
>
> Patient: Yes, I don't really like it. It's a bit gross.

Doctor: But it's not gross, it's good for your body.

Patient: Well, if it's good for my life (*para mi vida*), then I have to eat it.

Doctor: Very good. You're going to find that it will improve your health and you'll start to like it.

Example Two:

Doctor: You need to start eating less each day, okay? Do you regularly eat three tortillas with meals?

Patient: Yes, three tortillas.

Doctor: Well, you're going to need to start eating two.

Patient: Ah, okay.

Doctor: So, for example, if today I ate three tortillas with breakfast, well tomorrow I'm going to only eat two. We'll start with this, but we need to keep reducing the quantities, do you see? This way you can start losing weight.

Patient: Okay.

Doctor: And it is even better if you can substitute your tortillas with what is called whole-wheat bread. You get this at the supermarket.

Patient: It's expensive!

Doctor: It will be more expensive at first, but it will save you money over time. You see, the other kinds of bread have fat in them – they have oil and they have lard and these make you heavy. But whole-wheat bread will keep you healthy.

Patient: Okay, doctor.

Example Three:

Doctor: As for sweet breads, it would be better if you stopped eating them. Sweet breads and the regular breads you buy in the market, you should stop eating these. Instead, just whole-wheat bread. The kind you buy at the supermarket – this you can eat since it has fewer carbohydrates. It's also better if you stop eating cakes, French fries, fried yucca, *camote* (a kind of squash), or *malanga*

> (a tuber cooked like potatoes). None of these you should eat – they have too much sugar and when you eat them the sugar accumulates in your body where it produces fat, you see? It's also better that you don't eat anything canned – it will be better to eat fruits and vegetables instead.
>
> Patient: Okay, fruits and vegetables and whole-wheat bread. Anything else?
>
> Doctor: When you eat fruits, it's preferable that they are not prepared with sugar. Also no juices – only drink juices that are natural. You need to do this so that you don't gain weight. This will help you to be healthy. And you need to do more exercise. This will also help control the fat in your body. When you do more exercise, then you can eat more fats, sugars and carbohydrates.

In the examples above, we see how variegated possible valuations of eating become compressed into clear-cut dietary guidelines (good and bad), which themselves frame food and bodies in mechanistic terms as the consolidation of discrete components (fats, sugars, or carbohydrates). The doctors' advice is premised upon the idea that foods and bodies can be broken into parts that can then be summed together, much like currency, to make a whole (either a whole food or a whole body). The quantitative underpinnings of this logic, with their strong resonance with market exchange, are also pervasive in how the doctors frame dietary motivations. Just as these doctors treated body weight as a function of energetic credit and debt – one bowl of cornflakes offset by half an hour on a treadmill – so did they expect their patients to balance the costs of immediate appetites against the benefits of potential future savings. Built into their counselling was the presumption that eating was commensurate with moving, each reducible to the unit of the calorie, akin to the way in which the value of present or future health could be converted into price. The bread they spoke of was abstracted into nutrient components – carbohydrates, fibres, fats – that would have a measurable and predictable impact upon the body. There was no space in their discussions for the bread held by my friend's brother – bread that was, for him, inextricable from kinship and community, bread that he smelled, felt, and tasted, bread that he related to a form of nourishment that had nothing to do with nutrients.

In an influential article on the process of 'nutritionism', sociologist Gyorgy Scrinis connects abstraction like that in the doctors' discussions to an 'increasingly functional approach to food and the body,' which 'obscures the broader cultural, geographic, and ecological contexts in which foods, diets, and bodily health are situated' (2008: 42, 44; see also Aphramor et al., this volume). Yet, a multitude of reasons exist for why diverse meanings become compressed into numerical units. One claim I heard often was that counting created accountability.[1] Whether it was counting calories on the part of people who were dieting, or counting dollars on

1 For further reading on the conflation of valuing and accounting see especially Dunn (2005), Garfinkel (1967), Maurer (2002) and Nelson (2010).

the part of public health institutions working to remain fiscally solvent, several doctors I spoke with suggested that translating behaviours into numbers would lead people to act more responsibly. Underlying the calorie counts that they advocated was the idea that when it came to eating, making quantities of consumption appear in the tangible form of measures would give people more control over what they ingested. The abstraction and externalization of the otherwise intimate and internalized process of eating would, according to this logic, make it easier for people to control the physiological drive to eat more than was in their long-term health interests.[2] Whereas they saw bodily knowledge and desires about hunger and satiety as too personal and too contextual to be trustworthy, they held that a reliance on numbers, which could be evaluated by disinterested parties, encouraged the discipline of responsibility (see also Porter 1995).

Another reason why doctors and health educators advocated for the translation of eating into numerical units was that global health programmes, which sponsored the existing locally-implemented metabolic-illness treatment plans, did not have resources to respond to nuanced frameworks of health and nourishment. When food is imagined to be composed of interchangeable and equivalent parts it becomes comparable across time and between diverse national and ethnic groups. Historian Nick Cullather (2007) shows that the adoption by the world health community of a 'universal currency' of calories and nutrients following World War II allowed relatively small organizations to cover tremendous international terrain. Indeed, the United Nations-affiliated Guatemala City-based Institute for Nutrition for Central America and Panama (INCAP) was founded in 1949, and has since disseminated uniform nutritional guidelines based upon numeric quantities throughout the country.[3] Given that public health infrastructure in Guatemala is dwarfed by need and that more than two-dozen linguistic and ethnic groups live within the country's mountainous topography, many nutrition programmes in Guatemala have consolidated otherwise diverse meanings of food and eating into easily comparable values such as calories or carbohydrates. This consolidation allows the newly formed values to travel, expanding beyond the relatively small

2 Philosopher Annemarie Mol (2011) writes that a dominant repertoire of nutrition science is one of moderation, or thrift, in which the human body is depicted as greedy, and where limiting caloric intake by counting calories is a necessary means through which to restrict this greed. She offers 'satisfaction' or 'pleasure' as another way to relate to eating, but suggests that this is often avoided because of Christian associations between pleasure and sin.

3 Sociologists Stefan Timmermans and Marc Berg write that techniques of standardization developed in the beginning of the twentieth century responded to a growing interest in scientific and technological progress. By the 1980s, standardization's appeal lay in 'the ideology of the free, global market', and standardization practices rode on the coat tails of a wave of interest in globalization (2003: 12). The vision of INCAP and the number of other international health centres that were developed following World War II complicates their timeline of the history of standardization as these centres combined a drive for scientific and technological progress with a vision of global unification.

and specific sites at which the calculations were initially generated and thereby affecting a much larger audience.

The same reliance on reductive metrics that I saw within public health programmes was also adopted by food companies in Guatemala as a strategy for selling their products. Just as Guatemalan public health programmes sought to disseminate their information broadly, multi-national food companies also aimed to distribute their goods across varied audiences. Rather than reflect local variations, they used generic slogans so that the same food and weight loss advertisements appeared in rural and urban settings (as well as across international boundaries). Several public health officials with whom I spoke during my fieldwork complained that these businesses were employing a focus on 'nutritional health' in their advertisements for reasons that they felt had little to do with nutrition. One senior INCAP employee summarized a shortcoming he saw in public health reductionism as follows: 'A problem with creating generalized obesity prevention programmes that cover an entire population is that we end up with ideas about health that can be easily co-opted and transformed by the food industry.' He noted that the standardization of heterogeneous eating practices into easily mobile nutritional guidelines might allow for underfunded public health centres to expand their reach. But simplistic nutritional advice – such as '50% fewer calories!', 'Less fat!', 'More nutrition!' – also creates effective slogans through which food companies can package and sell a wide-range of products as 'nutritionally healthy.' The concern that he and others expressed was that the association of health with the consumption of particular low-carb' or 'high nutrition' products would silence ideas about health that could not be easily reduced to the sound bites provided by these metrics (see also Yates-Doerr 2012a).

The INCAP employee quote above articulated a challenge implicit in much of the work and research about nutrition. On one hand, there exists the need to create knowledge about food that can be easily transported from one site to another, as seen in efforts to build strong public health programmes, expansive advertising campaigns, or scientific food standards. On the other hand are the intersecting, overlapping, and variegated ways in which people value food in their daily culinary practices. In order to expand beyond always-situated life engagements in order to create knowledge that was generalizable, nutrition scientists would employ techniques of abstraction where human eating interactions – the 'why', and 'how' people do what they do with food – were translated into measureable variables. By, at least temporarily, fixing information about food into a variable, eating interactions were ostensibly made transportable. A premise of this approach is that what people do with food can be accurately studied through analysis of these variables, as though shifts in contexts will not alter the variable itself. Another premise is that it is possible to account for enough aspects of human interaction with food to create meaningful models of health and weight loss. During fieldwork, when dietary approaches failed, it was surmised that certain variables of dietary life were not taken into consideration or adhered to, such as the patient had eaten three tortillas instead of two, or a food that should have been treated as unhealthy had been treated as healthy. The problem was seen to lie in the specificities of the guidelines and not the process of guideline-making itself. After all, the language of metrics and numbers was treated as value-free: it represented values, but it did not contain them.

It is useful to pause here to consider the argument made by philosopher of science Helen Verran that numbers serve as ordering devices that do not, on their own, have either universal effects or negative connotations (see especially 2001). She suggests that because there are many ways of counting, it would be a mistake to assume that numeracy necessarily connotes a particular way of valuing; indeed, a dangerous sleight-of-hand can occur when practices of ordering and practices of valuing are taken as equivalents (2011). Borrowing from her writings, I might argue that the counting of nutrients or calories is not itself a coherent practice and because this counting can take numerous forms, the resulting 'calories' or 'nutrients' have heterogeneous, non-uniform meanings. To an extent, this argument is fitting for my fieldwork as people tended to reinterpret and transform the ostensibly uniform quantitative dietary standards with which they were presented. Nonetheless, I saw that the equations underlying many international nutrition guidelines *did* have the effect of condensing diverse ways of relating to food and eating into uniform practices. Though number-based dietary health 'standards' would inevitably have variable and unpredictable effects when put into practice – effects that I will explore in more detail below – the advocacy of dietary standards also had homogenizing and harmful consequences.

In the following explanation of carbohydrates given by a health educator to a classroom filled with rural highland women, we see how diverse eating practices become framed through a language of energetic equivalency, where the effects that food has upon the body are both quantifiable and controllable. Also notable are the ways in which eating is situated through a system of measurements. During this lecture the educator stood in front of a diagram illustrating the 'seven steps for healthy eating', which gave indications of appropriate portion-sizes: meat – at least once a week; dairy – at least twice a day; and vegetables, fruits, grains, cereals, and potatoes – every day; sugars and fats – in moderation. She said:

> Some foods have what are called carbohydrates. Perhaps you aren't familiar with this term, but carbohydrates are our source of energy. If you don't eat them, you won't be able to get up. You won't be able to accomplish anything in the day. You won't be able to run. You won't be able to do your work. You won't be able to care for your children, even. So we need to eat these things in order to have energy.

In this narrative, food is reduced to one of its component parts – the carbohydrate – which is presented as having a predictable, measurable, and cumulative effect on the human body. To carry out daily life functions one must ingest appropriate amounts, with the failure to do so resulting in an inability to accomplish daily activities. The logic here is mechanistic; the body is imagined as a motor, with food playing the functional role of fuel. Historian Anson Rabinbach connects the prevalence of mechanical metaphors for describing the human body in the early twentieth century to a vision of labour in which the self was also understood to be 'objective, measureable, and above all, conquerable' resulting in increased efficiency and ultimately in increased economic productivity (1990: 21). Framing the body as a motor implies that people do not eat for joy or relaxation or because food tastes good or because eating is a means of relating with others. Instead people eat because there is a monetary payoff. As the nutritionist above instructs her audience, eating provides the energy to do one's work and to accomplish something with one's day.

From this quote alone we can know little about how people responded to this style of advice. In the case of this class, and others like it, the audiences mostly remained silent, the efficacy or effects of the instructor's training hidden from the anthropologist-observer. But I saw in my additional fieldwork in medical clinics and while living with Guatemalan families – some for several months at a time – that many people became confused and preoccupied with the project of learning to eat by balancing energy input with energy expended. I often had women with far more cooking experience than I had turn to me for advice on how to prepare foods for themselves and their families. They never asked me about how to make eating enjoyable; they wanted to know how to make properly calculated meals – not too much fat, or sugar, or carbohydrates, the right amounts and kinds of oil – that would result in 'healthy' foods, healthy bodies, and ultimately, productive lives. In the following conversation about eating bread that took place between a nutrition educator and a patient, we can see the insecurity generated by this focus on nutrients.

> Educator: Do you ever eat cereals?
>
> Student: No.
>
> Educator: Bread?
>
> Student: Because they said that sweet breads were bad for me, French breads are all that I buy.
>
> Educator: Now you only buy French bread?
>
> Student: Yes, only French.
>
> Educator: And how many times do you eat this during the day?
>
> Student: Just three times a week.
>
> Educator: And how much do you eat when you eat this?

Student: Just two little pieces.

Educator: Two?

Student: Yes, I'm afraid to eat this because I'm not sure if it's okay to eat or not.

Educator: It's sweet breads that you shouldn't be eating.

Student: Yes, I don't eat sweet breads.

Educator: Well, you really shouldn't eat French bread either. The best is whole-wheat bread. It has the fewest fats and carbohydrates. It also has the most fibre.

Student: Okay. I was afraid I was making a mistake.

Educator: Ask at the supermarket about this bread. You're going to see that it's going to help you feel better and have more energy.

Student: Did you write it down?

Educator: Yes.

Student: I'm afraid I'll forget. I want to do it correctly.

In the above conversation the educator consistently reframes correct dietary behaviour as a matter of quantities: times per week, slices per episode of consumption, and levels of nutrients per slice. Also apparent is the student's anxiety in the face of these standardized regulations. When she does eat bread, she remains unsure whether it is harmful and as a result she eats it with fear. Although she expresses a desire to comply with what is 'correct', her uncertainty about what the rules are and how to follow them is also apparent.

The following exchange between a nutritionist and a woman who has registered as overweight (on the basis of her weight and height measurements and global health guidelines for normal weights) illustrates how this uncertainty about how to best follow dietary standards can, in turn, shape dietary practice. In the woman's case she has responded to confusion by skipping meals and by attempting to reduce both her overall caloric intake and the foods with fat in her diet. Note that following the nutritionist's suggestion to focus less on the guidelines and more on finding changes to her habits that might be sustainable, the patient nonetheless reiterates the importance of the standards.

Nutritionist: What do you usually eat through the day?

Patient: Mostly fruits and vegetables. No chicken, nothing with fat. I love milk, but I can't afford low-fat milk so I haven't been drinking milk. I've dropped a little weight, thank God.

Nutritionist: What is your eating schedule like?

Patient: Usually I skip breakfast or have, maybe, just a bit of coffee –

Nutritionist: Oh, that's not good. It's better to eat about six times a day so that you stay satisfied throughout the day. That's what we're going to recommend.

Patient: Okay, but I don't want you to make me fat. I need to lose weight – twenty pounds.

Nutritionist: We can give you recommendations to help with this, but it's important that you eat during the day so that your energy levels don't drop.

Patient: But I want to lose more weight than I have. I want to be in the normal weight range.

Nutritionist: Try not to worry about your weight. It's more important to change your habits in a way that you can maintain over time.

Patient: Okay, but I have twenty pounds that I have to lose.

Nutritionist as Ethnographer

So far in this chapter I have demonstrated how lived experiences of eating and dieting become circumscribed into calculative, metabolic equations from which dietary standards and guidelines emerge. I have further illustrated that these equations can have harmful effects, circumscribing diverse valuations of food into confusing regulations and guidelines. The realm of cooking and eating, which has otherwise been a domain of women's expertise, becomes a realm of insecurity and anxiety. This is, however, not the end of the story. Although many of the people I met presumed that dietary equations would have predictable and controllable effects upon the body, I also encountered a failure of metabolic calculations to achieve their ostensible ends. I saw time and again that dietary practices could not be contained within the mathematical abstractions typical of weight loss protocols (fewer carbs! More exercise!). Instead, dietary practices contained an innumerable 'excess' that rendered these guidelines ineffective. The guidelines were there but people, including several nutritionists who might have otherwise enforced the guidelines, worked around them.

While standards can create both accountability and transportability, thereby expanding the reach of public health nutrition programmes, several of the educators I worked with were exasperated with standards. I have written about these educators elsewhere (see especially Yates-Doerr 2012b), but it is useful to clarify that many of Guatemala's nutritionists are closely (inter)connected to people suffering from metabolic illnesses. I worked with numerous nutritionists who were just a generation removed from the poverty of the rural countryside and who remained attuned to the difficult economic conditions faced by many of their patients (and their relatives). They were aware of the gap between public health protocol and what I have termed 'nutrition-in-action' in an attempt to underscore how different the lived experiences of nutrition are from the calculative rationality that seems to define it. The widespread failure of metric-based diet programmes – a failure determined by persistent illness and weight gain as well as by mounting frustration with treatment strategies and patients' subsequent disappearances from clinics – made many of the educators that I met sceptical about the value of these measures. One health worker explained this to me as follows:

> We have the relatively new concept of 'the diet', which is based entirely upon eating foods in correct portions and eating a certain number of calories a day. Today everyone is becoming familiar with the idea that 'to eat well is to follow a diet.' People think that if you can measure out a proper diet, then you know about eating well. But following a diet doesn't necessarily lead to healthy eating and we are focusing too much of our training on following diets. If a diet is failing, if someone isn't losing weight or if their blood sugar levels stay high, we try to adjust their portions and the calories in their meals. But eating food in proper portions is not synonymous with eating well. It seems easier to teach dieting in 30 minute sessions, and right now this is where a lot of our energy is directed. But a lot is missing when we do this. So many people leave nutrition consultations with diet sheets, but with nothing that will be useful in their lives.

In contrast to the mathematical underpinnings of caloric formulas, which abstracted bodies from place and time (as well as from familial engagements with eating), many nutritionists sought to develop treatment strategies that emplaced bodies in their surroundings, recognizing the myriad ways in which commensality – the socio-material practice of eating together – defied the calorie's pretense of commensurability.

For example, despite the financial limitations faced by one of the nutrition clinics where I carried out research, its director was working to develop nutrition services that extended well beyond the technical qualities of different foods. He held that nutritionists needed to know about financial matters, intra-family relations, dietary preferences, daily routines (be they routines of work, exercise, or leisure); they also needed to know intimate details of their patients' histories, about the communities where they lived and their religion, and about the various anxieties that they grappled with in their lives. He explained the need for expansive services to me saying, 'if you want to encourage good nutrition you can't just talk about what people are eating. You need to address entire social, political, and economic systems.' He held that this, in turn, required not a dependence upon the dietary surveys or body weight measurements that were typical of nutrition consultations, but the development of long-term rapport between nutritionist and patient. I should point out that this rapport was also necessary for learning about relationships with materials: did patients have enough burners and pots to make separate meals for themselves? How might they make coffee without adding sugar to the pot? Did they take buses to work, and how far were the stops from where they lived? Could they afford the Tupperware necessary to pack themselves food when they were away from their kitchen? What was their means of refrigeration at home, if any? Echoing the director's interest in far-reaching services, another nutritionist I spoke with emphasized the need to move away from 'the quick-solution mentality that pervades much of nutrition education.' She, along with several others I met, advocated treatment strategies that addressed the diverse contexts of eating, instead of those that focused on the mechanistic consumption of nutrients. Time and again I was told that in the realm of nutrition, simplistic solutions such as calorie counts or portion size guidelines did not only *not work*, but they made patients' lives worse by adding a feeling of failure to the frustrations of poor health and weight gain.

I was often struck during my fieldwork by the parallels between the role of the 'good nutritionist' and my own training in ethnography. When confronted with the heterogeneous, incoherent multiplicities of life-as-lived, sociologist John Law asks social scientists the following question: 'If this is an awful mess … then would something less messy make a mess of describing it?' (2007). Answering in the affirmative, he advocates research methods that aim not to create order out of observable mess, but to participate in and relate with the social world(s) that this research simultaneously helps to bring into being (see also Law and Urry 2003). This vision of research methods resonates with the performative and reflexive turn that anthropology took in the 1970s and 1980s (see especially Geertz 1973; Marcus and Clifford 1986; Martin 1987; Rosaldo 1989) and which continues to situate ethnography as a research technique that trades the shortcuts of surveys and statistical standards for the intimacies of relational commitments. As geographer Sarah Whatmore views it, ethnography does not seek to collect data about and then report on a world that remains 'out there'; it is, instead, always both an intervention in the world, as well as a means of 'working together with those whom we are researching' (2003: 90). The aim is not distanced or detached description; as with nutritionists, whose expertise was contingent upon the engaged participation of their patients, any expertise an ethnographer gains is done so in collaboration with those around her. Just as the most sensitive nutritionists I worked among did not seek to adopt their patients' lives in order to develop relationships with them, neither do we aim to become the people we work with. Instead, we strive for interference – increasingly recognized to be not only an unavoidable outcome but a *goal* of ethnography[4] – by bringing together, but not dissolving together, multiple voices, actors, forms of expertise, and fields of experience.

By the time I began my fieldwork, I had already spent seven summers in Guatemala. On the basis of preliminary research and the steady stream of global health reports and evaluations I was reading, I expected to find a field dominated and overwhelmed by metrics. But although metrics were everywhere, I began to see while living and working with people who were grappling with what had only recently become a problem of 'too-much' weight how often these metrics were reinterpreted or altogether ignored. Weight – a word whose meaning might seem inextricable from the abstraction of numbers – was in practice situational, context-based, at once in and beyond an individual body. While nutritional protocol and guidelines typically focused on the *Indice Masa Corporal* (Body Mass Index), the *masa* of concern outside clinical protocol was the soft corn dough of tortillas, which mattered in ways that quantitative representations of weight did not. As the field of global health that I was studying was seeking to make clearer standards, many of the people I lived among were advocating, often in quiet ways, the importance of complexities – of staying with them, rather than trying to reduce them away.

Conclusion

I began this chapter with the story of a gift to highlight the significance of that which cannot be precisely measured, even in the midst of a caloric system that appears to operate within the clean balance of numbers. When I delivered the bread in December 2008, the power of a market logic, based on a system of monetary credit and debt and which loomed large in the Manhattan skyline where my friend's brother and I stood, was also collapsing around us. It was encounters and contexts like this that helped me to understand, when I turned to write about my fieldwork, what was needed from my own analysis. To tell the story about nourishment and fatness in Guatemala that I wanted to tell, a story that would honour the experiences of those around me, I would need to stay close to those occurrences of social life that could not easily be spoken, let alone fixed into formulas. I would need to point towards the pleasures of eating in a way that would not explain these pleasures away. I would need to not just make room for what might be called the 'excess' of the tastes and sentiments of social life, but to illustrate that these supposed excesses were everywhere. Rather than highlight the moral economies of eating and dieting, I would have to show that the calculative

4 The notion of 'interference', as employed by Haraway (1992), comes from physics where it is used to talk about two sound or light waves that, as they come together, make a complex pattern. For more on the theme of 'making interferences', see the recent special issue in *Current Anthropology* on engaged anthropology, edited by Low and Merry (2010).

logic implicit in the idea of economy might hide more than reveal. Indeed, even when it came to something as apparently quantitative as weight, to speak of economies might not make any sense at all.

Dietary standards, which might have seemed transparent, neutral, and easy to follow from the quiet confines of a doctor's advice within the nutrition clinic, became much less straightforward when confronted with mobile, overlapping, and often-conflicting desires of eating in everyday life. After all, it is easy to care about calorie-amounts and carbohydrate-levels in a setting where weight-loss is an obvious goal. It is more difficult for these measures to remain meaningful when confronted with other ways of valuing, including the various obligations and pleasures of kinship, and the longings and satisfactions of eating that are so often connected to our kin.

Standards work by stabilizing. Operating as a recursive practice, they constrain active phenomenon while simultaneously dictating the behaviours necessary for achieving the narrowly-defined dimensions of these constraints (Star and Lampland 2009: 14). And yet life, in its varying socio-material forms, will not be stilled. In spite of the ways in which standards seem to make differences cohere – to smooth the rough edges of variation into generalizable guidelines and normative ideals – the apparent uniformity that results from these standards remains elusive. Although standards will travel globally, and though they might at first appear to be everywhere the same, they remain, much like the tastes of various kinds of bread, situated within the localities in which people come into contact with them, where they are put into practice.

All too often, discussions of 'why we eat' are predicated on the idea that a range of factors motivates behaviour: the food eaten is cheap; it tastes good or healthy, or decadent; the eater is hungry or anxious or tired or bored or celebrating; the eater wants to share in the consumption of substance with family; the eater wants to isolate from others; and so on. For any given situation, when analysed against one another – that is, when added together on a causal scale – some of these factors will emerge as strong while others will disappear as weak; what remains from this summation is an explanation for the question of 'why'. Yet as my work with nutritionists, and the orientation of ethnography itself, make increasingly apparent, any direct answer to the question of 'why we eat' would be misleading; we cannot sum various motivations and causations for eating together into an understanding of what people do with food. The reasons are not just multiple across varying contexts, but are multiple within any given context, and this multiplicity also illustrates that such a thing as 'any given context' also falls apart. My interactions with nutritionists and patients who were exhausted with the calculative, functionalistic logic of dietetics suggest that for reasons of both nutritional well-being and theoretical acumen we might 'abandon the idea that offering an explanation is good for your health' (Latour 1988: 161). Insofar as we seek to elaborate 'contours of eating', as the editors of this volume put it in the introduction, it would do us well to leave the terrain of causal explanations behind. Rather than imagine that we can ever straightforwardly answer the question of 'why', we might instead seek to make differences visible – to hold them, without holding them still or compressing them together in such a way that they appear to disappear.

It is fitting, given the field of anthropology's historic reliance upon the insights of those with whom we do fieldwork, that people around me were adept at valuing practices and knowledges that defied reduction and generalization. My friend's decision to send his brother a suitcase filled with bread indicated that he was well aware of the importance of tastes and pleasures that would never be stabilized into standards and indicators. For the field of global public health, where wellbeing is dominated by the numerical alchemy of measurement, the work of describing that which eludes measurement remains a challenge for (and an achievement of) ethnography.[5] This challenge of depicting aspects of life that will never be fixed is not small, but the value of complexity underscores not only the importance of ethnographic knowledge but also its importance well beyond the domain of the social sciences.

Acknowledgements

Emily Martin, Rayna Rapp, Sally Merry, Tom Abercrombie, and Renato Rosaldo gave me endless support in crafting this project. I thank Annemarie Mol and the members of the *Eating Bodies* team. Eugene Raikhel and Somatosphere provided the impetus for me to write this chapter. I am grateful to the people I lived and worked with in Guatemala, whose ideas and lives have shaped my own.

References

Cullather, N. 2007. The foreign policy of the calorie. *The American Historical Review* 112(2), 337–64.

Dunn, E. 2005. Standards and person-making in east central Europe, in *Global Assemblages: Technology, Politics, and Ethics as Anthropological Problems*, edited by A. Ong and S.J. Collier. Malden, MA: Blackwell Publishing, 173–93.

Garfinkel, H. 1967. *Studies in Ethnomethodology*. Englewood Cliffs, N.J.: Prentice-Hall.

Geertz, C. 1973. *The Interpretation of Cultures: Selected Essays*. New York: Basic Books.

Groeneveld, I.F., Solomons, N.W. and Doak, C.M. 2007. Nutritional status of urban schoolchildren of high and low socioeconomic status in Quetzaltenango, Guatemala. *Pan American Journal of Public Health* 22(3), 169–77.

---

5 For further reading on the idea that numeracy operates with the sleight-of-hand of alchemy see especially Merry (2011) and Poovey (1998).

Haraway, D. 1992. The promises of monsters: A regenerative politics for inappropriate/d others, in *Cultural Studies*, edited by P.A. Treichler, C. Nelson and L. Grossberg. New York: Routledge, 295–337.

Latour, B. 1988. The politics of explanation: An alternative, in *Knowledge and Reflexivity: New Frontiers in the Sociology of Knowledge*, edited by S. Woolgar. London: Sage, 155–76.

Law, J. 2007. Making a mess with method, in *The SAGE Handbook of Social Science Methodology*, edited by W. Outhwaite and S.P. Turner. Los Angeles: SAGE Publications, 595–606.

Law, J. and Urry, J. 2003. Enacting the social. *Economy and Society* 33(3), 390–410.

Low, S.M. and Merry, S.E. 2010. Engaged anthropology: Diversity and dilemmas: An introduction to supplement 2. *Current Anthropology* 51(S2), S203–S226.

Marcus, G.E. and Clifford, J. 1986. *Writing Culture: The Poetics and Politics of Ethnography*. Berkeley: University of California Press.

Martin, E. 1987. *The Woman in the Body: A Cultural Analysis of Reproduction*. Boston: Beacon Press.

Maurer, B. 2002. Anthropological and accounting knowledge in Islamic banking and finance: Rethinking critical accounts. *The Journal of the Royal Anthropological Institute* 8(4), 645–67.

Merry, S.E. 2011. Measuring the world: Indicators, human rights, and global governance: With CA comment by John M. Conley. *Current Anthropology* 52(S3), S83–S95.

Nelson, D.M. 2010. Reckoning the after/math of war in Guatemala. *Anthropological Theory* 10(1–2), 87–95.

Poovey, M. 1998. *A History of the Modern Fact: Problems of Knowledge in the Sciences of Wealth and Society*. Chicago: University of Chicago Press.

Porter, T. 1995. *Trust in Numbers: The Pursuit of Objectivity in Science and Public Life*. Princeton: Princeton University Press.

Rabinbach, A. 1990. *The Human Motor: Energy, Fatigue, and the Origins of Modernity*. New York: Basic Books.

Rosaldo, R. 1989. *Culture & Truth: The Remaking of Social Analysis*. Boston: Beacon Press.

Star, S.L. and Lampland, M. 2009. Reckoning with standards, in *Standards and Their Stories: How Quantifying, Classifying, and Formalizing Practices Shape Everyday Life*, edited by M. Lampland and S.L. Star. Ithaca: Cornell University Press, 3–24.

Timmermans, S. and Berg, M. 2003. *The Gold Standard: The Challenge of Evidence-Based Medicine and Standardization in Health Care*. Philadelphia: Temple University Press.

Verran, H. 2001. *Science and an African Logic*. Chicago: University of Chicago Press.

Verran, H. 2011. Number as generative device: Ordering and valuing our relations with nature, in *Inventive Methods: The Happening of the Social World*, edited by N. Wakeford and C. Lury. Routledge, 220–54.

Whatmore, S. 2003. Generating materials, in *Using Social Theory: Thinking Through Research*, edited by M. Pryke, G. Rose and S. Whatmore. London: SAGE in association with the Open University, 89–104.

Yates-Doerr, E. 2012a. The opacity of reduction: Nutritional black-boxing and the meanings of nourishment. *Food, Culture and Society: An International Journal of Multidisciplinary Research* 15(2), 293–313.

Yates-Doerr, E. 2012b. The weight of the self: Care and compassion in Guatemalan dietary choices. *Medical Anthropology Quarterly* 26(1), 136–58.

Interlude

Entanglements: Fish, Guts, and Bio-cultural Sustainability

Elspeth Probyn

Entanglement is a central part of eating. It is both a figurative way of considering eating but also what physically happens when we eat. Every time we bite into something, we become mired in more than we envision. Of course, many seek out entanglement as a rationale for eating. This ranges from the search for ever-more rarefied eateries (as a title of a recent seminar put it: 'OMG I'm going to Noma!'[1]), the increasing fervour of televized displays of cooking as 'spectator sport', to the proliferation of eating-related allergies.

Across a number of scales, eating ties us into global-local geo-political webs: the ecological landscape of food-human and non-humans. As Cook et al. put it, we are '(un)knowingly connected to each other through the international trade … that entangles a range of economic, political, social, cultural, agricultural and other processes' (2004: 642). Through personal, institutional, and structural attachments, 'our dietary intake and health are affected by practices far away from our direct experience, and in reverse, our food choices have effects that reach beyond our own domestic worlds' (Cook et al. 1998: 163).

The chapters in this section touch on all of these forms of entanglement, and it is the connections between bodies, place and metabolism that I will further pursue here. Arguments in this section flag new areas of research that are conjoining several disciplines and opening the social sciences to the important questions raised by what is now commonly called the anthroposcene. As Sarah Whatmore argues, trying to understand 'the vital connections between *geo* (earth) and the *bio* (life)' (2006: 601) is now squarely on the research agenda as well as widely commented upon in the public sphere. This is finally seriously shaking up the 'entrenchment' of disciplines quarantined as either 'social' or 'natural' sciences. It is particularly in the areas of food and health that we see in action what Michel Callon calls 'hot situations' and 'the intensification of the interface between "life" and "informatic" sciences and politics' (cited in Whatmore 2006: 601). As Whatmore suggests, there is a 'visceral vernacular of social anxiety relating to food and health [that has] become routinely conversational matters' (ibid.: 606).

---

1 This was the title of a presentation by Nancy Lee to the Department of Gender & Cultural Studies' Work-in-Progress sessions for research students at the University of Sydney on 11th May 2012. Nancy's thesis is on the cultures of celebrity chefs.

As I argued several years ago (Probyn 2000), food seems to have taken over from sex as the most worried-about subject in Western society – it is continuously presented to us through popular books and a plethora of journalistic accounts as an area of taboo and worry. Just as in the nineteenth century sexuality was seen as a deep problem for society and rarely as a source of pleasure (at least in Foucauldian readings), apart from the kitchen porn of Nigella there remains little evidence of eating as other than complicated, uneasy and resulting in pathological conditions.

In Carnal appetites (2000) I sought to shift the terrain of thinking about subjectivity away from the textuality of representation to questions about what eating and food connect us with. There I wondered at the ways we eat into history, emotions, race relations, gender and sexuality. One can hardly ignore the geo-political ties of history that tether sugar, for instance, to the bloody industry of colonialism although the present day sugar industry tries its very best to present a wholesomely 'natural' face (Hollander 2003). But it is, after all, the trade in foodstuff and slave bodies that made the black Atlantic (Gilroy 1992) both the map of imperial desires and the trade route of primary food commodities, and also the place of 'double-consciousness' of Black culture.

In Australia, as with other 'settler' nations, Aboriginal people were controlled and eradicated through feeding policies that destroyed souls by severing them from their connections to lands and dreaming, in which bush food remains central. White man's food also killed bodies – immediately through the poisoning of rations, and then ever onwards through the damage of 'white food/white death' (Rowse 1998) and the ingestion of nutrition-less food and grog.[2] And it is food that is the principle cause of the enormous discrepancy in Type 2 diabetes rates amongst Aboriginal and non-Indigenous Australians, and it is food that is causing the gulf between the health of poorer and more affluent people world-wide. While of course the connections between humans and natural environments, and to genetics and habitus that result in Type 2 diabetes are complex, nonetheless the separation of the consumption of food from where it is produced is part of the picture. For instance, in Australia getting Aboriginal people away from a diet of highly-processed food and connecting people with traditional country and practice has shown good results. As Rowley et al. (2000) argue, successful prevention of obesity in some outstation communities has been associated with greater physical activity, consumption of bush foods, and ownership of and access to traditional homelands. However the basis of this research has to be taken with a pinch of salt. Its framing bypasses the structural inequalities that render many remote Aboriginal

2 Tim Rowse's (1998) book, *White Flour, White Power* describes most powerfully the full extent of the dehumanization of Aboriginals through systems of rationing that forced people to eat in what were for them unnatural and white ways, and how food stuff became part of an overall strategy to reduce Aboriginal people in status as well as in sheer numbers – through genocide. See also my chapter, 'Eating in Black & White' (in Probyn 2000), which was highly influenced by his argument.

lands hostage to prohibitively expensive fresh produce. It also can imply that urban Aboriginal people's deprivation can be magically resolved by sending them all 'off bush' – back to the lands from which they were violently removed generations ago.[3] The effects of the dislocation of people from rural and local farming in the increasing global urbanization in developing nations is also wreaking havoc on bodies. The rising rate of obesity is often associated with the destruction of local crops such as ground nut in developing African nations through the mass import of cheaper cooking oil filled with transfats (Probyn 2010). It is undeniable that the political economies of who eats what where and with what effect are central to the worldwide and socio-economic distribution of bad health through food.

As Emily Yates-Doerr's ethnographic research within public health programmes in Guatemala reveals, programmes that seek to reverse 'bad' eating habits must go far beyond the model of nutritional science, with its attachment to counting calories and to moralistic judgment. She writes: 'Weight – a word whose meaning might seem inextricable from the abstraction of numbers – was in practice situational, context-based, at once in and beyond an individual body'. She advocates 'the importance of complexities – of staying with them, rather than trying to reduce them'. It is indeed astonishing that within such a complex area as food and eating the causal thinking and methods of much public-health-based research could even be countenanced, let alone implemented. Yates-Doerr's analysis echoes Anna Tsing's (2005) argument for an ethnographic understanding of complexity within the global mapping of connection. The role of gathering and telling stories, of trying to see from multiple perspectives, which extend to those of the nonhuman, is not to erase incompatibility. Rather, as Tsing writes, 'we need to find out where it [complexity] makes a difference' (2005: 259).

This volume's attention to the 'contours of eating' and its aim of reflecting on difference and complexity rather than sameness and causality promises to reveal a much more interesting field of food studies than one would gather from a cursory understanding of much of what passes in its name. Lest it seem that nutritionalism is the only target, I hasten to add that overly *cultural* analyses of food can produce equally, though differently, curtailed understandings. In the new projects I am engaged in, I try to open up the mainstream concerns about what we eat, where our food comes from, and who and what is touched by our actions. This is to analyse food in what Michel Callon calls 'a network of relationships in which social and natural entities mutually control who they are and what they want' (1986: 203). This is a challenge on many fronts.

---

3 For a much more considered and in-depth understanding of these complex issues, see Elizabeth Povinelli's work, especially *The Empire of Love* (2006).

Eat More Fish

To begin by outlining some of the elements within this challenge, I want to turn to fish.[4] I'll start with my favourite fish, canned tuna. Several years ago, my local pharmacist remarked on the spate of colds I was contracting. She brusquely inquired: 'how much fish are you eating?' And when I said not much, she glared at me and ordered me to eat fish at least four times a week, promising that it would strengthen my cells against the onslaught of cold viruses. Of course we now know that it cannot just be any fish but must be of the deep sea and oily variety. While I am fond of sardines, a fish to which I will return, tuna sandwiches became my medicinal meal of choice. Just to be sure I also began taking fish oil capsules, although the Mayo Clinic warns that 'the evidence is stronger for the benefits of eating fish rich in omega-3 fatty acids than for using supplements' (2010). Their advice is that omega-3 can give you a healthy heart, and indeed my father who knows about hearts and fish had us eating sardines, mackerel and pilchards from an early age. (He worked for the British Heart Foundation at the time, and was an avid fisherman).

It seems that fish of the omega-3 variety can help hearts, stave off colds and make you happier and smarter. The medical findings widely publicized in the media have set off an enormous public quest for omega-bearing fish, to which retailers and pharmaceutical companies alike have responded with alacrity. Given that the scientific advice about the nutritional value of fish is relatively recent, and only began to gather force in the early 1990s, it is astonishing that there are now nearly 14 million hits on Google about the benefits of fish as a prime source of omega-3.

One aspect that soon jumps out across some of the websites is the 'thrifty gene', and particularly how it is linked to fish consumption. As Rachael Kendrick explains in her chapter, James Neel discovered the 'thrifty gene' in the early 1960s. In simple terms the theory is that former hunter-gatherer communities of people (normally these are taken to be Indigenous peoples) have a propensity to store fat in times of feast to compensate for future famine. However when these individuals find themselves with a steady supply of food, the gene continues to make the body store fat as if a famine were ahead. While, as Kendrick argues, the theory has been roundly discredited in medical research, it continues to reverberate in studies that seek to gauge why Indigenous populations should have such high rates of obesity. For instance, in a 2006 article in the *American Journal of Agricultural Economics*, Timothy Richards and Paul Patterson introduce their study:

---

4 Much of what follows derives from my project on 'Taste & Place', which is funded by an Australian Research Council grant. In this project I examine the circulation of different locally-produced foods through to their distanced places of consumption. These have included kangaroo (Probyn 2012b), and most concertedly, fish (Probyn 2011, 2012a, forthcoming b).

> The most common explanation for the apparently dysfunctional diets of many Native American groups lies in the notion that they possess a 'thrifty gene' (Neel 1962). Over many generations, surviving native societies develop a means of sustaining themselves during times of famine by building up stores of energy (fat) acquired during times of relative prosperity (2006: 543).

With this hypothesis, the authors propose to 'investigate the possibility that food, or more specifically nutrient, choices among Native Americans – even apparently addictive, self-destructive choices – result from fully rational, forward-thinking decisions' (ibid.: 543). After a number of complicated statistical computations, they find that: 'Native consumers fail to internalize future costs' (ibid.: 546). In other words, the reason that First Nations people will not pay more to consume fish rather than chips is that they do not think about how their choices will affect them in the long term. As Jennifer Poudrier, a Métis researcher, bluntly states, 'the whole thrifty-gene idea seems to me not to capture the subtlety and complexity … of type 2 diabetes in First Nations communities' (cited in Abraham 2010). She is being rather kind given that the continued use of Neel's ideas seem to further pathologize the Indigenous as well as the poor in terms of what and how they eat, of which of course she is well aware.

Follow Fish

In my 20s I became a fish-eating vegetarian, or what is now commonly called a pescatarian. The thinking that drove my eating was that we could more easily feed the world if we ate legume-based proteins. For me then, like many, the oceans seemed to be a vast ever-bountiful provider of food for humans, and so eating fish fitted into my admittedly naïve idea. It was an easy whim, which carried me for many years until I moved to Australia. Arriving in this land of unfamiliar meats, I could not help myself from eating kangaroo and sampling emu – two of the animals on the national coat of arms. The die was cast, and contrary to the fortitude shown by Samantha Hurn as she continued her fieldwork in the hunts of Wales despite her rather opposing alimentary philosophy as a vegan, I immersed myself in food research and returned to meat-eating.[5]

5 This raises an interesting question analogous to one asked by Traci Warkentin in her recent article: Must every animal studies scholar be vegan? (2012). Or to reframe, must every food studies scholar be omnivorous? Warketin's question is posed in the context of a seemingly blanket policy/philosophy in animal studies to not eat what you are investigating. Her response is a critique of an unthinking presumption of equivalences, to which she advocates reflection on 'the messy areas in between the extremes of veganism, vegetarianism, and meat-eating and yet still [be] allied with the goals and values of animal advocacy in multi-faceted ways' (ibid.: np).

Gradually it became hard to dismiss the fears about fish. At first it was anecdotal: the place in northern Scotland where we had had summer holidays as children catching and eating abundant fish, no longer had any fish. Then in 1992, in the land of my birth, Canada, the entire far right-hand side of the country was closed down when the fishing bans on cod were announced. It is hard to underestimate the full extent of this tragedy for fish and humans alike. The unbelievable had happened: the seemingly bottomless flow of cod from the Grand Banks had been reduced to 1% of its biomass. While Newfoundlanders ('Newfies') are often the butt of jokes in parts west of the Maritime Provinces, much of the blame goes to the Federal mismanagement of the fisheries. The end of the cod was a death-knell for the Maritimes and effectively terminated the livelihoods of 35,000 people directly employed in the fisheries and processing industries. Centuries-old ways of life vanished as seemingly permanently as the fish (the cod fisheries are still closed). The culprits are mismanagement and also that seemingly human desire for more and for the capacity of technology to give us more: the boats that can go further out to sea, the sonar equipment and spotting planes that mercilessly chase down the ever-declining fish.

So much for my hopes that the oceans could easily provide humans with their protein. But what about farming fish? On the surface, the idea that the oceans can feed humanity seems great.[6] But what of the horror stories of salmon fish farms in Scotland and Norway? It is said that farms in Scotland produce as much faecal matter as towns with populations of 65,000. The unprocessed poop filled with nitrates goes directly into the streams and out to sea, and effectively burns the ocean's eco-system.[7]

Farmers of the Sea

These thoughts were in my mind as I headed into fieldwork in South Australia. It was actually oysters that got me going down the western side of the Spencer Gulf in the vast and sparsely populated outback of South Australia. The oysters of Coffin Bay are renowned in Asia in large part because they are produced in the pristine waters of the Gulf and the southern Pacific. 'Pristine' is somewhat overused in marketing fish and shellfish from this region, but for good reason – it is

---

6 I am here paraphrasing a question raised by Carlos Duarte et al. (2009), a group of marine scientists who are exploring whether and how agriculture could be moved to the oceans in a bid for the sustainability of food and the planet.

7 Given the dearth of fish in the oceans, Scotland along with Chile, competes with Norway to be the largest producer of farmed salmon. See Marianne Lien and John Law's interesting project on fish farming in Norway (2011).

true. There are few people, little industry, and most importantly no rivers to take the agricultural pesticides into the sea. At least, for the time being.[8]

Port Lincoln is the hub of the Eyre Peninsula, famous for being home to the highest percentage of millionaires per capita in the whole of Australia. Their money comes from Southern Bluefin Tuna. I won't repeat what I have written elsewhere (Probyn 2012a, forthcoming a) but the tale of how tuna became the most valuable fish is fascinating (Bestor 2004). Suffice to say, it is a tale of fish and humans and technology, and now one that looks to have a sad ending as humans aided by technology have fished the stock to near-endangered.[9] Federal and State fisheries management has responded by reducing the quota quite drastically over the last decade. As one of my informants told me,[10] there are few weapons fishery management can wield but one is to control how many fish are caught through the system of Individual Transferable Quotas. This system aims to reduce catch and it often results is reducing the size of the fishing fleet as people sell or lease their quota. There are now only about 22 boats fishing tuna in Australia, and many fishers claim they can catch their yearly quota in a matter of weeks, if not in days.[11]

With the writing on the wall, Dinko Lukin[12] and others came up with the idea of catching smaller fish – thus reducing the tonnage of their catch in line with the set quota – and then feeding them up. Effectively the tuna fishers are now farmers. The practice is to catch the smaller tuna, which are caught in the Great Bight, and then to carefully transfer them into sea cages that are very slowly towed back to the waters off Port Lincoln. Once there, the fish are again very carefully transferred to huge commercial pens where they are fed and pampered. Given a single fish can sell for tens of thousands of dollars at the Tsukiji fish market in Tokyo, there is to be no rough handling of the tuna. When their times comes, they are instantly killed with a spike down the spinal cord – which is apparently one of the fastest and least stressful ways for a fish to die. They are then flown to Tsukiji

8 This state is in peril because of the mining boom in South Australia, and elsewhere. The huge mine at Olympic Dam needs more water and plans are underway to install a desalination plant at the top of the Gulf, which would threaten the quality of water, as well as the breeding habit of Giant Cuttlefish. For more, see Probyn forthcoming b.

9 There is an important distinction between commercial and artisanal fishing; the latter is being pushed and fished out by the huge fleets that often sail under Flags of Convenience, and therefore are not liable to national or regional quotas and regulation.

10 My warm thanks to Rob who taught me a great deal about the regulation side of fisheries.

11 The South Australian agency Primary Industry Research South Australia states: 'In October 2011, the Commission for the Conservation of Southern Bluefin Tuna (CCSBT) – the body responsible for managing the global quota for Southern Bluefin Tuna (SBT) – agreed to increase Australia's national quota allocation from 4,015 tonne in 2011 to 4,528 tonne in 2012 and to 4,698 tonne in 2013.' www.pir.sa.gov.au/__data/.../PIR7012AquaScope_Issue1_web.pdf [accessed: 30 May 2012].

12 There is much more to be said about Dinko who sadly died last year. See Probyn 2011 for an interview with him.

and within 16 hours they can go from alive in a pen off Port Lincoln, to dead and on the plate as beautiful sushi. There is something seductive about this. The very high value of the fish means that it is never squandered, although conversely it does mean that the fishers will do their utmost (including apparently death threats) against those who wish to close down the industry.

The story gets even more appealing when I hear of the operations of *Cleanseas: sustainable fish.* I first come across their advertising at various seafood markets. A lot of thought has gone into the beautiful presentation, which features a striking woman is a low-cut ball gown on the prow of a boat with a fish in her hand. The tag line is: 'Australian Born, Japanese Quality, Culinary Excellence'. *Cleanseas* does both tuna and *hiramasa*, or kingfish. The description reads:

> Produced from wild brood stock, hatchery raised fingerlings are grown in an open water, low stress environment. The use of natural feeds, minimal stocking densities and site fallowing practices, delivers a fish which is totally sustainable using the world's best practices.[13]

Hagen Stehr, the man behind *Cleanseas* is a legend in Australia and Japan. The story is that he jumped ship from the Foreign Legion and fell in love with a South Australian girl. His first fortune was from something nefarious – guns are hinted at. Then he, with the other tuna barons, hit it rich when they accumulated the tuna quota and started making real money with the rising price of tuna in the early 1980s. Instead of putting his money into racehorses (as his fellow tuna millionaire, Tony Santic did) or real estate (another tuna baron built the swanky millionaire's row marina in Port Lincoln as well as the local up-market hotel), Stehr put his into science. With the help of Australian scientists at the Commonwealth Scientific and Industrial Research Organization (CSIRO), and Japanese researchers at Kinko University, Stehr has accomplished amazing results. The kingfish are fully bred on land and raised on land. Kingfish is a native species, so much of the trouble of importing exotic fish such as Atlantic salmon has been avoided. Stehr then turned his sights to Bluefin tuna, a much trickier fish to grow in captivity. He has managed to breed and grow fingerlings, which are then transferred to sea pens. This remains at the experimental stage at his research facility in Arno Bay, although the Japanese have succeeded in commercializing the 'Kindai' artificially bred and raised tuna, which comes with certificates of its birth and rearing.

Stehr does not seem to be an overly modest man. *Cleanseas*' website states: 'The exciting advances in the industry are reflected in the recovering wild stocks.' Fisheries management would not go as far as to say that wild stock is now healthy, although the quota was raised significantly in the last season. Here we find again that connection with omega-3. The section on 'healthy eating' states that: 'Research has shown that Omega 3 has several healthful properties including: Reducing the risk of heart disease and Type 2 diabetes' and it 'eases the symptoms in conditions

13 http://www.cleanseas.com.au/main/home.html [accessed: 30 May 2012].

such as: Rheumatoid arthritis, Depression, Attention Deficit Disorder, Psoriasis, Obesity'.[14]

Who is Eating Whom, Where?

There is no doubt that *Cleanseas* (and the other tuna farmers to a lesser extent) tell a seductive story about sustainability. Indeed *Cleanseas*' presentation of itself deliberately foregrounds and aetheticizes sustainability. But at the end of the day it is still fish being fed to fish. Tuna, and kingfish (as well as salmon) are all carnivores. In the wild they eat anything that swims (including their own young, making them beautiful cannibals). Indeed the only thing saving fishers in many parts of the world is that when the carnivores have been fished out the stock of lobster and crab explodes. The diet of tuna in captivity mainly consists of the herring family – sardines, pilchards and anchovies. In South Australia 98% of the Total Allowable Catch of the sardine fishery goes into feeding the tuna and kingfish. The sardines are frozen into great blocks that then are transported by crane and dropped in the tuna pens.

The South Australian sardine fishery is a drop in the ocean compared to the world's leading producer of fishmeal, Peru. The Peruvian anchoveta is the largest fishery in the world by volume, and globally it contributes 30% of the fish consumed by other fish, as well as by poultry and pigs. While the anchoveta is good for the Peruvian economy it is less than good for the inhabitants, especially those living near the enormous fishmeal factories in the north of the country. Compared to just freezing sardines and throwing them to the fish as they do in South Australia, the process of mulching them down to make them more easily transported around the world is complicated and produces a lot of environmental damage experienced by the sea and its human communities. It also threatens long-term devastation. Members of the herring family are a keystone species, absolutely crucial to the overall wellbeing of the pelagic ecosystem. From phyto-plankton into sardines, they feed sea birds and mammals as well as larger fish. When they are fished out, the ocean will truly become a desert with large populations of jellyfish.[15]

In Ben Coles's terms, 'to eat a food is to eat its geography'. In many ways, this connection motivated the response in Peru to the worsening degradation of the environment and of individuals' health due to the effects of the high-volume commercial exploitation off the Peruvian anchoveta. Patricia Majluf, the Director of the Centre for Environmental Sustainability at Cayetona Heredia University in Lima, has spearheaded a programme to encourage the human consumption of these little fish. A zoologist by training, Majluf was concerned at the lack

14 http://www.cleanseas.com.au/main/healthy-eating.html [accessed: 30 May 2012].

15 The experimental cook, Heston Blumenthal, has used jellyfish in recipes to highlight this (see Highfield 2010).

of availability for people to buy and consume anchovies, as well as their bad image. In an inspired move, she set about building a public campaign for 'la rica anchoveta', which included a major media blitz and the enlistment of Peru's top restaurants – Peru is now seen as a global top gastronomic destination. She also got food aid programmes to distribute the fish to the poor, and galvanized processors to produce canned and frozen anchovetas. These fish have an extremely high percentage of protein along with, of course, omega-3. Their stock went up (in a manner of speaking) when Oprah Winfrey declared sardines to be a superfood, saying: 'Wild-caught sardines are low in mercury (unlike some types of tuna) and high in vitamin D; a three-ounce serving has as much calcium as a cup of milk. Even better, they're one of the Monterey Bay Aquarium's top picks for sustainability' (Barbour and Mount 2010: np). The result of the Peruvian initiative has been a 285% increase in human consumption of anchoveta, which equals 2% of the total allowable catch. At the same time research began to formulate a value chain analysis 'from sea to plate'.

Fish, Guts and Bio-Cultural Sustainability

To bring to a (temporary) close this meandering tale of complexity, I will go back to the three tenets that would seem to motivate the chapters in this section: connections between bodies, place, and metabolism. These avenues are extremely generative in opening up the analysis of food beyond its often-narrow framing either in terms of the pathological gaze of the nutritionist counting calories and weighing bodies, or of the committed semiotician scouring the surface for meanings. Thinking about the geographical connections between food production and consumption is the first base in terms of bridging disciplinary divides that so often curtail the full interest of engaged scholarship. This is to bring bodies and places together in ways that understand food in terms of its capacity to spawn 'metabolic intimacies',[16] to use a phrase from John Law and Annemarie Mol's (2006) wonderful article on boiling pigswill.

Taking my lead from the chapters I have tried to trace some of the connections, bodies and places that are spurred on by the popularity and near obsession with omega-3. A metabolic aspect of fish – omega-3 – becomes popularized through media reports of scientific research, which set into motion different connections.

16 As Law and Mol explain, the traditional practice of giving the family pig the scraps from the table enabled a pig-human metabolic intimacy. The practice was enabled by boiling the scraps, however when this process wasn't followed, an epidemic of foot and mouth broke out in northern England with the result that pigswill was banned. Law and Mol write: 'When pigs eat swill they eat what people do not want to eat. And when they eat whey and skimmed milk they are not in direct competition with people either. But when pigs start to eat potatoes this is no longer the case. They are in metabolic competition with people' (2006: 140).

These take us to the Peruvian anchoveta and the campaigns that attempt to bring sardines and anchovies back into human bodies, to do their good work in helping us to feel better, be smarter, etc.; in sum, to be a living part of a fish-guts-sea entanglement, which may help further awareness of the absolute necessity of framing and enacting bio *and* human sustainability.

References

Abraham, C. 2011. How the diabetes-linked 'thrifty gene' triumphed with prejudice over proof. *Globe and Mail*, 25th February.

Barbour, C. and Mount, R. 2010. *25 Superfoods to Incorporate Into Your Diet Now* [Online]. Available at: http://www.oprah.com/food/Superfoods-Ingredients-and-Recipes-for-a-Healthy-Diet/18#ixzz1wKBjXCku [accessed: 30 May 2012].

Bestor, T.C. 2004. *Tsukiji: The Fish Market at the Center of the World.* Berkeley and Los Angeles: University of California Press.

Callon, M. 1986. Some elements of a sociology of translation: Domestification of the scallops and fishermen of St. Brieuc Bay, in *Power, Action and Belief: A New Sociology of Knowledge?*, edited by J. Law. London: Routledge & Kegan Paul, 196–233.

Cook, I. et al. 2004. Follow the thing: Papaya. *Antipode* 36(4), 642–64.

Cook, I., Crang, P. and Thorpe, M. 1998. Biographies and geographies: Consumer understandings of the origins of foods. *British Food Journal* 100(3), 162–7.

Duarte, C.M., Holmer, M., Olsen, Y., Soto, D., Marba, N., Guiu, J., Black, K. and Karakassis, I. 2009. Will the Oceans feed humanity? *BioScience* 59(11), 967–76.

Gilroy, P. 1992. *The Black Atlantic: Modernity and Double Consciousness*. Cambridge: Harvard University Press.

Highfield, R. 2010. Heston Blumenthal: Food's future is jellyfish and chips. *New Scientist*. [Online]. Available at: http://www.newscientist.com/article/mg20627635.600-heston-blumenthal-foods-future-is-jellyfish-and-chips.html [accessed: 27 May 2012].

Hollander, G.M. 2003. Re-naturalizing sugar: Narratives of place, production and consumption. *Social & Cultural Geography* 4(1), 59–74.

Law, J. and Mol, A. 2008. Globalisation in practice: On the politics of boiling pigswill. *Geoforum* 39(1), 133–43.

Lien, M. and Law, J. 2011. 'Emergent Aliens': On salmon, nature and their enactment. *Ethnos* 76(1), 65–87.

Mayo Clinic. No Date. Omega-3 in fish: How eating fish helps your heart. [Online]. Available at: http://www.mayoclinic.com/health/omega-3/HB00087 [accessed: 30 May 2012].

Povinelli, E. 2006. *The Empire of Love: Toward a Theory of Intimacy, Genealogy, and Carnality*. Durham: Duke University Press.

Probyn, E. 2000. *Carnal Appetites: FoodSexIdentities*. London: Routledge.

Probyn, E. 2010. Feeding the world: Towards a messy ethics of eating, in *A Critical Introduction to Consumption*, edited by E. Potter and T. Lewis. London: Palgrave.

Probyn, E. 2011. Swimming with tuna: Human-ocean entanglements, *Australian Humanities Review* 51, 97–114.

Probyn, E. 2012a. In the interests of taste & place: Economies of attachment, in *The Global Intimate*, edited by G. Pratt and V. Rosner. New York: Columbia University Press.

Probyn, E. 2012b. Moving food: Of roo. *New Formations* 74, 33–45.

Probyn, E. Forthcoming a. Women following fish in a more-than-human world. *Gender, Culture & Place*.

Probyn, E. Forthcoming b. Sustaining fish-human communities? A more-than-human question, in *Contemporary Ethnography across Disciplines*, edited by C. Pope and B. Reinhart. Amsterdam: Springer.

Richards, T.J. and Patterson, P. 2006. Native American obesity: An economic model of the 'Thrifty Gene' theory. *American Journal of Agricultural Economics* 88(3), 542–60.

Rowley, K.G., Gault, A., McDermott, R., Knight, S., McLeay, T. and O'Dea, K. 2000. Reduced prevalence of impaired glucose tolerance and no change in prevalence of diabetes despite increasing BMI among Aboriginal people from a group of remote homeland communities. *Diabetes Care* 23(7), 898–904.

Rowse, T. 1998. *White Flour, White Power: From Rations to Citizenship in Central Australia*. Cambridge: Cambridge University Press.

Tsing, A. 2005. *Friction: An Ethnography of Global Connection.* Princeton: Princeton University Press.

Warkentin, T. 2012. *Must Every Animal Studies Scholar be Vegan? Hypatia.* [Online]. Available at: http://onlinelibrary.wiley.com.ezproxy1.library.usyd.edu.au/journal/10.1111/(ISSN)1527-2001/ [accessed: 30 May 2012].

Whatmore, S. 2006. Materialist returns: Practising cultural geography in and for a more-than-human world. *Cultural Geographies* 13(4), 600–609.

Index